Green Chemistry for the Development of Eco-Friendly Products

Kavita Shakya Chahal
Government College, Bichhua, India

Twinkle Solanki
Government College, Bichhua, India

A volume in the Advances in
Industrial Ecology (AIE) Book Series

IGI Global
PUBLISHER of TIMELY KNOWLEDGE

Published in the United States of America by
IGI Global
Engineering Science Reference (an imprint of IGI Global)
701 E. Chocolate Avenue
Hershey PA, USA 17033
Tel: 717-533-8845
Fax: 717-533-8661
E-mail: cust@igi-global.com
Web site: http://www.igi-global.com

Library of Congress Cataloging-in-Publication Data

Names: Chahal, Kavita, 1979- editor. | Solanki, Twinkle, 1993- editor.
Title: Green chemistry for the development of eco-friendly products / Kavita Chahal and Twinkle Solanki, editor.
Description: Hershey, PA : Engineering Science Reference, an imprint of IGI Global, [2022] | Includes bibliographical references and index. | Summary: "This book covers green chemistry, which is the design of chemical products and processes that reduce or eliminate the use or generation of hazardous substances, demonstrating the life cycle of a chemical product, including its design, manufacture, use, and ultimate disposal"-- Provided by publisher.
Identifiers: LCCN 2021051135 (print) | LCCN 2021051136 (ebook) | ISBN 9781799898511 (h/c) | ISBN 9781799898528 (s/c) | ISBN 9781799898535 (ebook)
Subjects: LCSH: Green chemistry. | Green products.
Classification: LCC TP155.2.E58 G737 2022 (print) | LCC TP155.2.E58 (ebook) | DDC 660.028/6--dc23/eng/20211230
LC record available at https://lccn.loc.gov/2021051135
LC ebook record available at https://lccn.loc.gov/2021051136

This book is published in the IGI Global book series Advances in Industrial Ecology (AIE) (ISSN: pending; eISSN: pending)

British Cataloguing in Publication Data
A Cataloguing in Publication record for this book is available from the British Library.

For electronic access to this publication, please contact: eresources@igi-global.com.

Advances in Industrial Ecology (AIE) Book Series

Syed Abdul Rehman Khan
Xuzhou University of Technology, China

ISSN:pending
EISSN:pending

MISSION

The **Advances in Industrial Ecology (AIE) Book Series** examines current, state-of-the-art studies in the areas of Industrial ecology and Circular economy. Industrial ecology is a rapidly growing field that systematically examines local, regional, and global materials along with energy uses and flows in products, processes, industrial sectors and economies. It focuses on the potential role of industry in reducing environmental burdens throughout the product life cycle from the extraction of raw materials, to the production of goods, to the use of those goods and to the management of the resulting wastes. Industrial ecology is ecological in that it (1) places human activity -- industry in the very broadest sense -- in the larger context of the biophysical environment from which we obtain resources and into which we place our wastes, and (2) looks to the natural world for models of highly efficient use of resources, energy, and byproducts. By selectively applying these models, the environmental performance of industry can be improved. Industrial ecology sees corporate entities as key players in the protection of the environment, particularly where technological innovation is an avenue for environmental improvement. As repositories of technological expertise in our society, corporations provide crucial leverage in attacking environmental problems through product and process design.

The **Advances in Industrial Ecology (AIE) Book Series** encompasses empirical, analytical, theoretical, reviews, communications, and conceptual research, and comprehensive reviews of relevant research case studies of effective applications in the areas of Industrial ecology and circular economy. It provides a practical and comprehensive forum for exchanging scientific knowledge and ideas which bridge the technological advancement and sustainable practices with industrial ecological management. The goal of this series is to bring together the expertise of professionals who are working in the field of industrial ecology and sustainability across the globe.

COVERAGE

- Waste
- Urban Metabolism
- Recycling
- Material Efficiency
- Waste Management
- Environmental Science
- Technological Advancement in Industrial Ecology
- Circular Economy
- Green Practices
- Industrial Ecology

The Advances in Industrial Ecology (AIE) Book Series (ISSN pending) is published by IGI Global, 701 E. Chocolate Avenue, Hershey, PA 17033-1240, USA, www.igi-global.com. This series is composed of titles available for purchase individually; each title is edited to be contextually exclusive from any other title within the series. For pricing and ordering information please visit http://www.igi-global.com/book-series/advances-industrial-ecology/270446. Postmaster: Send all address changes to above address.

Titles in this Series

For a list of additional titles in this series, please visit: http://www.igi-global.com/book-series/

Innovative Nanocomposites for the Remediation and Decontamination of Wastewater
Azad Kumar (M.L.K. P.G. College, India)
Engineering Science Reference • © 2022 • 335pp • H/C (ISBN: 9781668445532) • US $240.00

Handbook of Research on SDGs for Economic Development, Social Development, and Environmental Protection
Cristina Raluca Gh. Popescu (University of Bucharest, Romania & The Bucharest University of Economic Studies, Romania)
Engineering Science Reference • © 2022 • 531pp • H/C (ISBN: 9781668451137) • US $325.00

Integrating Blockchain Technology Into the Circular Economy
Syed Abdul Rehman Khan (Xuzhou University of Technology, China)
Business Science Reference • © 2022 • 307pp • H/C (ISBN: 9781799876427) • US $250.00

IGI Global
PUBLISHER of TIMELY KNOWLEDGE

701 East Chocolate Avenue, Hershey, PA 17033, USA
Tel: 717-533-8845 x100 • Fax: 717-533-8661
E-Mail: cust@igi-global.com • www.igi-global.com

Table of Contents

Foreword xiv

Preface xv

Acknowledgment xxiv

Chapter 1
Effective Utilization of Thermal Power Plant Waste Fly Ash for Value Addition of Plastic Products: A Conceptual Use of Fly Ash in Polymeric Materials 1
Omdeo Kishorrao Gohatre, Central Institute of Petrochemicals Engineering and Technology, India
Subhaprada Sahoo, Central Institute of Petrochemicals Engineering and Technology, India
Kashmira Majhi, Central Institute of Petrochemicals Engineering and Technology, India
Sunil S. Suresh, Independent Researcher, India
Jaidev K., Laboratory for Advanced Research in Polymeric Materials (LARPM), Central Institute of Petrochemicals Engineering and Technology, India

Chapter 2
HyLaw or Hydrogen Law: A Regulation for Removal of Legal Barriers to the Deployment of Fuel Cells and Hydrogen Applications 24
Nima Norouzi, Bournemouth University, UK

Chapter 3
Evaluating Wastes as Low-Cost Substrates for Use in Microbial Citric Acid Production 43
Bilge Sayın Börekçi, Ardahan University, Turkey

Chapter 4
Green Polymer Chemistry and Bio-Based Plastic 77
Ankur Khapre, University of Hyderabad, India
Payal Panwar, BITS Pilani, India

Chapter 5
Green Solvents for Sustainable Organic Synthesis: An Overview 104
Anuradha Singh, Department of Chemistry, S. S. Khanna Girls Degree College, University of Allahabad, Prayagraj, India
Kavya Singh, S. S. Khanna Girls Degree College, University of Allahabad, Prayagraj, India
Swastika Singh, S. S. Khanna Girls Degree College, University of Allahabad, Prayagraj, India

Chapter 6
Recycled Raw Banana Peels for Bioplastics 129
Suchismita Satapathy, KIIT University, India
Divyalok Samal, KIIT University, India
Partha Pratim Sarmah, KIIT University, India
Tushar Mandal, KIIT University, India
Abhinav Sinha, KIIT University, India
Simranpreet Singh, KIIT University, India

Chapter 7
Scientomeric Evaluation and Visual Aanalytics of the Scientific Literature Production on Bioplastics 142
Anchal Rana, DAV College, Chandigarh, India
Nidhi Gupta, Panjab University, Chandigarh, India
Anjali Sharma, DAV College, Chandigarh, India
Rupak Chakravarty, Panjab University, Chandigarh, India

Chapter 8
Magnetic Nanoparticles for Environmental Management 174
Mustafa Akgün, Ardahan University, Turkey

Chapter 9
An Overview to Green Nanotechnology for Environmental Issues 190
Naveen Kumar Chourasia, Government College, Bichhua, India

Chapter 10
Crop Improvement Technology With *Lawsonia inermis*....................................200
Shweta Saxena, School of Studies in Botany, Jiwaji University, Gwalior, India
Sweta Prakash, Department of Botany, Rajmata Scindia Government Girl's P.G. College, Chhindwara, India
Sadhna Panday, Department of Botany, Government KRG P.G. College, Gwalior, India

Compilation of References 211

Related References.. 249

About the Contributors 272

Index.. 275

Detailed Table of Contents

Foreword xiv

Preface xv

Acknowledgment xxiv

Chapter 1
Effective Utilization of Thermal Power Plant Waste Fly Ash for Value Addition of Plastic Products: A Conceptual Use of Fly Ash in Polymeric Materials 1

Omdeo Kishorrao Gohatre, Central Institute of Petrochemicals Engineering and Technology, India
Subhaprada Sahoo, Central Institute of Petrochemicals Engineering and Technology, India
Kashmira Majhi, Central Institute of Petrochemicals Engineering and Technology, India
Sunil S. Suresh, Independent Researcher, India
Jaidev K., Laboratory for Advanced Research in Polymeric Materials (LARPM), Central Institute of Petrochemicals Engineering and Technology, India

Over recent decades, most plastic materials processed by incorporating some fillers in it. In this regard, many researchers have continued to close attention towards the study of various fillers and their effects on the advanced polymeric material. Similarly, these fillers and additives incorporate within the polymer matrix to meet various applications and to enhance its mechanical properties, strength, durability, etc. The trends have been significantly arise to the use of waste material as filler owing to its advanced properties in plastic material, ease of availability, and its low cost. In general, fly ash (FA) is a waste by-product generated from most thermal power plants and contains varieties of different elemental particles. It is utilized as filler material in a variety of polymeric materials to make sports equipment/devices, household products, construction industries, and in many other engineering applications. Hence, this section particularly focused on fly ash filler and its composites preparation using different polymer matrix.

Chapter 2
HyLaw or Hydrogen Law: A Regulation for Removal of Legal Barriers to the Deployment of Fuel Cells and Hydrogen Applications 24
Nima Norouzi, Bournemouth University, UK

In recent years, hydrogen technology has been at the forefront of environmental discussions to meet increasingly tough climate protection goals and particularly low emissions targets in the transportation sector. Like any major change, a transition to hydrogen energy faced challenges in many countries, which caused several problems in the growth of the hydrogen share of the total energy supply portfolio. In 2018, Hydrogen Law (Hylaw) was introduced, which removes the legal barriers to the deployment of fuel cells and hydrogen applications. It is a flagship project aimed at boosting the market uptake of hydrogen and fuel cell technologies providing market developers with a clear view of the applicable regulations while calling policymakers' attention to legal barriers to be removed. This chapter introduces a consistent framework for the Hylaw regulations that makes is a clear and precise statement and an interconnection between law and energy management policies.

Chapter 3
Evaluating Wastes as Low-Cost Substrates for Use in Microbial Citric Acid Production 43
Bilge Sayın Börekçi, Ardahan University, Turkey

With the increasing population, developing technology, and industry, the importance given to waste control/effective assessment studies continue with increasing momentum. The use of wastes in the production of biotechnological products is preferred due to its advantages in reducing environmental pollution, preventing nutrient and biomass losses, recycling, and decreasing costs. Citric acid (CA) is an intermediate product formed by the oxidation of carbohydrates to carbon dioxide in the Krebs cycle. This organic acid is used in many industrial areas such as pharmaceuticals and cosmetics. It is also an important organic acid in the food industry and is used as an acidifier, a stabilizer, an antioxidant, a flavor enhancer, and a preservative. Today, CA production is produced by microorganisms through fermentation. In addition, some wastes, such as molasses, glycerol, whey, olive mill wastewater, and various fruit wastes can be evaluated for use in the production of CA. This study reviewed the microbial production of CA using various wastes and some factors affecting the production.

Chapter 4
Green Polymer Chemistry and Bio-Based Plastic 77
Ankur Khapre, University of Hyderabad, India
Payal Panwar, BITS Pilani, India

The overview briefly surveys the practice of green chemistry in polymer science. The conservation and sustainable utilization of polymers is achieved through documentation of indigenous knowledge through the ideology for syntheses of green polymers. The economic importance of polymers includes the study of the relationship between people and polymers. The chapter offers a systematic approach to study the variety of materials. It intersects many fields including the need for green polymer chemistry, environmentally friendly methods for syntheses of green polymers, biodegradable polymers, recycling of polymers, and many more topics. This study aims to provide an updated survey of the green synthesised polymers in order to obtain recyclable materials for various industrial and indoor applications. A number of barriers that hinder the adoption of newer technology have also been discussed. All of these areas are experiencing an increase in research activity with the development of new tools and technologies. Examples are given of recent developments in green polymer chemistry.

Chapter 5
Green Solvents for Sustainable Organic Synthesis: An Overview 104
Anuradha Singh, Department of Chemistry, S. S. Khanna Girls Degree College, University of Allahabad, Prayagraj, India
Kavya Singh, S. S. Khanna Girls Degree College, University of Allahabad, Prayagraj, India
Swastika Singh, S. S. Khanna Girls Degree College, University of Allahabad, Prayagraj, India

Green or sustainable chemistry is now recognised as an important scientific subject, and research in this area has led to the creation of cleaner and safer chemical processes, with many new advances occurring each year. Researchers are very interested in unconventional or alternative solvents for chemical synthesis. Chemical solvents make up around 80% of all chemical compounds utilised in a variety of essential chemical processes, notably in the chemical industry. Unfortunately, many of these solvents are volatile organic compounds (VOCs) produced from petrochemicals, which pose a variety of health and environmental hazards. In this chapter, different environmentally benign solvent alternatives have been discussed based on their green matrix in chemical processes.

Chapter 6
Recycled Raw Banana Peels for Bioplastics 129
Suchismita Satapathy, KIIT University, India
Divyalok Samal, KIIT University, India
Partha Pratim Sarmah, KIIT University, India
Tushar Mandal, KIIT University, India
Abhinav Sinha, KIIT University, India
Simranpreet Singh, KIIT University, India

Petroleum-based plastics may not be sustainable in the long run as crude oil will get scarce. Hence, it is imperative to find alternative sources of material from which plastics can be prepared. Starch, found abundantly in raw banana peel, which is a well-known kitchen waste, can be used to prepare biodegradable plastics. The preparation and some of its properties including its biodegradability have been studied here. A homemade DIY bioplastic sample has been prepared and its biodegradability tested. The world today is suffering an unprecedented menace of plastic pollution. The ubiquitous presence of plastic has sent alarm bells all across the globe. So bioplastics may be the solution.

Chapter 7
Scientomeric Evaluation and Visual Aanalytics of the Scientific Literature Production on Bioplastics 142
Anchal Rana, DAV College, Chandigarh, India
Nidhi Gupta, Panjab University, Chandigarh, India
Anjali Sharma, DAV College, Chandigarh, India
Rupak Chakravarty, Panjab University, Chandigarh, India

Incessant use of plastics in the past decades has been a serious issue of environmental concern. The aim of this study is to examine the current status of the bioplastic research field by identifying the key publications or scientific productions, the most productive researchers in terms of publications and citations, latest hotspots (keywords), the countries with the most active research, collaborations between the countries, highly cited journals, highly contributing organizations, and co-cited references with their cluster analysis. This chapter considers whole literature study of bioplastics (i.e., documents published from 1894 to 2021). One thousand eight hundred sixty-three documents were identified using the Scopus database. The largest number of papers published were during the year 2020 with 308 records. Bibliometric analysis shows author "Guerrero, A." was the highly productive author and USA was the most productive region of the world. The Journal of Polymers and the Environment was the most common outlet for publication.

Chapter 8
Magnetic Nanoparticles for Environmental Management 174
Mustafa Akgün, Ardahan University, Turkey

Magnetic nanoparticles are an emerging technique that has attracted attention in recent years in nanotechnology, biomedical, electronics, environmental science, and engineering applications. Nanoparticles have optical, electrical, catalytic, and thermal properties with their supermagnetic properties, large surface area, and biocompatibility. The major benefit of using nanoparticles is that due to their size, they can be accurately oriented and can be targeted and interacted with a specific biological entity or marker. In addition, it is easy to separate the magnetic property from the aqueous solution with the application of an external magnetic field. From an environmental perspective, MNPs have been used as catalysts in the purification of whey; removal of heavy toxic metals such as Arsenic (As), Lead (Pb), and toxic pollutants such as Fluoride (F) from contaminated water; and photocatalytic degradation of dyes and pollutants in water. In this study, types of magnetic nanoparticles, synthesis methods, properties, and environmental science and engineering applications are included.

Chapter 9
An Overview to Green Nanotechnology for Environmental Issues.................. 190
Naveen Kumar Chourasia, Government College, Bichhua, India

Environmental pollution is a serious issue that has a negative impact on human health. The situation is exacerbated by the gradual development of industrialization and urbanization. The removal of toxins from the environment is characterized as environmental remediation. Environmental remediation refers to all of the approaches used to lessen the risks of pollution in the environment. Nanotechnology has the potential to significantly reduce existing environmental issues. Green nanotechnology is the science and technology for developing nanoparticles using environmentally benign approaches. Green nanotechnology makes use of nature's biological qualities (through a variety of activities). Another area where nanotechnology can play a critical role is in environmental remediation because the properties of nanoparticles can be designed by designing their size and shape. Due to their huge surface area and, as a result, strong reactivity, nanoparticles perform significantly better than bulk materials for environmental remediation.

Chapter 10
Crop Improvement Technology With *Lawsonia inermis*................................. 200
Shweta Saxena, School of Studies in Botany, Jiwaji University, Gwalior, India
Sweta Prakash, Department of Botany, Rajmata Scindia Government Girl's P.G. College, Chhindwara, India
Sadhna Panday, Department of Botany, Government KRG P.G. College, Gwalior, India

Lawsonia inermis L. (family Lythraceae), often known as henna, is a dye-producing plant cultivated in various parts of the world for cosmetic purposes. Since time immemorial, leaf powder made from this little tree has been used to beautify skin, hair, fingernails, leather, silk, and wool. The plant's leaf, which contains an active dye (red orange pigment), lawsone, is widely employed in cosmetic and pharmaceutical industries (2-hydroxy-1,4 naphthoquinone). Lawsone, an essential secondary metabolite, accumulates in the plant's aerial parts, with the highest concentration of 1.0–1.4% in the young leaf petiole. Analgesic, hypoglycemic, hepatoprotective, immune stimulant, anti-inflammatory, antibacterial, anti-dermatophytic, protein glycation inhibition, anti-sickling, antioxidant, anti-fertility, tuberculostatic, wound healing, anticomplimentary, and anticancer properties have all been reported for the plant. Henna is currently recognised as a valuable source of unique natural ingredients for the creation of medications and commercial products for a variety of ailments.

Compilation of References 211

Related References 249

About the Contributors 272

Index 275

Foreword

Green chemistry is the development of chemical products and processes that limit the usage of hazardous compounds and their creation. It also includes the development and marketing of economically viable industrial processes that lessen the risk to human health and the environment. the importance of chemistry in everyday life and explains the societal benefits of expanding the use of green chemistry. It contains an update from the field's frontiers.

This book is an essay on green chemistry and uses a straight forwarded approach to providing a broad but thorough overview of the subject, beginning with traditional green chemistry topics and moving on to less commonly covered topics such as chemistry of longer wear and population, as well as environmental chemistry. Topics like these highlight the importance of chemistry in everyday life and demonstrate the tangible benefits that increased green chemistry use can provide to society.

Practice in Green Chemistry brings together the hands-on experience of an interdisciplinary team from academia and industry to offer a unique perspective on the practicality of implementing green chemistry concepts in support of a worldwide push toward a more sustainable future. This book will appeal to industrial chemists, chemical engineers, students, teachers, researchers and anyone interested in learning more about green chemistry. It is primarily intended for use by students and lecturers, but it will also appeal to anybody interested in learning more about green chemistry.

For environmental sustainability, it is critical to educate the next generation of chemists on green chemistry challenges such as waste minimization and clean synthesis, and hence, following research trends in the field of Green Chemistry is critical for directing future research. Green chemistry professionals from around the world provide chapters with experience teaching at various academic levels, providing a cohesive overview of possible approaches to incorporating green chemistry into existing curricula.

Nitin Nema
Vedica College of B Pharmacy, India

Preface

AN OVERVIEW OF THE SUBJECT MATTER

Green chemistry is the design of chemical products and processes that reduce or eliminate the use or generation of hazardous substances. Green chemistry applies across the life cycle of a chemical product, including its design, manufacture, use, and ultimate disposal. Green chemistry is also known as sustainable chemistry.

Green chemistry is the key to sustainable development, as it will direct the scientific community to the more persuasive and innovative solutions for the existing problems and will present opportunities for new processes and safer products in an economically competitive manner.

The main objective of green chemistry is to redefine and modify the manufacturing processes and products in order to decrease hazards to human health. This would be achieved by making better use of available resources for the development of a chemical process, as well as it will aim to reduce waste generated in any preparation or handling of chemicals.

Green chemistry's 12 principles that demonstrates the breadth of the concept of green chemistry:

1. **Prevent waste**: Design chemical syntheses to prevent waste. Leave no waste to treat or clean up.
2. **Maximize atom economy**: Design syntheses so that the final product contains the maximum proportion of the starting materials. Waste few or no atoms.
3. **Design less hazardous chemical syntheses**: Design syntheses to use and generate substances with little or no toxicity to either humans or the environment.
4. **Design safer chemicals and products**: Design chemical products that are fully effective yet have little or no toxicity.
5. **Use safer solvents and reaction conditions**: Avoid using solvents, separation agents, or other auxiliary chemicals. If you must use these chemicals, use safer ones.

6. **Increase energy efficiency**: Run chemical reactions at room temperature and pressure whenever possible.
7. **Use renewable feedstocks**: Use starting materials (also known as feedstocks) that are renewable rather than depletable. The source of renewable feedstocks is often agricultural products or the wastes of other processes; the source of depletable feedstocks is often fossil fuels (petroleum, natural gas, or coal) or mining operations.
8. **Avoid chemical derivatives**: Avoid using blocking or protecting groups or any temporary modifications if possible. Derivatives use additional reagents and generate waste.
9. **Use catalysts, not stoichiometric reagents**: Minimize waste by using catalytic reactions. Catalysts are effective in small amounts and can carry out a single reaction many times. They are preferable to stoichiometric reagents, which are used in excess and carry out a reaction only once.
10. **Design chemicals and products to degrade after use**: Design chemical products to break down to innocuous substances after use so that they do not accumulate in the environment.
11. **Analyze in real time to prevent pollution**: Include in-process, real-time monitoring and control during syntheses to minimize or eliminate the formation of byproducts.
12. **Minimize the potential for accidents**: Design chemicals and their physical forms (solid, liquid, or gas) to minimize the potential for chemical accidents including explosions, fires, and releases to the environment.

A DESCRIPTION OF WHERE THIS TOPIC FITS IN THE WORLD TODAY

Governments and scientific communities throughout the world recognize that the practice of green chemistry and engineering not only leads to a cleaner and more sustainable earth, but also is economically beneficial with many positive social impacts. These benefits encourage businesses and governments to support the development of sustainable products and processes. The United States, desiring to reward and celebrate significant achievements in Green Chemistry, has given out an annual award since 1996, the Presidential Green Chemistry Challenge Award.

BENEFITS OF GREEN CHEMISTRY TO THE WHOLE WORLD TODAY

Human Health

- Cleaner air: Less release of hazardous chemicals to air leading to less damage to lungs
- Cleaner water: less release of hazardous chemical wastes to water leading to cleaner drinking and recreational water
- Increased safety for workers in the chemical industry; less use of toxic materials; less personal protective equipment required; less potential for accidents (e.g., fires or explosions)
- Safer consumer products of all types: new, safer products will become available for purchase; some products (e.g., drugs) will be made with less waste; some products (i.e., pesticides, cleaning products) will be replacements for less safe products
- Safer food: elimination of persistent toxic chemicals that can enter the food chain; safer pesticides that are toxic only to specific pests and degrade rapidly after use
- Less exposure to such toxic chemicals as endocrine disruptors

Environment

- Many chemicals end up in the environment by intentional release during use (e.g., pesticides), by unintended releases (including emissions during manufacturing), or by disposal. Green chemicals either degrade to innocuous products or are recovered for further use
- Plants and animals suffer less harm from toxic chemicals in the environment
- Lower potential for global warming, ozone depletion, and smog formation
- Less chemical disruption of ecosystems
- Less use of landfills, especially hazardous waste landfills

Economy and Business

- Higher yields for chemical reactions, consuming smaller amounts of feedstock to obtain the same amount of product
- Fewer synthetic steps, often allowing faster manufacturing of products, increasing plant capacity, and saving energy and water
- Reduced waste, eliminating costly remediation, hazardous waste disposal, and end-of-the-pipe treatments

- Allow replacement of a purchased feedstock by a waste product
- Better performance so that less product is needed to achieve the same function
- Reduced use of petroleum products, slowing their depletion and avoiding their hazards and price fluctuations
- Reduced manufacturing plant size or footprint through increased throughput
- Increased consumer sales by earning and displaying a safer-product label (e.g., Safer Choice labeling)
- Improved competitiveness of chemical manufacturers and their customers

A DESCRIPTION OF THE TARGET AUDIENCE

The book would be very much appropriate for Students pursuing chemistry at school, undergraduate and postgraduate course. Also, Business personals relying on chemical-based industries would need to understand the potential of green chemistry in the environment. Environmentalists and other laypersons should also need to know about green chemistry.

It is open for all academic, teaching, and industry professionals, including environmental scientists and law and policy makers, who constitute the green chemistry community. Articles, perspective and feature articles, reports on recent innovations, as well as educational resources from those working in related disciplines across chemistry, chemical engineering, political science, jurisprudence, and economics are also welcomed.

IMPORTANCE OF EACH CHAPTER

Chapter 1: Effective Utilization of Thermal Power Plant Waste Fly Ash for Value Addition of Plastic Products – A Conceptual Use of Fly Ash in Polymeric Materials

Over recent decades, most plastic materials processed by incorporating some fillers in it. In this regard, many researchers have continued to close attention towards the study of various fillers and their effects on the advanced polymeric material. Similarly, these fillers and additives incorporate within the polymer matrix to meet various applications and to enhance its mechanical properties, strength, durability, etc. The trends have been significantly arise to the use of waste material as filler owing to its advanced properties in plastic material, ease of availability, and its low cost. In general, fly ash (FA) is a waste by-product generated from most thermal power plants and contains varieties of different elemental particles. It is utilized as filler material in a variety of polymeric materials to make sports equipment/devices, household products, construction industries, and in many other engineering applications. Hence, this section particularly focused on fly ash filler and its composites preparation using different polymer matrix.

Chapter 2: HyLaw or Hydrogen Law – A Regulation for Removal of Legal Barriers to the Deployment of Fuel Cells and Hydrogen Applications

In recent years, hydrogen technology has been at the forefront of environmental discussions to meet increasingly tough climate protection goals and particularly low emissions targets in the transportation sector. Like any major change, a transition to hydrogen energy faced challenges in many countries, which caused several problems in the growth of the hydrogen share of the total energy supply portfolio. In 2018 Hydrogen Law (Hylaw) was introduced, which stands for Hydrogen Law and removing the legal barriers to the deployment of fuel cells and hydrogen applications. It is a flagship project aimed at boosting the market uptake of hydrogen and fuel cell technologies providing market developers with a clear view of the applicable regulations while calling policymakers' attention to legal barriers to be removed. This chapter is aimed to introduce a consistent framework for the Hylaw regulations and make it a clear and precise statement as an interconnection between law and energy management policies.

Chapter 3: Evaluating Wastes as Low-Cost Substrate for Use in Microbial Citric Acid Production – Evaluating Wastes for Use in Citric Acid Production

With the increasing population, developing technology, and industry, the importance given to waste control/effective assessment studies continue with increasing momentum. The use of wastes in the production of biotechnological products is preferred due to its advantages in reducing environmental pollution, preventing nutrient and biomass losses, recycling and decreasing costs. Citric acid (CA) is an intermediate product formed by the oxidation of carbohydrates to carbon dioxide in the Krebs cycle. This organic acid is used in many industrial areas such as pharmaceuticals and cosmetics. It is also an important organic acid in the food industry and is used as an acidifier, a stabilizer, an antioxidant, a flavor enhancer, and a preservative. Today, CA production is produced by microorganisms through fermentation. In addition, some wastes, such as molasses, glycerol, whey, olive mill wastewater, and various fruit wastes can be evaluated for use in the production of CA. This study reviewed the microbial production of CA using various wastes and some factors affecting the production.

Chapter 4: Green Polymer Chemistry and Bio-Based Plastic

The overview briefly surveys the practice of green chemistry in polymer science. The conservation and sustainable utilization of polymer is achieved through documentation of indigenous knowledge through the ideology of green syntheses polymers. The economic importance of polymers includes the study of the relationship between people and polymer. The chapter offers a systematic approach to the wide variety of materials in polymerization. It intersects many fields including the need for green polymer chemistry, environmentally friendly methods for syntheses green polymer, biodegradable polymer, recycling of polymers and many more topics. This study aims to provide an updated survey of the main green synthesised polymers in order to obtain recyclable materials for various industrial and indoor applications. a number of barriers that hinder the adoption of newer technology have also been discussed. All of these areas are experiencing an increase in research activity with the development of new tools and technologies .example are given of recent development in green polymer chemistry.

Chapter 5: Green Solvents for Sustainable Organic Synthesis

Green or sustainable chemistry is now recognised as an important scientific subject, and research in this area has led to the creation of cleaner and safer chemical processes, with many new advances occurring each year. Researchers are very interested in unconventional or alternative solvents for chemical synthesis. Chemical solvents make up around 80% of all chemical compounds utilised in a variety of essential chemical processes, notably in the chemical industry. Unfortunately, many of these solvents are volatile organic compounds (VOCs) produced from petrochemicals, which pose a variety of health and environmental hazards. In this chapter, different environmentally benign solvent alternatives have been discussed based on their green matrix in chemical processes.

Chapter 6: Recycled Raw Banana Peels for Bioplastics

Petroleum-based plastics may not be sustainable in the long run as crude oil will get scarce. Hence, it is imperative to find alternative sources of material from which plastics can be prepared. Starch, found abundantly in raw banana peel which is a well-known kitchen waste can be used to prepare biodegradable plastics. The preparation and some of its properties including its biodegradability have been attempted to study here. A homemade DIY bioplastic sample has been prepared and its biodegradability tested. The world today is suffering an unprecedented menace of plastic pollution. The ubiquitous presence of plastic has sent alarm bells all across the globe. So bioplastics may be the solution to it is tested.

Chapter 7: Scientometric Evaluation and Visual Analytics of the Scientific Literature Production on Bioplastics

Incessant use of plastics in the past decades has been a serious issue of environmental concern. The aim of this study is to examine the current status of bioplastic research field by identifying the key publications or scientific productions, the most productive researchers in terms of publications and citations, latest hotspots (keywords), the countries with the most active research, collaborations between the countries, highly cited journals, highly contributing organizations and co-cited references with their cluster analysis. This paper considers whole literature study of bioplastics i.e., documents published from 1894 to 2021. 1863 documents were identified using the Scopus database. Highest number of papers published were during the year 2020 with 308 records. Bibliometric analysis shows author "Guerrero, A" was the highly productive author and USA was the most productive region of the world. The Journal of Polymers and the Environment was the most common outlet for publication.

Chapter 8: Magnetic Nanoparticles for Environmental Management

Magnetic nanoparticles is an emerging technique that has attracted attention in recent years in nanotechnology, biomedical, electronics, environmental science and engineering applications. Nanoparticles have optical, electrical, catalytic and thermal properties with their supermagnetic properties, large surface area and biocompatibility. The major benefit of using nanoparticles is that due to their size, they can be accurately oriented and can be targeted and interacted with a specific biological entity or marker. In addition, it is easy to separate the magnetic property from the aqueous solution with the application of an external magnetic field. From an environmental perspective, MNPs have been used as catalysts in the purification of whey, removal of heavy toxic metals such as Arsenic (As), Lead (Pb) and toxic pollutants such as Fluoride (F) from contaminated water, and photocatalytic degradation of dyes and pollutants in water. In this study, types of magnetic nanoparticles, synthesis methods, properties and environmental science and engineering applications are included.

Chapter 9: An Overview to Green Nanotechnology for Environmental Issues

Environmental pollution is a serious issue that has a negative impact on human health. The situation is exacerbated by the gradual development of industrialization and urbanization. The removal of toxins from the environment is characterized as environmental remediation. Environmental remediation refers to all of the approaches used to lessen the risks of pollution in the environment. Nanotechnology has the potential to significantly reduce existing environmental issues. Green nanotechnology is the science and technology for developing nanoparticles using environmentally benign approaches. Green nanotechnology makes use of nature's biological qualities (through a variety of activities). Another area where nanotechnology can play a critical role is in environmental remediation because the properties of nanoparticles can be designed by designing their size and shape. Due to their huge surface area and, as a result, strong reactivity, nanoparticles perform significantly better than bulk materials for environmental remediation.

Chapter 10: Crop Improvement Technology with *Lawsonia inermis*

Lawsonia inermis L. (family Lythraceae), often known as henna, is a dye-producing plant cultivated in various parts of the world for cosmetic purposes. Since time immemorial, leaf powder made from this little tree has been used to beautify skin, hair, fingernails, leather, silk, and wool. The plant's leaf, which contains an active dye (red orange pigment), lawsone, is widely employed in cosmetic and pharmaceutical industries (2-hydroxy-1,4 naphthoquinone). Lawsone, an essential secondary metabolite, accumulates in the plant's aerial parts, with the highest concentration of 1.0–1.4 percent in the young leaf petiole. Analgesic, hypoglycemic, hepatoprotective, immune stimulant, anti-inflammatory, antibacterial, anti-dermatophytic, protein glycation inhibition, anti-sickling, antioxidant, anti-fertility, tuberculostatic, wound healing, anticomplimentary, and anticancer properties have all been reported for the plant. Henna is currently recognised as a valuable source of unique natural ingredients for the creation of medications and commercial products for a variety of ailments.

CONCLUSION

Green chemistry, which was established about two decades ago, has attracted much attention. It reflects the efforts of academia and industry to address the challenges related to sustainable development of the chemical industry, and continuous progress is being made, both in academia and industry. Briefly, green chemistry is the utilization of a set of principles to reduce or eliminate the use or generation of hazardous substances in the design, manufacture and applications of chemical products. Green chemistry is a multidisciplinary field and covers areas such as synthesis, solvents, catalysis, raw materials, products, and efficient processes.

Kavita Shakya Chahal
Government College, Buchhua, India

Twinkle Solanki
Government College, Bichhua, India

Acknowledgment

The editors would like to acknowledge the help of all the people involved in this project and, more specifically, to the authors and reviewers that took part in the review process. Without their support, this book would not have become a reality.

First, the editors would like to thank each one of the authors for their contributions. Our sincere gratitude goes to the chapter's authors who contributed their time and expertise to this book.

Second, the editors wish to acknowledge the valuable contributions of the reviewers regarding the improvement of quality, coherence, and content presentation of chapters. Most of the authors also served as referees; we highly appreciate their double task.

Kavita Shakya Chahal
Government College, Bichhua, India

Twinkle Solanki
Government College, Bichhua, India

Chapter 1
Effective Utilization of Thermal Power Plant Waste Fly Ash for Value Addition of Plastic Products:
A Conceptual Use of Fly Ash in Polymeric Materials

Omdeo Kishorrao Gohatre
https://orcid.org/0000-0003-1430-8826
Central Institute of Petrochemicals Engineering and Technology, India

Subhaprada Sahoo
Central Institute of Petrochemicals Engineering and Technology, India

Kashmira Majhi
Central Institute of Petrochemicals Engineering and Technology, India

Sunil S. Suresh
Independent Researcher, India

Jaidev K.
Laboratory for Advanced Research in Polymeric Materials (LARPM), Central Institute of Petrochemicals Engineering and Technology, India

DOI: 10.4018/978-1-7998-9851-1.ch001

ABSTRACT

Over recent decades, most plastic materials processed by incorporating some fillers in it. In this regard, many researchers have continued to close attention towards the study of various fillers and their effects on the advanced polymeric material. Similarly, these fillers and additives incorporate within the polymer matrix to meet various applications and to enhance its mechanical properties, strength, durability, etc. The trends have been significantly arise to the use of waste material as filler owing to its advanced properties in plastic material, ease of availability, and its low cost. In general, fly ash (FA) is a waste by-product generated from most thermal power plants and contains varieties of different elemental particles. It is utilized as filler material in a variety of polymeric materials to make sports equipment/devices, household products, construction industries, and in many other engineering applications. Hence, this section particularly focused on fly ash filler and its composites preparation using different polymer matrix..

INTRODUCTION

Fly ash is the residue formed during coal combustion known by the names such as coal ash and pulverized flue ash. It is generally have light tan in colour and has a fine particulate structure along with its not having virtual odour and is considered as industrial waste pollution. In recent years about million tonnes of coal is burnt to produce electricity so that a huge amount of fly ash is being generated which possesses environmental pollution and many other serious issues as mention in **Figure 1.** Moreover, India generates about 120 to 150 million tons of fly ash from burning coal in approximately 120 thermal power plants than any other country (Akhtar & Tarannum, 2018). It is also forecasted that the global fly ash market will be expected valued will be USD 11.0 billion by the end of 2024 (Globe News Wire, 2018). The major market for fly ash is within Asia pacific region including countries such as China, Japan, India, South Korea, and rest of the Asia pacific region.

The coal is major source for energy generation and being available abundantly worldwide. During energy production, the coal is used to burn and during combustion some volatile material as well as carbon also burn off. In meantime the coal impurities such as quartz, clays, feldspar etc. are hugely fused and stayed in suspension form. Thereafter, these fused particles are getting passed using flue gas. At the end these flue gas approaches to low temperature, these fused substances cause to coagulate and form spherical particles. So, these coagulated spherical form of particles called as "Fly ash". Similarly, the remaining matter are going to agglomerate as well as settle down at the bottom of the furnace are referred as "Bottom ash". It is also expected that fly ash market will reach US$ 64,761.9 Mn by 2022 which was US$ 39,548.1 Mn in 2015. The growing rate of this fly ash market is reported at a CAGR of 7.3% during the period of 2016-2022 (forecasted period) (Surabhi, 2017).

Figure 1. The most important issues to be analyzed from the generated waste fly ash

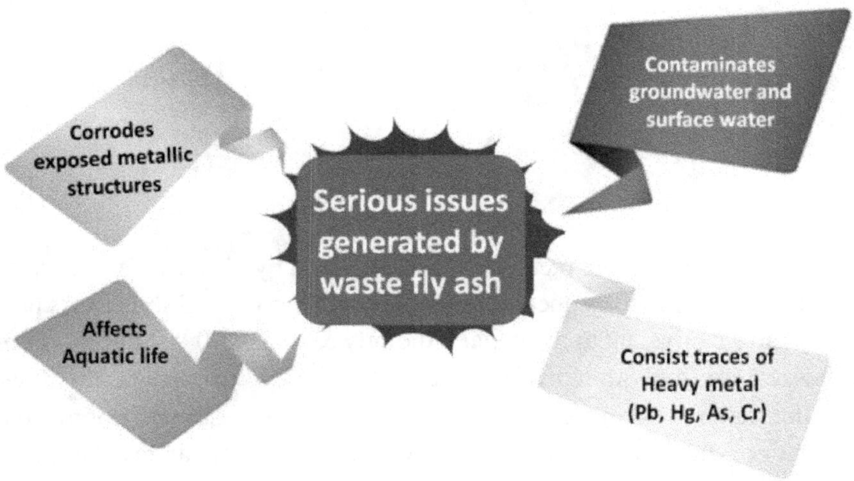

The another report marked the global fly ash market sized was valued at USD 8.54 billion in 2019 and expected to reach USD 13.33 billion by 2027, (with 5.83%CAGR) as illustrated in **Figure 2**. The report is basically analysed including both type of fly ash (Class F & Class C) within the study period 2016-2026. The most of the report indicates that there is rise in the fly ash market by > 4% CAGR. A very large amounts of this particles is formed by different thermal power plants, and their disposal generates major environmental challenges (Fly ash market size, 2020).

However, such generated waste fly ash is mainly composed of SiO_2 with low-lying contents of Al_2O_3, Na_2O, Fe_2O_3, K_2O, CaO, MgO, SO_3, etc. Accordingly, fly ash categorized as class-F as well as class-C type depends upon the percentage of calcium and silicon present in it. In a current situation, such waste fly ash finds utilization in most of the applications such as in brick manufacturing, road, and building construction, cementing industries, mineral filler, structural fills, road base/sub-base, soil modification, and others (Attarde et al., 2014). Nowadays, fly ash finds a place in plastic industries due to its desired properties of the final plastic products after the incorporation of fly ash in it. Similarly, fly ash having a low cost, ease of availability causes rising popularity in plastic industries.

Figure 2. Fly ash market report (Fly ash market-growth, n.d.)

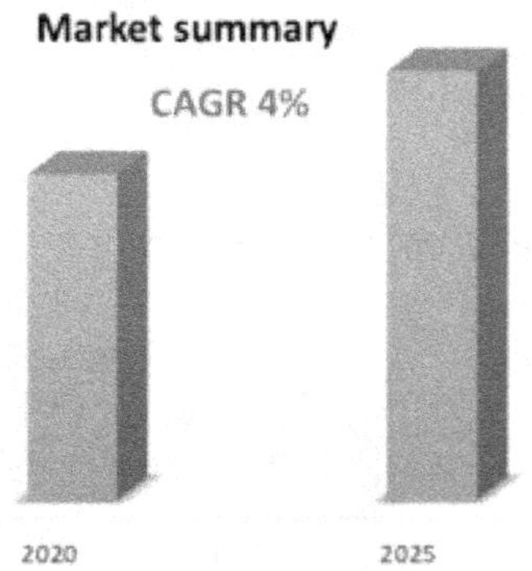

Fastest Growing market	Asia Pacific
Largest market	Asia Pacific ➤ China ➤ India ➤ Japan ➤ South Korea ➤ Rest of Asia-Pacific
CAGR	> 4%
Study Period	2016-2026
Type	Class F Class C

It is well known that the metal is getting replaced by plastics material due to its lightweight, corrosion resistance, heat-insulating material, and sometimes provides strength that is equivalent to the metal (Andrady & Neal, 2009). Majority of plastic materials required high strength to weight ratio and hence, to overcome this drawback many filler material incorporate within polymer matrix results in improvement in the strength of material. However, sometimes fly ash shows an excellent property when incorporated in the polymer matrix. The filler consumption causes to lower in consumption of costlier polymer matrix and also help to alternate the overall composite properties (Shubham & Tiwari, 2013). For this reason, polymer industries search for a low-cost and lightweight material that shows excellent properties with better processability. Therefore, fly ash waste is the better alternatives with its cheaper cost and ease of availability.

FLY ASH GENERATION AND ITS PHYSICAL PROPERTIES

The fly ash physically occurs as very fine particles that having minute average diameter, and that has high surface area and light texture. On considering chemical nature of fly ash is considered as amorphous and mixture of Ferro-alluminosilicate minerals. These both chemical and physical appearance of fly ash depends upon the type of coal used, the method of combustion adopted, set temperature during combustion as well as collection method. Fly ash is the form of ash generated during the separation of flue gas of a power station burning pulverized coal. It is precipitated as a fine particulate material from the stack of gases by the combustion of solid fuels such as coal in an industrial furnace (Swanepoel & Strydom, 2002). Generally, the amount of fly ash is >70% of the entire amount of coal ash produced within power plants only. In general, fly ash can be found in its fineness and globular particles form. The average particle size of fly ash generally lying ranging in the size of <1 to 100 μm (Rossow, 2003). Some of the globular, solid spheres in shape, and some in hollow cenosphere. The present spherical/globular particles improve the fluidity and workability whenever used with other material.

COMPOSITION BASED CLASSIFICATION OF FLY ASH

The fly ash is primarily composed of various elements such as SiO_2 with low contents Al_2O_3, Na_2O, Fe_2O_3, K_2O, MgO, etc., and is listed in **Figure 3** and **Figure 4**. Accordingly, the classification of fly ash can be done broadly on its present composition. In the current situation, there are two types of fly ash waste is procured from most of the thermal power plants. Based on its chemical composition the fly ash is classified as either Type- C or Type-F fly ash as seen in earlier section. The first type i.e. Type 'C' fly ash which referred to a high quantity of calcium (lime) present in it. Whereas, in Type 'F' fly ash low in calcium less than 10%, and consists of more than 60% of silane (SiO_2) content (Federal Highway Administration, 2017; Rossow, 2003). Class C fly ash is another category for fly ash and has high cementing abilities as formed by burning of sub-bituminous coal. The presence of lime content (24%) in class C fly ash in large amount. Similarly, Class (F) type fly ash are generated from the combustion of bituminous as well as anthracite coals. This Class (F) type fly ash have less than 10% lime content.

Figure 3. Classification of fly ash based on its composition

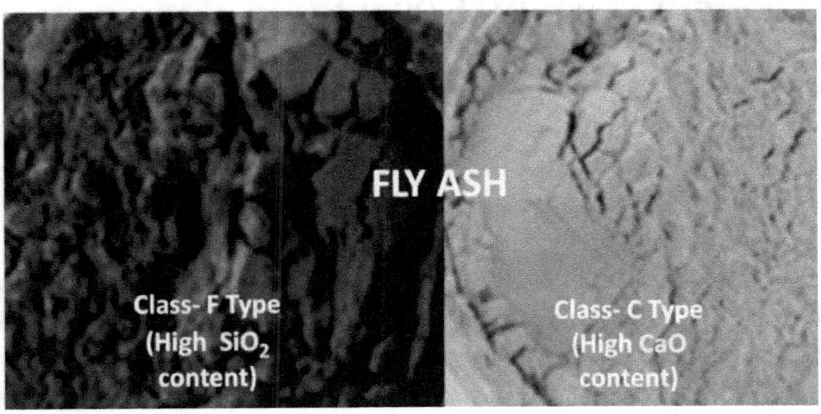

Figure 4. Different compounds present in class-C and class-F type fly ash

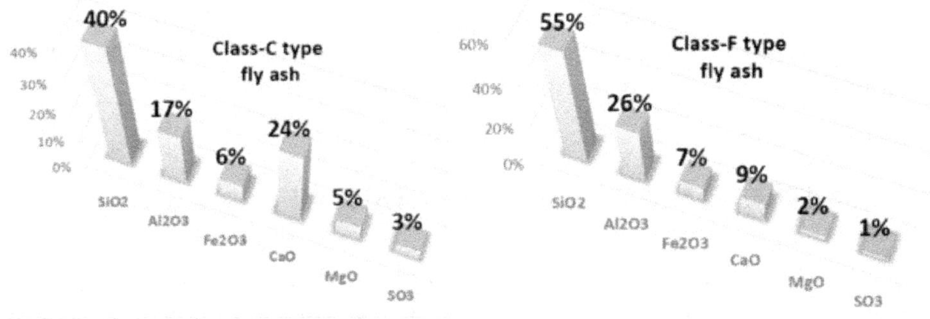

The chemical and mineral constituents determine the colour of fly ash **(Figure 3)**. As the iron content increased in fly ash gives brownish colour whereas the gray and black colour of fly ash depends upon the existence of unburned carbon content in it. Usually, the diameter of fly ash is less than 53μ. From several studies, it is found that the consistency of fly ash is floury and some are also found in fine granular texture. The fine texture of fly ash indicates the low water permeability because of its maximum dry density capacity. However, the bulk density of fly ash is low so, it can be a considered as superior material for manufacturing lightweight building blocks. But, fly ash particles results in increasing the potential of dust generation and creates problems during the transportation as well as storage of dry fly ash.

EFFECT OF WASTE FLY ASH ON THE ENVIRONMENT AND HUMAN HEALTH

The utilization of fly ash gives some environmental benefits, especially this fly ash waste in many applications such as in concrete, no need to dispose into landfills which ultimately conserve other materials and natural resources. Fly ash is dumped into landfills, lakes, rivers, and ponds. Since it is small in size and light in weight so that it can easily emit into the air by loading or unloading and transport and wind. In the air, it remains as fugitive dust. It can pollute the environment and threatens human health if it is discharged directly into the atmosphere from the power plant chimneys without any further treatment of filtration. There is also the presence of many other toxic elements in fly ash such as arsenic, mercury, Pb (lead), uranium, carcinogenic chromium (VI) compound, molybdenum (Mo), nickel (Ni), boron, Cd (cadmium), Cu (copper), and others. This results in various respiratory diseases like asthma, bronchitis, and lung cancer due to long-term inhalation of fly ash (Herndon

& Whiteside, 2017; Valavanidis et al., 2008). The particles of fly ash can also move to the heart and increase the risks of cardiac arrest. They travel to the brain via the nerves of the nose and cause neurological diseases.

A significant amount of silica is present within fly ash in both form like crystalline and amorphous forms (Singh & Subramaniam, 2018). The presence of silica element in waste fly ash is also considered as a human carcinogen. Crystalline silica can be respirable and lodges in the lungs. This causes scarring of lung tissue leads to fatal lung disease like silicosis (Golsby, 2015). Fly ash can also create pollution in lakes, rivers, and streams as well as it also increases the alkalinity of the soil and ultimately affects the pH level. Similarly, whenever it is disposed into large earthen dams it can cause many waterborne diseases. The present metals in fly ash can accumulate with lakes, rivers, and ponds which is highly toxic to water living organisms. The metal traces present in fly ash not only cause damage to the health of aquatic animals but also affects those who eat them. Immune suppression, reproductive failure, and acute poisoning will be caused by drinking the water containing the particles of fly ash (Vengosh et al., 2019).

Therefore, to overcome and to reduce the environmental damage as well as human health concern, there is a necessity to create a fly ash processing facility to utilize this material in any other form of applications. The fly ash filled polymer composites has economic and commercial importance. On considering its fruitful advantages and improvement in the properties of plastic products the fly ash could be utilized effectively as partial replacement of many other fillers.

CRITERIA AND SOLUTION FOR INTERFACIAL INTERACTION OF SILICA/POLYMER MATRIX

The important criteria for the improvement in the mechanical strength of polymer products is formation of proper interfacial interaction between fly ash particles and polymer matrix. There is chances lowering the mechanical properties of plastic products if there is weak or absence of interfacial interaction between the polymer matrix and fly ash filler (Sharma & Mahanwar, 2010). As in the fly ash, consists of different inorganic elements. Among them, the silica particles play a crucial role in interfacial interaction within the polymer matrix. In which the interaction mechanisms can occur through hydrogen bonding. And also, these interfacial interaction network shows an additional functional group of silica. The interaction process of a hydroxyl group in a silica molecule causes hydrogen bonding with the functional group of the polymer matrix. X. Y. Zhuang et al., also predicts in their study that there is a hydrogen bonding between -OH together with poly (phenylene oxide) functional group (Zhuang et al., 2016). Such silica in fly ash having a hydroxyl group also called hydrophilic silica that simply reacts to polymer matrix through the bonding process. The interfacial-interaction of silica particles with polymer matrix through the hydrogen bonding is represented in **Figure 5.**

Figure 5. An interfacial-interaction within the fly ash particles surface together with the polymer matrix

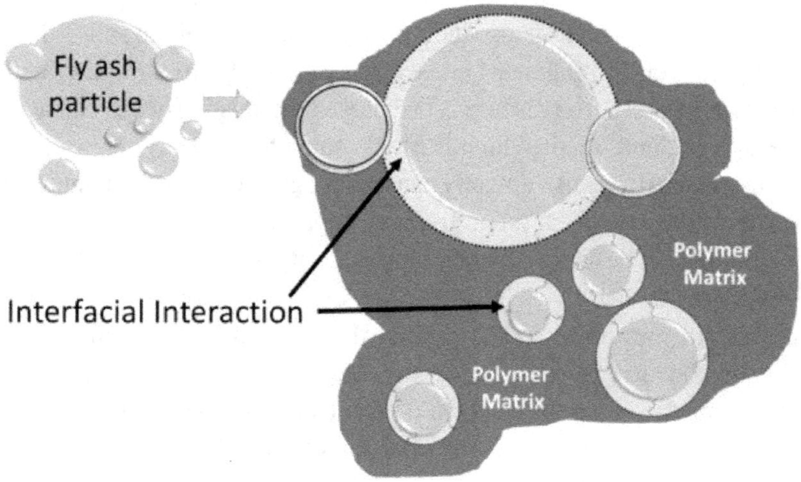

Based on the interfacial interaction mechanism, the fly ash particles are most widely used with several polymers and represent their marginal properties. Sometimes many researchers and industrialist also add some other additives to improve the interfacial interaction mechanism in between filler particles and polymer matrix. Among them, vegetable oil also has been used to create an excellent compatibility between polymer matrix and fly ash particles (Saumya et al., 2016). The surface modification of filler particles plays a crucial role in making filler more compatible with polymer matrix.

As per Y. F. Fen et al., there is lack of improvement in the strength of polymer composite as there is lower interfacial bonding between the untreated fly ash the polymer. In this regards, it is necessary to provide proper surface modification to fly ash particles using either surfactants or by proper coupling agent that leads to improvement in interfacial bonding with polymer (Yang et al., 2006). Similarly, S. Bonda et al., and O. K. Gohatre et. al., also reported that the surface treatment of fly ash particles using bis (3-triethoxysilylpropyl) tetrasulfane (Si69) i.e. silane coupling agent improved the interfacial bonding in between the polymer matrix and fly ash particles. This improvement in interfacial adhesion observed in improvement in their mechanical properties (Bonda et al., 2012; Gohatre et al., 2021). Also, the use of titanate based coupling agent (LICA 38) modified fly ash within the matrix not only improve the interfacial bonding between fly ash particles and the matrix but also enhancement in the mechanical and thermal properties of prepared polymer/ fly ash composites (Kulkarni & Mahanwar, 2016).

There are some reported coupling agent such as titanate (Parvaiz et al., 2011), silane (Atikler et al., 2006; Chand et al., 2010; Das et al., 2010; Goh et al., 2016) and surfactant (lauryl sulfate- SLS, stearic acid) (Kamal et al., 2012; Wang et al., 2011; Yao et al., 2013) that results in the better filler-matrix physical bonding with an improvement in mechanical, thermal and many other properties.

DEVELOPMENT OF POLYMERIC BASED COMPOSITES USING FLY ASH FILLER

In the current scenario, various researchers have been considering the beneficial use of fly ash with a polymer matrix, whereas the disposal of fly ash causes environmental pollution. Fly ash consists of several beads of different sizes ranging from 0.1 to 200μm. These beads can be used as fillers in many polymer composites as seen earlier section. There are many other studies shows, where the fly ash is generally applied in various polymers such as Polyvinyl chloride (PVC) (Khoshnoud & Abu-Zahra, 2019; Sushma & Kumar, 2014), High density polyethylene (HDPE) (), Epoxy (Kulkarni & Kishore, 2002) Polypropylene (PP) (Nath et al., 2009; Satapathy et al., 2011; Sengupta et al., 2011) Acryloniltrile butadiene styrene (ABS) (Kulkarni et al., 2014), Polyether ether ketone (PEEK) (Parvaiz et al., 2011) for the improvement in the properties of the same. Some of the instances regarding the effects of fly ash on different polymeric materials have been described as below.

Polypropylene/Fly Ash (PP/FA) Composites

Polypropylene (PP) is another type of thermoplastic consist tough and flexible properties and being widely used in variety of applications such as stationery products, household containers, some automotive components etc. This is because of its low price and ease of availability. Similarly PP can recycled and hence it has less impact on the environment. The study of PP/fly ash based composites exhibits good properties with low cost on the incorporation of fly ash in an appropriate amount within a polymer matrix. As we know the PP polymer shows ductile nature as used in its virgin stage. However, when the fly ash filler loading increases to PP polymer that reaches its material stiffness, more brittle and behaves like weaker material. This phenomenon of lack in properties is due to less interfacial interaction between fly ash PP. In some studies the PP composites using silane treatment containing amine and a vinyl-based functional group which produces changes in high strength as well as more stiffness to that of non-silane modified fly ash-based composites. In general, the coupling agent used as a surface modification for the fly ash particles increases the more interfacial interaction within the polymer matrix. The hardness value of composites also depends upon the interaction within the filler and matrix.

As per SEM analysis performed for PP/fly ash composite by K.Y. Wong and R.W Truss was observed from surface analysis that the bonding between the modified fly ash with polymer matrix is higher in comparison to the composites prepared by untreated fly ash particles (Wong & Truss, 1994). The PP/fly ash based composite has been intensively studied by J. Gummadi et al., and suggested that fly ash is good filler material for polypropylene matrix. From the PP/fly ash composite study it is also indicate that there is drastic change the properties of composites as fly ash particles variation occurred. (Gummadi et al., 2012). For instance, M. Ajorloo et al., suggested a feasible study of recycled PP and fly ash particles formulation for automotive parts. The formulation consist of recycled polypropylene, fly ash, rubber and some compatibilizer for better bonding achievement within polymer matrix and filler particles. From the recycling study, it was positively examined that the composite consist 20-40 wt.% of recycled PP, some mechanical properties can achieved which meets the requirement0 of some auto parts. (Ajorloo et al., 2021). In a PP/fly ash composite study performed by I. Kastiawan concluded that the smaller particle size of fly ash greatly influenced the strength and flexural properties of the composites. (Gummadi et al., 2012; Kastiawan et al., 2020). This improvement in the strength, hardness and tensile strength of composites caused by the perfectly distributed fly ash particles within the polymer matrix (Mishra & Shimpi, 2005; Saleh & Mustafa, 2011).

High-Density Polyethylene/Fly Ash (HDPE/FA) Composites

In the current situation, well-known commodity polymer i.e. polyethylene grouped material HDPE is most widely used in a domestic application. The previous study made by N. T. Chinh et al. mentioned in their study that there is a difference in the chemical and structural nature of HDPE and fly ash. However, this structural difference creates phase separation as well as incompatibility in between the HDPE/ fly ash. Therefore there is a necessity to make fly ash surface-modified so that it will impart compatibility with HDPE polymer matrix (Chinh et al., 2016). Similarly, S. Baglari also reported that there is an effective enhancement in thermal stability of HDPE/fly ash composites, as the fly ash filler is embedded within the HDPE polymer matrix. Such HDPE/fly ash composites-based material can also be used in the electronic packaging application where the heat dissipation creates a big problem (Baglari et al., 2011). I. Ahmad et al., also notified that the strength of the HDPE/fly ash composites can also be increased by using finer particle size of fly ash. The ductile properties also retain to some extend as the less than 3 μm particle size fly ash used within the HDPE matrix (Ahmad & Mahanwar, 2010). However, the loading capacity of fly ash within polymer matrix was also successfully studied and make some utilized into building materials such as lumber and roof shakes. The study made on loading up with 50-80% of fly ash within recycled polyethylene to make some special purpose building materials. By compounding of 50% of fly ash adds the stiffness and ultimately reduce the cost. Though the injection molded car trim also made by compounding of 15-30% of fly ash within polymer matrix (PT Online, n.d.).

Poly (Vinyl Chloride)/Fly Ash (PVC/FA) Composites

Poly (vinyl chloride) PVC is second most thermoplastic material commonly used in various applications such as pipe, sheet and cable/wire sheathing due to its advantageous properties. PVC can be used in its flexible as well as rigid form as it depends upon the presence and absence of plasticizer in it respectively. The plasticizer added to PVC, it is known as flexible or plasticized poly (vinyl chloride) (PPVC) and whenever plasticizer not added to PVC, called as un-plasticized poly(vinyl chloride) (UPVC). It is nowadays growing research topic for many researchers and industrialist because of its excellent fire retardency as well as low cost with chemically stable nature (Eaves, 2004; Rabinovitch et al., 1997; Thomas et al., 2004). However, apart from its fire and flame resistant properties, it has excellent water resistant property and has been used widely for making raincoats and shower curtains (Sushma & Kumar, 2014). The observation made by P. Khosnoud et al., on interfacial interaction between rigid-PVC/FA and concluded that it improved the mechanical properties of the composites. Such improvement in the mechanical properties of the composites by incorporating fly ash particles up to a certain ratio. Similarly, the observation made on the composites, reveals that the particle size of silicon dioxide and calcium carbonate within a polymer matrix plays a vital role in improving the properties of the composites. The composites prepared with the incorporation of smaller particles had higher surface area so they showed better mechanical and viscoelastic performance and microstructural properties. Since larger particles fly ash contains iron oxide so the composites reinforced with larger particles were stiffer than composites incorporated with smaller particles (Khoshnoud & Abu-Zahra, 2019). From B. Pukanszky's study, it was investigated that the composites with smaller particles had lower adhesion factors and higher interfacial parameters. Higher concentration (Si) was consistent on an interface because –OH groups significantly enhanced the interfacial adhesion and introduced good adhesion of fly ash with the PVC matrix (Pukánszky & Móczó, 2004).

Similarly, the incorporation of fly ash within waste PVC recovered from waste electrical wires and cables insulation has been studied by O. K. Gohatre et al. Generally, during waste wire and cable insulation manufacturing process, plasticizer is added to make it more flexible as well as for ease of application. Therefore, by considering the readily available plasticizer within waste PVC insulation will create additional benefits to proper wetting of fly ash within the same. From the study it has been observed that incorporation of fly ash without modification will achieved the mechanical properties of waste polymer matrix up to some extent. Whereas, there is drastic change with improvement in the same properties in very attractive manner as silane treated fly ash loaded within polymer matrix (Gohatre et al., 2020; Gohatre et al., 2021). The same study of fly ash added PVC composites has been recently suggested by many authors (Kumar et al., 2019; Van der Merwe et al., 2014; Xue et al., 2021; Zhou et al., 2012). Apart from, use of fly ash within flexible PVC, the addition of fly ash fillers will also influences the mechanical properties of rigid PVC/fly ash composites. There are many researcher apart A. W. Nugroho et al., have studied the impact and tensile strength of PVC/fly ash composites. The composites shows promising properties with improved strength and rigidity of composites with proper loading of fly ash filler. It was concluded that, the rigidity of polymer composites has been achieved due to well dispersed and excellent interaction surface area of fly ash filler within rigid PVC (Nugroho et al., 2020).

Acrylonitrile Butadiene Styrene (ABS)/ Fly Ash Composites

Acrylonitrile butadiene styrene (ABS) is a thermoplastic and consist of two phases i.e. one is amorphous phase like styrene acrylonitrile (SAN) copolymer and another phase with rubber content of polybutadiene. ABS is generally used in many engineering application due to its excellent mechanical and dimensional properties. However it is having some of its restriction in properties such as weak UV resistant and flame resistant property (Ez-Zahraoui et al., 2021). Therefore, whenever ABS is used in electrical and equipment product will add some amount of flame retardant (FR) additives to overcome this problem. Similarly, on considering the negative impact on environment of these flame retardant based ABS material after its end of life, many researchers have focusing to alter this additives with fly ash fillers to overcome this problem. The author Y. Xuan et al., also focused on ABS/FA based composite especially for three dimensional (3D) printing used to make prototype model of craft. From the study it is indicated that the use of surface modified (silane treated) fly ash within ABS matrix improves its melt flow rate as well as tensile strength but there is down in its impact strength (Xuan et al., 2019). There are very few research has been made study on ABS/fly ash composite till now. The author, S. Bonda et al. briefly studied the viscoelastic, thermo-mechanical, and behavior of fly ash incorporated ABS composites. To construct these composites ABS as a polymer matrix, fly ash as filler were used. The Bis(3-triethoxysilylpropyl) tetrasulfane and also called (Si69) was used for the surface modification of fly ash. Fly ash was undergone various treatments like surface treatment, chemical treatment, thermal treatment, and hybrid treatment.

The two different types of composites were prepared using treated and untreated fly ash polymer matrix such as ABS. From the tensile test result, it was observed that the composites showed improvement in tensile strength by 1.3 times greater with the untreated fly ash loading. As the untreated fly ash particles were incompatible with the matrix so the tensile strength reduced with the fly ash loading from 10 to 50wt%. Fly ash is a spherically shaped particle with a low aspect ratio and has little friction with the polymer chains. Therefore, it contributes to the reduction in properties. However, it was also noticed that, as the treated fly ash incorporated into the polymer matrix, the composites showed optimum thermal stability (Bonda et al., 2012).

Epoxy/ Fly Ash Composites

Epoxy polymer is widely used in various engineering applications because of its excellent heat resistant, chemical resistant, moisture resistant and it good adhesion to many substrates. Therefore, in order to improve its additional properties and with lowering the cost, fly ash filler introduce. It is also considered as the properties of fly ash filled epoxy composites affects by the characteristics of the fillers size, shape and volume fraction. T. Chaowasakoo and N. Sombatsompop studied the effect of fly ash on epoxy polymer matrix (Chaowasakoo & Sombatsompop, 2007). The basic inorganic content in fly ash that have high specific surface area, active functional groups as well as a balanced dispersion of fly ash particles makes it nano-sized additives in epoxy resin to improve mechanical and thermal properties of epoxy/fly ash based composites. It is observed from the study, the effective amount of silane coupling agent can used for getting better results in mechanical properties. Otherwise, beyond the ratio, there is remarkable reduction in the mechanical properties of prepared composites. It is also expected that thermoset based hydrophobic epoxy material with hydrophilic nature fly ash filler shows somewhat poor interfacial bonding. Therefore, the surface modified fly ash generally used within polymer matrix to get extraordinary properties of prepared composites. It is also observed that silane coupling agent causes adhesion between thermoset matrix and fly ash particles (Guhanathan et al., 2001; Şen & Nugay, 2000). Also, epoxy is generally cure by using mostly hardener with other fillers. For instance, J. Yin et al., also mentioned in their study that when FA particles added to epoxy resin, that greatly affects the curing characteristics of epoxy resin (Yin et al., 2020). It is likely to be mentioned here that the improvement in the resistance to degradation (acidic or basic) condition can also achieved in fly ash filler filled epoxy resin as that of unfilled epoxy resin. The presence of void and density of composites is minimized with increase in fly ash filler content in epoxy resin (Nagendiran et al., 2016).

FEASIBLE APPLICATION OF POLYMER AND FLY ASH FILLER BASED COMPOSITES MATERIAL

The fly ash mixed polymer products are most widely used in our daily routine. Some useful products such as plastic crates, some municipality dust bins, boxes. The fly ash has been used with different types of thermoplastic. The schematic representation of the proper mixing of fly ash and polymer matrix is as shown in **Figure 6**. During fly ash incorporation the size and shape of fly ash particles mostly consider. The size and shape of fly ash particles determine the physical properties of the respective polymer matrix. It has been also observed that such inorganic-based filler causes poor interfacial interaction and adhesion with a polymer matrix. However, to overcome such poor adhesion of fly ash within a polymer matrix, the fly ash is mostly surface modified or treated with some silane-based coupling agent. This coupling agent created a bridging effect within that fly ash surface to that polymer matrix by providing proper interfacial interaction within them. The proper fly ash to polymer matrix determined the accurate mixing and interfacial interaction of polymer matrix with fly ash filler. There is a decrease in melt viscosity with more brittle properties as the concentration of fly ash filler increases within the polymer matrix (Pardo et al., 2010). The fly ash incorporated polymer composites have better physical and mechanical properties. So it can be used in various construction industries. As fly ash is already a scrap material so its cost is very less. So that it can be easier to be utilized fillers in the manufacturing of tiles floorings in the different polymeric sectors. Products made from fly ash incorporated polymer composites will be more strain resistant and more durable.

Similarly, the current craze of utilization of fly ash within polymer matrix increase nowadays to manufacturing tiles. Tiles manufacturing using waste fly ash considered as cost-effective as compared to clay tiles and also helps to irrigable topsoil. Accordingly, plastic products should show promising properties as per the demand or necessity. In such case, many of the plastic product manufacturing industries use additives and fillers in plastic products to take advantage of plastic product. In such cases many costly and some hazardous material used as fillers and additives in plastic which ultimately creates a negative impact on the environment as this plastic products gets outdated. So, the waste fly ash material is the best option for such costs as well as hazardous fillers. Many of the polymer researchers get attention towards the use of fly ash.

However, the incorporation of fly ash waste as filler provides a promisable improvement in the mechanical properties of polymer material. The enhancement in mechanical properties such as tensile strength and modulus of plastic products gets enhanced by using the proper incorporation of fly ash within a polymer matrix.

Figure 6. Current use of fly ash in plastic industries to get desired properties products

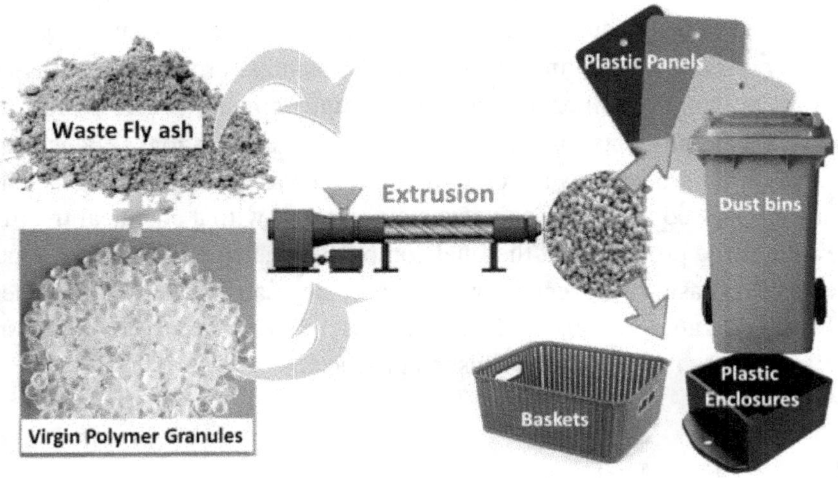

ADVANTAGES AND DISADVANTAGES OF UTILIZATION OF FLY ASH WITHIN POLYMER MATRIX

On considering the properties of fly ash, is much similar to the calcium carbonate with significant difference whenever it is used in polymer matrix. Fly ash consist of fine powder of small balls without any sharp edges those are present in calcium carbonate. However, fly ash is less dense on comparing calcium carbonate and thus it is usually replace by fly ash. These fine sphere particles provides stiffness to the plastic material but at same time increased the stiffness when excess in quantity. Similarly, it is also found that during processing these fly ash spheres are acts as ball bearing and starts improvement in the throughput.

Also, on considering advantages of fly ash usage, it is also having some drawback. The colour is the major drawback of fly ash whenever used in plastic products. The fly ash found in dirty gray and tan in colour whereas, $CaCO_3$ is much brighter white in colour than fly ash. Consequently, the availability of fly ash is not in proper particle size and can make cost twice when it is refined and also there is presence of some undesirable residues such as unburned carbon causes absorption of oil and resin.

CONCLUSION

On considering an increasing demand for energy all around the world, it looks that coal-based electricity generation will increase in tremendous manner in upcoming years. This improvement in electricity generation will increase in developing countries. Thus by keeping in view the issues generated by fly ash as well as for proper disposal. Although the fly ash material has several advantages over its some disadvantages and can be used due to its improvement in mechanical properties as it is utilised in various polymers as filler material. Such improvement in several properties of prepared plastic products makes its use in different applications. Fly ash can also play an important role as raw material for different plastics applications. Such waste fly ash can be used according to its particle size and with a chemical treatment to achieved desired properties of the final composite material. In the view of positive side of waste fly ash whenever used within a polymer matrix that ultimately increases the fly ash utilization and also reduces the consumption of many mineral resources as well as reduce the cost of the finished products.

ACKNOWLEDGMENT

This research received no specific grant from any funding agency in the public, commercial, or not-for-profit sectors.

REFERENCES

Ahmad, I., & Mahanwar, P. A. (2010). Mechanical properties of fly ash filled high density polyethylene. *Journal of Minerals & Materials Characterization & Engineering*, *9*(03), 183–198. doi:10.4236/jmmce.2010.93016

Ajorloo, M., Ghodrat, M., & Kang, W. H. (2021). Incorporation of recycled polypropylene and fly ash in polypropylene-based composites for automotive applications. *Journal of Polymers and the Environment*, *29*(4), 1298–1309. doi:10.100710924-020-01961-y

Akhtar, M. N., & Tarannum, N. (2018). Flyash as a resource material in construction industry: A clean approach to environment management. In *Sustainable construction and building materials*. IntechOpen.

Andrady, A. L., & Neal, M. A. (2009). Applications and societal benefits of plastics. *Philosophical Transactions of the Royal Society of London. Series B, Biological Sciences*, *364*(1526), 1977–1984. doi:10.1098/rstb.2008.0304 PMID:19528050

Atikler, U., Basalp, D., & Tihminlioğlu, F. (2006). Mechanical and morphological properties of recycled high-density polyethylene, filled with calcium carbonate and fly ash. *Journal of Applied Polymer Science*, *102*(5), 4460–4467. doi:10.1002/app.24772

Attarde, S., Marathe, S., & Sil, A. (2014). Utilization of fly ash in construction industries for environment management. *International Journal of Environmental*, *3*(2), 117–121.

Baglari, S., Kole, M., & Dey, T. K. (2011). Effective thermal conductivity and coefficient of linear thermal expansion of high-density polyethylene—Fly ash composites. *Indian Journal of Physics*, *85*(4), 559–573. doi:10.100712648-011-0059-x

Bonda, S., Mohanty, S., & Nayak, S. K. (2012). Viscoelastic, mechanical, and thermal characterization of fly ash-filled ABS composites and comparison of fly ash surface treatments. *Polymer Composites*, *33*(1), 22–34. doi:10.1002/pc.21242

Chand, N., Sharma, P., & Fahim, M. (2010). Correlation of mechanical and tribological properties of organosilane modified cenosphere filled high density polyethylene. *Materials Science and Engineering A*, *527*(21-22), 5873–5878. doi:10.1016/j.msea.2010.06.022

Chaowasakoo, T., & Sombatsompop, N. (2007). Mechanical and morphological properties of fly ash/epoxy composites using conventional thermal and microwave curing methods. *Composites Science and Technology, 67*(11-12), 2282-2291.

Chinh, N. T., Mai, T. T., Trang, N. T. T., Giang, N. V., Trung, T. H., Huong, N. T. T., & Hoang, T. (2016). Tensile, electrical properties and morphology of polyethylene/modified fly ash composites using ultraflow. *Vietnam Journal of Chemistry*, *54*(6), 776.

Das, K., Ray, D., Adhikary, K., Bandyopadhyay, N. R., Mohanty, A. K., & Misra, M. (2010). Development of recycled polypropylene matrix composites reinforced with fly ash. *Journal of Reinforced Plastics and Composites*, *29*(4), 510–517. doi:10.1177/0731684408099415

Deepthi, M. V., Sharma, M., Sailaja, R. R. N., Anantha, P., Sampathkumaran, P., & Seetharamu, S. (2010). Mechanical and thermal characteristics of high density polyethylene–fly ash cenospheres composites. *Materials & Design*, *31*(4), 2051–2060. doi:10.1016/j.matdes.2009.10.014

Eaves, D. (2004). *Handbook of polymer foams*. Rapra Technology Ltd.

Ez-Zahraoui, S., Kassab, Z., Ablouh, E. H., Sehaqui, H., Bouhfid, R., Alami, J., & Qaiss, A. E. K. (2021). Effect of fly ash and coupling agent on the structural, morphological, thermal, and mechanical properties of polyamide 6/acrylonitrile-butadiene-styrene blend. *Polymer Composites*.

Federal Highway Administration. (2017). *Department of Transportation's Federal Highway Administration Privacy Policy*. https://www.google.nl/ url?sa = t&source = web & rct = j&url = https:// www.fhwa.dot.gov/ pavement/ recycling/ fach01. Cfm & ved = 2ahU KEwj okpa 7wZn uAhW s7XM BHVQ SAWo QFjA jegQ IKhAB & usg = AOvV aw3p G8U4 gmdq U8iK NvRH 6xR6

Fly ash market-growth, trends, covid-19 impact and forecasts 2021-2026. (n.d.). https://www.mordorintelligence.com/industry-reports/fly-ash-market

Fly ash market size, share & industry analysis, by type (class F and Class C), by application (cement & concretes fills & embarkments, waste stabilization, mining, oil field service and road stabilization and others) and regional forecast, 2020-2027. (2020). https://www.fortunebusinessinsights.com/industry-reports/fly-ash-market-101087-

Globe News Wire. (2018). https://www.globenewswire.com/ news- release/ 2018/ 10/ 04/ 1601143/ 0/en/ Global- Fly- Ash- Market- Will- Reach- USD- 11-0- Billion-by- 2024- Zion- Market- Research.html

Goh, C. K., Valavan, S. E., Low, T. K., & Tang, L. H. (2016). Effects of different surface modification and contents on municipal solid waste incineration fly ash/ epoxy composites. *Waste Management, 58*, 309-315.

Gohatre, O. K., Biswal, M., Mohanty, S., & Nayak, S. K. (2020). Study on thermal, mechanical and morphological properties of recycled poly (vinyl chloride)/fly ash composites. *Polymer International*, *69*(6), 552–563. doi:10.1002/pi.5988

Gohatre, O. K., Biswal, M., Mohanty, S., & Nayak, S. K. (2021). Effect of silane treated fly ash on physico-mechanical, morphological, and thermal properties of recycled poly (vinyl chloride) composites. *Journal of Applied Polymer Science*, *138*(19), 50387. doi:10.1002/app.50387

Golsby, A. (2015). *New Generation Polymer Technology Available to be Used in Plaster and Cementitious Fibre and Non-Fibre Spray On and Grout Systems.* Academic Press.

Guhanathan, S., Devi, M. S., & Murugesan, V. (2001). Effect of coupling agents on the mechanical properties of fly ash/polyester particulate composites. *Journal of Applied Polymer Science*, *82*(7), 1755–1760. doi:10.1002/app.2017

Gummadi, J., Kumar, G. V., & Rajesh, G. (2012). Evaluation of flexural properties of fly ash filled polypropylene composites. *International Journal of Modern Engineering Research*, *2*(4), 2584–2590.

Herndon, J. M., & Whiteside, M. (2017). Further evidence of coal fly ash utilization in tropospheric geoengineering: Implications on human and environmental health. *J Geog Environ Earth Sci Intn*, *9*(1), 1–8. doi:10.9734/JGEESI/2017/31417

Kamal, M., Sharma, C. S., Upadhyaya, P., Verma, V., Pandey, K. N., Kumar, V., & Agrawal, D. D. (2012). Calcium carbonate (CaCO3) nanoparticle filled polypropylene: Effect of particle surface treatment on mechanical, thermal, and morphological performance of composites. *Journal of Applied Polymer Science*, *124*(4), 2649–2656. doi:10.1002/app.35319

Kastiawan, I., Sutantra, I. N., & Sutikno. (2020). Correlation of Holding Time and Bottom Ash Particle Size to Mechanical Properties of Polypropylene Composite. *Key Engineering Materials*, *867*, 172–181. doi:10.4028/www.scientific.net/KEM.867.172

Khoshnoud, P., & Abu-Zahra, N. (2019). The effect of particle size of fly ash (FA) on the interfacial interaction and performance of PVC/FA composites. *Journal of Vinyl and Additive Technology*, *25*(2), 134–143. doi:10.1002/vnl.21633

Kulkarni, M. B., Bambole, V. A., & Mahanwar, P. A. (2014). Effect of particle size of fly ash cenospheres on the properties of acrylonitrile butadiene styrene-filled composites. *Journal of Thermoplastic Composite Materials*, *27*(2), 251–267. doi:10.1177/0892705712443253

Kulkarni, M. B., & Mahanwar, P. A. (2016). Studies on effect of titanate-coupling agent (0.5, 1.5, and 2.5%) on the mechanical, thermal, and morphological properties of fly ash–filled polypropylene composites. *Journal of Thermoplastic Composite Materials*, *29*(3), 344–365. doi:10.1177/0892705713518795

Kulkarni, S. M., & Kishore. (2002). Effects of surface treatments and size of fly ash particles on the compressive properties of epoxy based particulate composites. *Journal of Materials Science*, *37*(20), 4321–4326. doi:10.1023/A:1020648418233

Kumar, P. N., Nagappan, V., & Karthikeyan, C. (2019). Effects of fly ash, calcium carbonate fillers on mechanical, moisture absorption properties in poly vinyl chloride resin. *Materials Today: Proceedings, 16*, 1219–1225. doi:10.1016/j.matpr.2019.05.217

Mishra, S., & Shimpi, N. G. (2005). *Comparison of nano CaCO 3 and flyash filled with styrene butadiene rubber on mechanical and thermal properties*. Academic Press.

Nagendiran, S., Badghaish, A., Hussein, I. A., Shuaib, A. N., Furquan, S. A., & Al-Mehthel, M. H. (2016). Epoxy/oil fly ash composites prepared through in situ polymerization: Enhancement of thermal and mechanical properties. *Polymer Composites, 37*(2), 512–522. doi:10.1002/pc.23207

Nath, D. C. D., Bandyopadhyay, S., Yu, A., Zeng, Q., Das, T., Blackburn, D., & White, C. (2009). Structure–property interface correlation of fly ash–isotactic polypropylene composites. *Journal of Materials Science, 44*(22), 6078–6089. doi:10.100710853-009-3839-3

Nugroho, A. W., Prasetyo, M. K. P., & Budiyantoro, C. (2020). Effect of Fly Ash on the Mechanical Properties of Polyvinyl Chloride-Fly Ash Composite. In *Proceedings of the 6th International Conference and Exhibition on Sustainable Energy and Advanced Materials* (pp. 667-674). Springer. 10.1007/978-981-15-4481-1_63

Pardo, S. G., Bernal, C., Ares, A., Abad, M. J., & Cano, J. (2010). Rheological, thermal, and mechanical characterization of fly ash-thermoplastic composites with different coupling agents. *Polymer Composites, 31*(10), 1722–1730. doi:10.1002/pc.20962

Parvaiz, M. R., Mohanty, S., Nayak, S. K., & Mahanwar, P. A. (2011). Effect of surface modification of fly ash on the mechanical, thermal, electrical and morphological properties of polyetheretherketone composites. *Materials Science and Engineering A, 528*(13-14), 4277–4286. doi:10.1016/j.msea.2011.01.026

PT Online. (n.d.). https://www.ptonline.com/articles/fly-ash-filler-stages-a-comeback

Pukánszky, B., & Móczó, J. (2004, August). Morphology and properties of particulate filled polymers. *Macromolecular Symposia, 214*(1), 115–134. doi:10.1002/masy.200451009

Rabinovitch, E. B., Isner, J. D., Sidor, J. A., & Wiedl, D. J. (1997). Effect of extrusion conditions on rigid PVC foam. *Journal of Vinyl and Additive Technology, 3*(3), 210–215. doi:10.1002/vnl.10193

Rossow, M. (2003). *Fly Ash Facts for Highway Engineers*. New York NY Continuing Education and Development.

Saleh, N. J., & Mustafa, S. N. (2011). A study of some mechanical, thermal and physical properties of polymer blend with Iraqi kaolin filler. *Eng. & Tech. Journal*, *29*(11), 2114–2132.

Satapathy, B. K., Das, A., & Patnaik, A. (2011). Ductile-to-brittle transition in cenosphere-filled polypropylene composites. *Journal of Materials Science*, *46*(6), 1963–1974. doi:10.100710853-010-5032-0

Saumya, K. M., Alam, M., & Hussain, A. (2016). Effect of fly ash particle and vegetable oil on the mechanical properties of fly ash-vegetable oil reinforced hard PVC plastic. *Int Res J Eng Technol*, *3*, 833–839.

Şen, S., & Nugay, N. (2000). Uncured and cured state properties of fly ash filled unsaturated polyester composites. *Journal of Applied Polymer Science*, *77*(5), 1128–1136. doi:10.1002/1097-4628(20000801)77:5<1128::AID-APP21>3.0.CO;2-D

Sengupta, S., Pal, K., Ray, D., & Mukhopadhyay, A. (2011). Furfuryl palmitate coated fly ash used as filler in recycled polypropylene matrix composites. *Composites. Part B, Engineering*, *42*(7), 1834–1839. doi:10.1016/j.compositesb.2011.06.021

Sharma, A. K., & Mahanwar, P. A. (2010). Effect of particle size of fly ash on recycled poly (ethylene terephthalate)/fly ash composites. *International Journal of Plastics Technology*, *14*(1), 53–64. doi:10.100712588-010-0006-2

Shubham, P., & Tiwari, S. K. (2013). Effect of fly ash concentration and its surface modification on fiber reinforced epoxy composite's mechanical properties. *IJSER*, *4*, 1173–1180.

Singh, G. B., & Subramaniam, K. V. (2018). Characterization of Indian fly ashes using different experimental techniques. *Indian Concrete Journal*, *92*(3), 10–23.

Surabhi, S. (2017). Fly ash in India: Generation vis-a-vis utilization and global perspective. *Int J Appl Chem*, *13*(1), 29–52.

Sushma, S. P., & Kumar, A. K. (2014). Hardness and tensile testing of PVC and fly ash composite. *Advance research and innovations in mechanical, material science, industrial engineering and management-ICARMMIEM*, 273-277.

Swanepoel, J. C., & Strydom, C. A. (2002). Utilisation of fly ash in a geopolymeric material. *Applied Geochemistry*, *17*(8), 1143–1148. doi:10.1016/S0883-2927(02)00005-7

Thomas, N. L., Eastupy, J. R., & Quirk, P. (2004). *Rigid PVC foam. Formulating for sustainability, Blowing agent and foaming process*. Academic Press.

Valavanidis, A., Fiotakis, K., & Vlachogianni, T. (2008). Airborne particulate matter and human health: Toxicological assessment and importance of size and composition of particles for oxidative damage and carcinogenic mechanisms. *Journal of Environmental Science and Health. Part C, Environmental Carcinogenesis & Ecotoxicology Reviews*, *26*(4), 339–362. doi:10.1080/10590500802494538 PMID:19034792

Van der Merwe, E. M., Mathebula, C. L., & Prinsloo, L. C. (2014). Characterization of the surface and physical properties of South African coal fly ash modified by sodium lauryl sulphate (SLS) for applications in PVC composites. *Powder Technology*, *266*, 70–78. doi:10.1016/j.powtec.2014.06.008

Vengosh, A., Cowan, E. A., Coyte, R. M., Kondash, A. J., Wang, Z., Brandt, J. E., & Dwyer, G. S. (2019). Evidence for unmonitored coal ash spills in Sutton Lake, North Carolina: Implications for contamination of lake ecosystems. *The Science of the Total Environment*, *686*, 1090–1103. doi:10.1016/j.scitotenv.2019.05.188 PMID:31200305

Wang, Z. H., Lu, Y. L., Liu, J., Dang, Z. M., Zhang, L. Q., & Wang, W. (2011). Preparation of nanoalumina/EPDM composites with good performance in thermal conductivity and mechanical properties. *Polymers for Advanced Technologies*, *22*(12), 2302–2310. doi:10.1002/pat.1761

Wong, K. Y., & Truss, R. W. (1994). Effect of flyash content and coupling agent on the mechanical properties of flyash-filled polypropylene. *Composites Science and Technology*, *52*(3), 361–368. doi:10.1016/0266-3538(94)90170-8

Xuan, Y., Huajin, T., & Zhiwen, Q. (2019, October). Effect of mixing time of coupling agent on properties of 3D printing prototype of fly ash filled ABS. *IOP Conference Series. Materials Science and Engineering*, *631*(3), 032025. doi:10.1088/1757-899X/631/3/032025

Xue, C., Nan, H., Lu, Y., Chen, H., Zhao, C., & Xu, S. (2021). Effects of inorganic-organic surface modification on the mechanical and thermal properties of poly (vinyl chloride) composites reinforced with fly-ash. *Polymer Composites*, *42*(4), 1867–1877. doi:10.1002/pc.25942

Yang, Y. F., Gai, G. S., Cai, Z. F., & Chen, Q. R. (2006). Surface modification of purified fly ash and application in polymer. *Journal of Hazardous Materials*, *133*(1-3), 276–282. doi:10.1016/j.jhazmat.2005.10.028 PMID:16314041

Yao, N., Zhang, P., Song, L., Kang, M., Lu, Z., & Zheng, R. (2013). Stearic acid coating on circulating fluidized bed combustion fly ashes and its effect on the mechanical performance of polymer composites. *Applied Surface Science*, *279*, 109–115. doi:10.1016/j.apsusc.2013.04.045

Yin, J., Teng, Y., Meng, X., Ge, Y., & Zhang, X. (2020). Influence of fly ash on the curing characteristics of an epoxy resin. *Polymers & Polymer Composites*, *28*(8-9), 579–588. doi:10.1177/0967391119894083

Zhou, Y. S., Liu, S., Liu, J. H., Sun, J. L., & Zhou, M. J. (2012). Rheological and Mechanical Properties of Fly Ash Microspheres Modified PVC. *Plastics*.

Zhuang, X. Y., Chen, L., Komarneni, S., Zhou, C. H., Tong, D. S., Yang, H. M., Yu, W. H., & Wang, H. (2016). Fly ash-based geopolymer: Clean production, properties and applications. *Journal of Cleaner Production*, *125*, 253–267. doi:10.1016/j.jclepro.2016.03.019

Chapter 2
HyLaw or Hydrogen Law: A Regulation for Removal of Legal Barriers to the Deployment of Fuel Cells and Hydrogen Applications

Nima Norouzi
https://orcid.org/0000-0002-2546-4288
Bournemouth University, UK

ABSTRACT

In recent years, hydrogen technology has been at the forefront of environmental discussions to meet increasingly tough climate protection goals and particularly low emissions targets in the transportation sector. Like any major change, a transition to hydrogen energy faced challenges in many countries, which caused several problems in the growth of the hydrogen share of the total energy supply portfolio. In 2018, Hydrogen Law (Hylaw) was introduced, which removes the legal barriers to the deployment of fuel cells and hydrogen applications. It is a flagship project aimed at boosting the market uptake of hydrogen and fuel cell technologies providing market developers with a clear view of the applicable regulations while calling policymakers' attention to legal barriers to be removed. This chapter introduces a consistent framework for the Hylaw regulations that makes is a clear and precise statement and an interconnection between law and energy management policies.

DOI: 10.4018/978-1-7998-9851-1.ch002

INTRODUCTION

A consensus is fast emerging that hydrogen will play a key role as an energy vector and a pillar in the ongoing energy transition. It promises to accelerate transformative changes across many sectors, most notably energy and transport. This chapter draws together some of the most experienced global energy experts' insight to provide a timely and insightful perspective on how hydrogen projects may proceed and the sector develops worldwide (Abánades et al., 2013; Abbasi & Abbasi, 2011).

Energy lawyers are accustomed to the emergence of new technologies(Ajanovic, 2008; Alazemi & Andrews, 2015). Nevertheless, each emergent technology's unique characteristics need to be respected. It would be complacent to think that hydrogen can be treated like natural gas or other energy sources for legal and regulatory frameworks, investment cases, financing structures, operational requirements, revenue stream arrangements, and the panoply of other elements that need to be considered to formulate an effective commercialization model (Andress et al., 2009; Arnason & Sigfusson, 2000).

The term "hydrogen economy" is not new, but the role that hydrogen can, and is expected, to play in the economies of many of the jurisdictions covered in this chapter demonstrates the revitalized ambitions of this subsector. But this chapter also highlights the fact that progress is not equal in all places. An emerging suite of technologies and an immature web of policy and regulatory frameworks in some jurisdictions are developing quickly into a supportive system ready to welcome private sector investment in other countries. What is clear is that the promise of hydrogen developments and uses is rapidly evolving as governments and market players are waking up to its benefits and potential. With many countries committing to having major low-carbon hydrogen projects underway by 2030 and committing to achieving net-zero targets, investors have to take a truly global perspective on the sector (Baker, 1980; Balat, 2008). This chapter sets out the ease (or otherwise) of developing hydrogen projects across the jurisdictions covered – highlighting the status of hydrogen developments in each country; considering the market prospects and opportunities ahead that are key for our clients who are seeking to enter or expand in this sector; what challenges need to be overcome to reach national and international goals and how the national and international specific legislation and regulations in each jurisdiction facilitate this growing sector (Ball & Weeda, 2015).

Environmental Policy

Supranational policies and frameworks helpfully chapter the longer-term direction and developments at national levels. In this case, the supranational commitment is made through the Paris Agreement by 189 countries, representing 97% of global emissions (Barclay, 2006). All of the countries covered in this chapter are signatories to the Paris Agreement (albeit the US has notified the United Nations of its intention to withdraw from the Paris Agreement). The Paris Agreement is one of the most ambitious international agreements within the United National Framework Convention on Climate Change ("UNFCC") (Bargigli et al., 2004). It commits signatories to respond to the threat of climate change by keeping any global temperature rise this century to well below 2 degrees Celsius above pre-industrial levels, and better still, to pursue efforts to limit the temperature increase to 1.5 degrees Celsius above those levels. Since then, several countries have adopted legally binding targets to reach "net-zero" in their greenhouse gas emissions by 2050(Beghi, 1983). Alongside national governments, similar commitments are being made by major businesses and investors, who are also seeking to decarbonize their products and processes (Bertel et al., 2004).

The countries covered in this chapter have or are in the process of creating legal frameworks to support their vision. For many, this vision includes hydrogen playing a key role in achieving their Paris Agreement climate change ambitions and net-zero targets in several sectors, notably transportation, heating, and industry (Bhandari et al., 2014).

Figure 1. Framework of hydrogen law (HyLaw) (adopted from cms law)

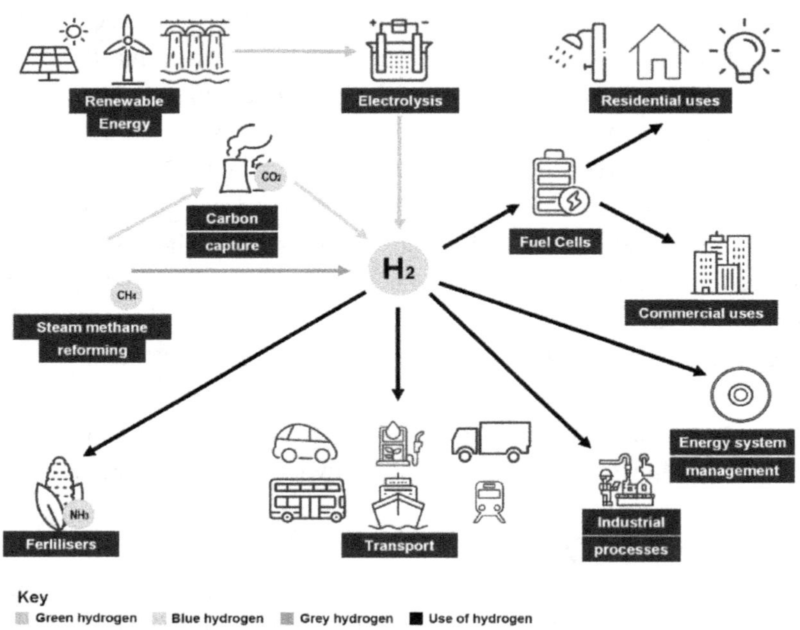

DIFFERENT LEGAL REGIMES FOR DIFFERENT "TYPES" OF HYDROGEN

From investors' perspective in the sector, it is important to recognize that hydrogen is classified according to its production while the molecules are indistinguishable. References to the following colors are for hydrogen produced in turn: "grey" from methane gas, "black" from coal, or "brown" from lignite. Around 95% of the world's hydrogen is grey or brown, 76% from natural gas, and 23% from coal(Bose & Malbrunot, 2007). "Blue" hydrogen is produced using methane gas with carbon capture and storage technology. This type of hydrogen is seen as carbon neutral but not strictly renewable. "Green" hydrogen is produced through electrolysis, which is the process of splitting up water (H_2O) into hydrogen and oxygen, using renewable energy (for example, wind or solar energy). Several other "colors" of hydrogen are less relevant from an investor's perspective and are not covered in this chapter (Chiesa et al., 2005; Cipriani et al., 2014; Cigolotti et al., 2018).

It is understood that a move away from grey hydrogen will be necessary to meet the decarbonization targets many countries have set out. However, blue hydrogen's attraction comes from the scale of production that it offers and the ability to use carbon capture storage technologies to prevent carbon dioxide emissions into the atmosphere. Many of the jurisdictions covered in this chapter focus on the advancement of green and blue hydrogen(Cohce et al., 2010).

Table 1. Hydrogen regulation type

Green Hydrogen	Blue Hydrogen
Green hydrogen is typically made through electrolysis - which in its simplest terms requires an electrolyzer to break down water (H2O) into hydrogen and water using renewable sources such as wind, solar and hydro to generate the electricity used for the process (Cox & Williamson, 1997).	Blue hydrogen is seen by many as an enabler in the commercial development of low-carbon hydrogen projects. Its success depends on the role of carbon capture usage and storage (CCUS), which suits some countries (such as the UK, Netherlands, Norway, and parts of the US) better than others that do not already have storage infrastructure exploit.
An example is the Austrian Hydrogen mobility AG, which announced green hydrogen produced from wood gasification at previously unattainable production and energy efficiency.	Several hydrogen plus CCUS projects and business models are under consideration and covered in the chapter. For example, the UK specifically develops business models for hydrogen and CCUS projects over the coming years(Dalebrook et al., 2013).
While many on-station projects are being developed; as a result, just 2% of global hydrogen production is currently produced by electrolysis. However, green hydrogen is perceived as the ultimate goal, and there are new national and supranational policies emerging that promote the use of electrolyzers (Iordache et al., 2018).	The UK is not alone in having spotted the blue hydrogen opportunity. In the Netherlands, led by a consortium comprising state-owned companies, the Porthos project focuses on capturing CO_2 within Rotterdam's port from existing hydrogen production facilities to produce large-scale blue hydrogen and reduce emissions 2030. The captured CO_2 will be transported and stored in a depleted gas field in the North Sea(Kreutz et al., 2005).

The taxonomy may seem a geeky subject, but classification and legal precision are crucial. In particular, the lack of harmonization and classification is a common complaint whether from financial institutions, developers, or investors seeking to understand and compare opportunities in the sector. There are some developments in this area. For example, the EU Taxonomy 3, published in March 2020, considers what should be treated as "low carbon" or "renewable." In this case, for the manufacture of hydrogen, hydrogen is considered renewable where (Kohl, 2018):

- The level of direct CO_2 emissions from the manufacturing of hydrogen is 5.8 tCO_2e/t;
- Electricity use for hydrogen produced by electrolysis is at or lower than 58 MWh/t; and
- The average carbon intensity of the electricity produced for hydrogen manufacturing is at or below 100 gCO_2e/kWh.

In practice, investors, developers, financiers, and advisors seeking to invest in renewable hydrogen projects should note that this narrows the options to "green" and "blue" hydrogen."

Legal System in Transportation

Hydrogen has the potential to play a significant role in the transportation sector. The International Energy Agency ("IEA") estimates while batteries are seen as a viable technology for passenger vehicles, as seen through the uptake of electric vehicles ("EV") globally, hydrogen-based mobility is a complimentary option. Hydrogen can be applied in various sectors such as local public passenger transport, heavy-duty road transport, commercial vehicles, marine, rail, and even aviation. The chapter has highlighted several jurisdictions (notably Korea and Japan) where fuel cell technology ("FCEV") shows significant prospects for use in road and rail transport (McCarty et al., 2018).

This raises several legal issues for companies seeking to enter the hydrogen mobility sector. Despite the first hydrogen refueling station installed over ten years ago, only a few of the countries we surveyed have specific hydrogen refueling legislation. In this uneven landscape, while in Germany, Denmark, the UK, and the Netherlands, authorities can look to existing rules. Several different authorities may need to develop the legal and regulatory framework for hydrogen in other countries. Further, it is more difficult to navigate the permitting and regulatory regimes without existing rules, with the process slowed down if the authorities are less familiar with the issues (Movahedian et al., 2021).

The opportunity to educate, create and harmonize some standards may be afforded by focusing on hydrogen buses' roll-out. Many of the countries we surveyed focus on public transport as the key area for using hydrogen in the mobility sector. The UK has already seen the roll-out of zero-emission hydrogen buses and has committed to introducing 3,000 buses by 2024. Similarly, Japan is aiming to have 1,200 fuel-cell buses on the road by 2030. More than decarbonization purposes, hydrogen-fuelled public transport is gaining popularity due to the added benefits of improving local air quality and bringing public health benefits (Norouzi, 2021a).

Given that contracting for public transport and associated infrastructure often involves a governmental authority as the counterparty (whether for the land rights, the leases of the transport, etc.), buses could be the catalyst for the administrative knowledge build-up processes. With internments are For example, if state funding is available to encourage individuals to purchase hydrogen-fuelled vehicles, state aid and procurement issues will need to be carefully considered and managed. Assuming this can be done purposefully and effectively, hydrogen fuel transport could mirror the rapid uptake seen in EVs in some countries. Though our experts agree that significant changes to the existing practices would be needed in these areas at an earlier stage of development, hydrogen technologies could also be used in shipping and aviation (Norouzi et al., 2021).

Legal System in Industry

Unsurprisingly the chapter underscores the significant hydrogen demand in industrial processes. The IEA estimates that the total global demand for hydrogen will be around 40 million tonnes per year over the coming decade(Norouzi & Ataei, 2021a). This includes using hydrogen in various industrial processes, including oil refining, ammonia production, and steel production. The hydrogen used in these industries is generally "grey," As such, there is a large opportunity to decarbonize this sector by transitioning to blue or green hydrogen. One such example is the HyNet project in North West England, which proposes to develop a hydrogen cluster in which ten large industrial sites would be converted to use 100% green hydrogen (Norouzi & Ataei, 2021b).

The challenge with adapting and retrofitting existing infrastructure to produce low carbon hydrogen is not without its challenges. As with much else in this developing sector, first-of-its-kind projects act as pathfinders in the administrative, permitting, and regulatory processes. The time and complexity are a burden that needs to be addressed to facilitate further developments and attract outside investments and financings. As it is already used in the manufacturing process, there are fewer novel technical barriers to consider. Nonetheless, based on current technology, the production cost is inevitably higher, which, combined with the economic and regulatory challenges that require thorough understanding and review, are delaying unlocking a sector suited for this transformation(Norouzi, 2021b).

Legal System in Heating

Several countries have identified hydrogen as part of their plan for decarbonizing heating. In the UK, for example, the gas network provides natural gas to over 80% of residential homes and commercial buildings for heating(Norouzi, 2021c). Decarbonizing this network could be paramount to achieving net-zero aims. Countries like Portugal, Germany, and France, have initiated pilot projects to test hydrogen blending into the gas grid. Some, such as the Netherlands, intended to eliminate natural gas use in the built environment by 2050 (Pandev et al., 2017). Several demonstration projects currently test the blending of up to 20% of hydrogen into the existing gas grid from a technical and engineering perspective. What is less clear is how the technical and safety tests about the limitations of hydrogen blending link with the current legal frameworks. At present, while there is a harmonized legal regime for ownership and operation of the gas networks (though the EU Third Package Directive), the legislation's design is firmly centered around methane gas and gas quality standard (based on calorific value; Wobbe Index) (Schitea et al., 2018). However, blending hydrogen changes the calorific value of gas carried in the grid. The existing ways of regulating gas transmission, the payment terms for the various entities involved, and questions of gas quality standards impacted by blending and deblending hydrogen into/out of the grid require further consideration. As identified in our chapter, in several jurisdictions, the gas transmission rules were drafted into law decades ago, well before the opportunities for heating with hydrogen, let alone having a 100% hydrogen gas network, were contemplated. If the ambition is to enhance the prospects of using hydrogen networks for heating, further work and studies are needed to link the safety and technical integrity requirements in all parts of the heating chain, including the end-user interface, with the current regulatory frameworks. The final picture is also likely to be hugely dependent on the country, and region specifics may vary according to their seasonality of demand and available infrastructure levels (Pandev, 2017).

DISCUSSION

Lack of administrative practice and guidance adds to the costs and complexity of authorizations; while some jurisdictions, such as Austria, South Korea, and Japan are centralizing the permitting processes and simplifying the requirements on operators seeking to develop hydrogen projects, in many other jurisdictions, the complexities in the permitting regimes are yet to be addressed. As we have seen with other emerging technologies, the lack of past projects for guidance can result in inconsistencies in the responsible authorities' approach and lead to complex, protracted discussions before authorities can interpret the existing rules and grant the necessary permits (Chiesa et al., 2005). The investors' burden to engage stakeholders in the process is thus extended to capacity building of the public and the competent authorities, which adds to the cost and time involved in developing hydrogen projects (Chiesa et al., 2005; Cipriani et al., 2014; Cigolotti et al., 2018).

Faster deployment can facilitate a reduction in costs; Following on from the above, a key challenge in the hydrogen market today is cost and the need to create a sufficiently large marketplace to achieve economies of scale. While electrolysis using electricity from renewable energy sources may be the more environmentally sustainable method of producing hydrogen, it can currently be two to three times more expensive than hydrogen produced with natural gas or fossil fuels (without CCUS). As we have learned from other new technologies, costs fall when enough projects can form a critical investment mass (Kohl, 2018). National policies are key for attracting more projects to be developed and thus driving down costs. Recognizing this, the European Commission published an ambitious strategy in July 2020. This seeks critical mass investment to make hydrogen more cost-effective. Today, estimated costs for fossil-based hydrogen are around 1.5 EUR/kg for the EU, fossil-based hydrogen with carbon capture and storage around 2 €/kg, and renewable hydrogen 2.5-5.5 EUR/kg. The strategy notes that carbon prices in the range of EUR 55-90 per tonne of CO2 would be needed to make fossil-based hydrogen with carbon capture competitive with fossil-based hydrogen today. The costs of renewable hydrogen are going down, however. For example, by nearly 60% in the last ten years, electrolyzer costs have already been reduced and are expected to halve in 2030 compared to today's economies of scale. Electrolyzer costs are predicted to decline from €900/kW to €450/KW or less in the period after 2030 and €180/kW after 2040 (Iordache et al., 2018).

In the absence of a facilitative legislative framework and the need for reform, reflective of the nascent yet developing array of hydrogen technologies, legislative frameworks have not always caught up with development ambitions. Another key challenge in this chapter is the lack of a clear legal and regulatory framework for hydrogen. The majority of countries rely on their existing gas regulations to regulate hydrogen (Abánades, 2013). Due to the different nature and use of hydrogen, these frameworks are not always appropriate, and market players would benefit from introducing a clear regulatory framework to encourage the development of a hydrogen economy. From a lack of clear legislation, the counterbalance is not suited to new ways of working. However, most hydrogen is produced and consumed on the same site or transported short distances by road or pipeline (Abánades, 2013; Abbasi

& Abbasi, 2011), for transporting hydrogen over longer distances is restricted. The flammable characteristics of hydrogen require extreme care when handling (and transporting it). Hydrogen is a colorless, odorless, and flammable gas, and its large-scale use has commonly been perceived as risky because of how easily it may leak and ignite in relatively low temperatures. Much of how hydrogen storage and transport is treated has developed over time from industrial uses of hydrogen. Therefore, the understanding of hydrogen as a fuel source by relevant stakeholders and authorities is important in deciding whether to authorize the activity in these newer contexts.

The European Hydrogen Economy

Several countries have adopted national hydrogen strategies, but the EU has set out a transnational hydrogen strategy applicable across all member states, many of which are covered in the chapter. The long-awaited Hydrogen Strategy for a Climate-Neutral Europe (the "Strategy") sets out the EU-wide vision for various decarbonizing sectors across Europe. The strategy estimates that hydrogen is expected to provide at least 13% of the final energy mix by 2050 in Europe. This Europe-wide strategy is expected to encourage a hydrogen economy's development across Europe, following several European national hydrogen strategies. On the back of this EU-wide framework, some other jurisdictions explored in this chapter await the pending publication of a roadmap or strategy for implementing hydrogen in their countries. These strategies will help set a clear path for accelerating a hydrogen economy's deployment in the respective countries (Ball & Weeda, 2015). As we have seen in other sectors (e.g., renewables), the EU-wide direction can stimulate the national legislative systems to create a further local policy and thus investors greater certainty in venturing into this sector. This is the expectation for countries, such as Poland, where the Polish Ministry of Climate intends to publish the Polish Hydrogen Strategy setting out its vision for developing hydrogen in Poland (Norouzi, 2021c).

Market Prospects of Hydrogen

Several ambitious hydrogen strategies are being released worldwide, sending signals to investors regarding such countries' openness to developing hydrogen projects and stimulating public awareness and acceptance. In countries where we have seen new technologies like hydrogen take off, the strategies need to be supported by capital availability. At present, while projects may fall below the capital requirements of large banks and institutional investors, many of the projects rely on public support measures and government-level financing initiatives that encourage the uptake of hydrogen. For example, the French Government intends to include support measures for hydrogen projects in the French economic recovery plan presented in Autumn 2020. The Minister of Economics has indicated a possible investment increase of up to several billion euros in hydrogen (Norouzi, 2021b). In early 2020 the UK government announced a £90 million fund to tackle emissions from homes and heavy industry. £70 million of this includes funding for two of Europe's first-ever large-scale, low-carbon hydrogen production plants: on the River Mersey and in Aberdeen, and developing technology to harness offshore wind the Grimsby coast to power electrolysis and produce hydrogen. With state-sponsored support, it is important to consider the state aid requirements to achieve the right balance of support and retain value to consumers. Further, given that state entities and local authorities manage many public grants, appropriate public procurement steps are key to success. Indeed, navigating the pathway of successfully bidding for, securing, and maintaining government support and funding has been identified as a key challenge by many countries listed in this chapter(Dalebrook, 2013).

CONCLUSION

The last 12 months have seen a flurry of announcements and plans being published. A plethora of hydrogen applications is poised to emerge to help businesses decarbonize a range of sectors and unlock new market opportunities. Beyond the energy, transport, and heating opportunities, countries are looking to low-carbon hydrogen to improve air quality in cities, produce local jobs, improve energy security, and provide much-needed grid stability services to offset some of the issues associated with increased levels intermittent generation. Hydrogen is a platform that spans several industries and legal disciplines, being an energy vector. As energy lawyers working across the globe, we know that the reasons why a given hydrogen project will succeed in a given location will be uniquely dependent on the most important factors in that scenario: clean air or backup power. The demand for hydrogen has grown more than threefold since 1975 and continues to rise. While this is almost entirely supplied from fossil fuels at present, the R&D and demonstration efforts are finding and refining yet more ways of producing low-carbon hydrogen that can help decarbonize a range of sectors. For example, during 2019, the fuel cell electric vehicle market almost doubled due to expansion in Japan and China. The same trajectory is true for other modes of transport: passenger vehicles, heavy-duty vehicles, public transport, and even railways. Beyond transport, the grey hydrogen demand in the industry presents a significant opportunity to decarbonize this sector. The pipeline networks used by blending different hydrogen levels into the gas networks that were not used on-site. These are just some of the examples our experts highlight in this chapter. From Saudi Arabia to Japan to California, the range of legal frameworks, business and financing models, operational requirements, and many other factors are being tested as we write. Hydrogen has an important and increasing role in the global, national and local economies. It remains to be seen which countries will lead the way in uncovering the promise of the hydrogen opportunity. From our experience, countries that support commercially sound business cases and establish enabling regulatory frameworks to support the development of this technology are the likely early winners in this race.

Further, as the costs of electrolysis decrease and a symbiotic relationship developed between renewables and hydrogen production and offtake, new business models will emerge, and new products will be created. The consensus across the chapter's contributors is that hydrogen will play a key role as an energy vector and a core pillar in the energy transition we are experiencing. This may be the beginning.

ACKNOWLEDGMENT

This research received no specific grant from any funding agency in the public, commercial, or not-for-profit sectors.

REFERENCES

Abánades, A., Rubbia, C., & Salmieri, D. (2013). Thermal cracking of methane into hydrogen for a CO2-free utilization of natural gas. *International Journal of Hydrogen Energy*, *38*(20), 8491–8496. doi:10.1016/j.ijhydene.2012.08.138

Abbasi, T., & Abbasi, S. A. (2011). 'Renewable' hydrogen: Prospects and challenges. *Renewable & Sustainable Energy Reviews*, *15*(6), 3034–3040. doi:10.1016/j.rser.2011.02.026

Ajanovic, A. (2008). On the economics of hydrogen from renewable energy sources as an alternative fuel in transport sector in Austria. *International Journal of Hydrogen Energy*, *33*(16), 4223–4234. doi:10.1016/j.ijhydene.2008.06.041

Alazemi, J., & Andrews, J. (2015). Automotive hydrogen fuelling stations: An international review. *Renewable & Sustainable Energy Reviews*, *48*, 483–499. doi:10.1016/j.rser.2015.03.085

Andress, R. J., Huang, X., Bequette, B. W., & Martin, L. L. (2009). A systematic methodology for the evaluation of alternative thermochemical cycles for hydrogen production. *International Journal of Hydrogen Energy*, *34*(9), 4146–4154. doi:10.1016/j.ijhydene.2008.11.118

Arnason, B., & Sigfusson, T. I. (2000). Iceland—A future hydrogen economy. *International Journal of Hydrogen Energy*, *25*(5), 389–394. doi:10.1016/S0360-3199(99)00077-4

Baker, C. R. (1980). *Production of hydrogen for the commercial market: current and future trends*. Academic Press.

Balat, M. (2008). Potential importance of hydrogen as a future solution to environmental and transportation problems. *International Journal of Hydrogen Energy*, *33*(15), 4013–4029. doi:10.1016/j.ijhydene.2008.05.047

Ball, M., & Weeda, M. (2015). The hydrogen economy–vision or reality? *International Journal of Hydrogen Energy*, *40*(25), 7903–7919. doi:10.1016/j.ijhydene.2015.04.032

Barclay, F. J. (2006). *Fuel cells, engines and hydrogen: an exergy approach*. John Wiley & Sons. doi:10.1002/9780470030264

Bargigli, S., Raugei, M., & Ulgiati, S. (2004). Comparison of thermodynamic and environmental indexes of natural gas, syngas and hydrogen production processes. *Energy*, *29*(12-15), 2145–2159. doi:10.1016/j.energy.2004.03.018

Beghi, G. (1983). An introduction to hydrogen technology. In *Hydrogen-energy vector of the future* (pp. 5–12). Conference.

Bertel, E., Lee, K. S., & Nordborg, C. (2004). Nuclear energy in the hydrogen economy. *NEA News*, 10-13.

Bhandari, R., Trudewind, C. A., & Zapp, P. (2014). Life cycle assessment of hydrogen production via electrolysis–a review. *Journal of Cleaner Production*, *85*, 151–163. doi:10.1016/j.jclepro.2013.07.048

Bose, T., & Malbrunot, P. (2007). *Hydrogen: facing the energy challenges of the 21st century*. John Libbey Eurotext.

Chiesa, P., Consonni, S., Kreutz, T., & Williams, R. (2005). Co-production of hydrogen, electricity and CO2 from coal with commercially ready technology. Part A: Performance and emissions. *International Journal of Hydrogen Energy*, *30*(7), 747–767. doi:10.1016/j.ijhydene.2004.08.002

Cigolotti, V., Mcphail, S. J., & Tommasino, M. C. (2018). *HyLAW National Policy Chapter—Italy*. HyLaw Project.

Cipriani, G., Di Dio, V., Genduso, F., La Cascia, D., Liga, R., Miceli, R., & Galluzzo, G. R. (2014). Perspective on hydrogen energy carrier and its automotive applications. *International Journal of Hydrogen Energy, 39*(16), 8482–8494. doi:10.1016/j.ijhydene.2014.03.174

Cohce, M. K., Dincer, I., & Rosen, M. A. (2010). Thermodynamic analysis of hydrogen production from biomass gasification. *International Journal of Hydrogen Energy, 3*(10), 4970–4980. doi:10.1016/j.ijhydene.2009.08.066

Cox, K. E., & Williamson, K. D. (Eds.). (1997). *Hydrogen, Its Technology and Implications: Implication of Hydrogen Energy*. CRC Press LLC.

Dalebrook, A. F., Gan, W., Grasemann, M., Moret, S., & Laurenczy, G. (2013). Hydrogen storage: Beyond conventional methods. *Chemical Communications, 49*(78), 8735–8751. doi:10.1039/c3cc43836h PMID:23964360

Iordache, I., Schitea, D., Floristean, A., Stefănescu, I., & Brahy, N. (2018). INC-DTCI-ICSI RM. Vâlcea participation in hylaw project. *Smart Energy and Sustainable Environment, 21*(1), 75–82.

Kohl, B. E. (2018). *H2-Herstellung; voestalpine Beitrag zum HyLaw Workshop*. Voestalpine AG.

Kreutz, T., Williams, R., Consonni, S., & Chiesa, P. (2005). Co-production of hydrogen, electricity and CO2 from coal with commercially ready technology. Part B: Economic analysis. *International Journal of Hydrogen Energy, 30*(7), 769–784. doi:10.1016/j.ijhydene.2004.08.001

McCarty, R. D., Cox, K. E., & Williamson, K. D. (2019). Hydrogen: Its Technology and Implications: Hydrogen Properties. *International Journal of Innovative Research in the Humanities, 1*(1), 22–28.

Norouzi, N. (2021a). *Analysis of legal considerations and challenges of wastewater recycling management.* 5th International Congress of Developing Agriculture, Natural Resources, Environment and Tourism of Iran, Shiraz, Iran.

Norouzi, N. (2021b). Post-COVID-19 and globalization of oil and natural gas trade: Challenges, opportunities, lessons, regulations, and strategies. *International Journal of Energy Research, 45*(10), 14338–14356. doi:10.1002/er.6762 PMID:34219899

Norouzi, N. (2021c). An Overview on the renewable hydrogen market. *International Journal of Energy Studies, 6*(1), 67–94.

Norouzi, N., & Ataei, E. (2021a). Covid-19 Crisis and Environmental law: Opportunities and challenges. *Hasanuddin Law Review*, *7*(1), 46–60. doi:10.20956/halrev.v7i1.2772

Norouzi, N., & Ataei, E. (2021b). Environmental Protection Regulations in the Light of Public Law and Social Obligations. *Research Journal of Ecology and Environmental Sciences*, *1*(1), 1–16.

Norouzi, N., Khanmohammadi, H. U., & Ataei, E. (2021). The Law in the Face of the COVID-19 Pandemic: Early Lessons from Uruguay. *Hasanuddin Law Review*, *7*(2), 75–88. doi:10.20956/halrev.v7i2.2827

Pandev, M., Lucchese, P., Mansilla, C., Duigou, A. L., Abrashev, B., & Vladikova, D. (2017). Hydrogen Economy: The future for a sustainable and green society. *Izvestiia po Himiia*, *49*, 84–92.

Schitea, D., Floristean, A., & Brahy, N. (2018). INC-DTCI-ICSI RM. Vâlcea participation in hylaw project. *Progress of Cryogenics & Isotopes Separation*, *21*(1).

ADDITIONAL READING

Alhajji, A. F. (2008). What is energy security? Definitions and concepts. *Oil, Gas & Energy Law*, *6*(3).

Cigolotti, V., Mcphail, S. J., & Tommasino, M. C. (2018). *HyLAW National Policy Chapter—Italy*. HyLaw Project.

Floristean, A. (2019). *List of Legal Barriers*. HyLaw Project.

Heffron, R. J. (2015). *Energy law: an introduction*. Springer. doi:10.1007/978-3-319-14191-6_1

Kohl, B. E. (2018). *H2-Herstellung; voestalpine Beitrag zum HyLaw Workshop*. Voestalpine AG.

Schuman, S., & Lin, A. (2012). China's Renewable Energy Law and its impact on renewable power in China: Progress, challenges and recommendations for improving implementation. *Energy Policy*, *51*, 89–109. doi:10.1016/j.enpol.2012.06.066

Talus, K. (2013). *EU energy law and policy: a critical account*. Oxford University Press. doi:10.1093/acprof:oso/9780199686391.001.0001

Williams, I. (Ed.). (1930). *Pedeir keinc y Mabinogi*. Gwasg Prifysgol Cymru.

KEY TERMS AND DEFINITIONS

Hydrogen Economy: The hydrogen economy is an envisioned future in which hydrogen is used as a fuel for heat and hydrogen vehicles, for energy storage, and for long distance transport of energy. In order to phase out fossil fuels and limit global warming, hydrogen can be created from water using intermittent renewal sources such as wind and solar, and its combustion only releases water vapor to the atmosphere. Hydrogen is a powerful fuel, and a frequent component in rocket fuel, but numerous technical challenges prevent the creation of a large-scale hydrogen economy. These include the difficulty of developing long-term storage, pipelines and engine equipment; a relative lack of off-the-shelf engine technology that can currently run safely on hydrogen; safety concerns due to the high reactivity of hydrogen fuel with environmental oxygen in the air; the expense of producing it by electrolysis; and a lack of efficient photochemical water splitting technology. Hydrogen can also be the fuel in a fuel cell, which produces electricity with high efficiency in a process which is the reverse of the electrolysis of water. The hydrogen economy is nevertheless slowly developing as a small part of the low-carbon economy. As of 2019, hydrogen is mainly used as an industrial feedstock, primarily for the production of ammonia and methanol, and in petroleum refining. Although initially hydrogen gas was thought not to occur naturally in convenient reservoirs, it is now demonstrated that this is not the case; a hydrogen system is currently being exploited in the region of Bourakebougou, Mali, producing electricity for the surrounding villages. More discoveries of naturally occurring hydrogen in continental, on-shore geological environments have been made in recent years and open the way to the novel field of natural or native hydrogen, supporting energy transition efforts. As of 2019, almost all (95%) of the world's 70 million tons of hydrogen consumed yearly in industrial processing are produced by steam methane reforming (SMR) that also releases the greenhouse gas carbon dioxide. A possible less-polluting alternative is the newer technology methane pyrolysis, though SMR with carbon capture also has much reduced carbon emissions. Small amounts of hydrogen (5%) are produced by the dedicated production of hydrogen from water, usually as a byproduct of the process of generating chlorine from seawater. As of 2018 there is not enough cheap clean electricity (renewable and nuclear) for this hydrogen to become a significant part of the low-carbon economy, and carbon dioxide is a by-product of the SMR process, but it can be captured and stored. As a more cost-effective alternative to hydrogen economy mainly the methanol economy is discussed.

Hydrogen Production: Hydrogen production is the family of industrial methods for generating hydrogen gas. As of 2020, the majority of hydrogen (~95%) is produced from fossil fuels by steam reforming of natural gas, partial oxidation of methane, and coal gasification. Other methods of hydrogen production include biomass gasification, no CO2 emissions methane pyrolysis, and electrolysis of water. The latter processes, methane pyrolysis as well as water electrolysis can be done directly with any source of electricity, such as solar power. The production of hydrogen plays a key role in any industrialized society, since hydrogen is required for many essential chemical processes. In 2020, roughly 87 million tons of hydrogen was produced worldwide for various uses, such as oil refining, and in the production of ammonia (through the Haber process) and methanol (through reduction of carbon monoxide), and also as a fuel in transportation. The hydrogen generation market was expected to be valued at US$115.25 billion in 2017.

Hydrogen Safety: Hydrogen has one of the widest explosive/ignition mix range with air of all the gases with few exceptions such as acetylene, silane, and ethylene oxide. This means that whatever the mix proportion between air and hydrogen, when ignited in an enclosed space a hydrogen leak will most likely lead to an explosion, not a mere flame. This makes the use of hydrogen particularly dangerous in enclosed areas such as tunnels or underground parking. Pure hydrogen-oxygen flames burn in the ultraviolet color range and are nearly invisible to the naked eye, so a flame detector is needed to detect if a hydrogen leak is burning. Like natural gas, hydrogen is odorless and leaks cannot be detected by smell. This is the reason odorant chemical is injected into the natural gas to deliver the rotten-egg odor. Hydrogen codes and standards are codes and standards for hydrogen fuel cell vehicles, stationary fuel cell applications and portable fuel cell applications. There are codes and standards for the safe handling and storage of hydrogen, for example the standard for the installation of stationary fuel cell power systems from the National Fire Protection Association. Codes and standards have repeatedly been identified as a major institutional barrier to deploying hydrogen technologies and developing a hydrogen economy. As of 2019 international standards are needed for the transport, storage and traceability of environmental impact. One of the measures on the roadmap is to implement higher safety standards like early leak detection with hydrogen sensors. The Canadian Hydrogen Safety Program concluded that hydrogen fueling is as safe as, or safer than, compressed natural gas (CNG) fueling. The European Commission has funded the first higher educational program in the world in hydrogen safety engineering at the University of Ulster. It is expected that the general public will be able to use hydrogen technologies in everyday life with at least the same level of safety and comfort as with today's fossil fuels.

Hydrogen Storage: Hydrogen storage is a term used for any of several methods for storing hydrogen for later use. These methods encompass mechanical approaches such as high pressures and low temperatures, or chemical compounds that release H_2 upon demand. While large amounts of hydrogen is produced, it is mostly consumed at the site of production, notably for the synthesis of ammonia. For many years hydrogen has been stored as compressed gas or cryogenic liquid, and transported as such in cylinders, tubes, and cryogenic tanks for use in industry or as propellant in space programs. Interest in using hydrogen for on-board storage of energy in zero-emissions vehicles is motivating the development of new methods of storage, more adapted to this new application. The overarching challenge is the very low boiling point of H_2: it boils around 20.268 K (−252.882 °C or −423.188 °F). Achieving such low temperatures requires significant energy.

Hydrogen Technologies: Hydrogen technologies are technologies that relate to the production and use of hydrogen. Hydrogen technologies are applicable for many uses. Some hydrogen technologies are carbon neutral and could have a role in preventing climate change and a possible future hydrogen economy. Hydrogen is a chemical widely used in various applications including ammonia production, oil refining and energy. Hydrogen is not a primary energy source, because it is not naturally occurring as a fuel. It is, however, widely regarded as an ideal energy storage medium, due to the ease with which electric power can convert water into its hydrogen and oxygen components through electrolysis and can be converted back to electrical power using a fuel cell. There are a wide number of different types of fuel and electrolysis cells. The potential environmental impact depends primarily on the methods used to generate the hydrogen fuel.

Hylaw: HyLaw stands for Hydrogen Law and removal of legal barriers to the deployment of fuel cells and hydrogen applications. It is a flagship project aimed at boosting the market uptake of hydrogen and fuel cell technologies providing market developers with a clear view of the applicable regulations whilst calling the attention of policy makers on legal barriers to be removed. The project brings together 23 partners from Austria, Belgium, Bulgaria, Denmark, Finland, France, Germany, Hungary, Italy, Latvia, Norway, Poland, Romania, Spain, Sweden, Portugal, the Netherlands and United Kingdom and is coordinated by Hydrogen Europe. The HyLaw partners will first identify the legislation and regulations relevant to fuel cell and hydrogen applications and legal barriers to their commercialisation. They will then provide public authorities with country specific benchmarks and recommendations on how to remove these barriers. HyLaw main outputs will be – • An online and publicly available database compiling legal and administrative processes applicable to hydrogen and fuel cell technologies in 18 countries across Europe; • National policy chapters describing each legal and administrative process, highlighting best practices, legal barriers and providing policy recommendations; • A pan-European policy chapter targeted towards European decision makers; • National and European workshops for dissemination of the findings and convincing public authorities to remove barriers. – HyLaw started in January 2017 and will run until December 2018. The database will be maintained by Hydrogen Europe for minimum three years after the end of the project.

Public Perception of Hydrogen: Public understanding of hydrogen will have a tremendous impact on current as well as future policy initiatives for vehicle as well as portable and stationary applications. Numerous studies have been performed to analyze the public's current perception and understanding of hydrogen. Most energy producing technologies have an attached combination of positive and negative stigmas and means of understanding by the general public. For example, individuals who are aware of the environmental effects of a possible nuclear meltdown may deem nuclear power as a negative entity. On the contrary, individuals who are aware of the quantities of carbon emissions being reduced by using one less fossil fuel driven power plant may find nuclear power as a positive entity. A combination of scientific understanding with common associated social themes anchored by pre-existing knowledge will have a significant impact on the future hydrogen policy. Many studies have been carried out on the topic of Hydrogen Public Perception and degree of acceptance. The Institute for Social, Cultural and Public Policy Research at the University of Salford, evaluated a variety of survey based studies and performed a critical analysis of these selected findings. The article emphasizes that public perception is largely formed on an overall uneducated or misinformed hydrogen knowledge base. Dr. Miriam Ricci states within the article that "Providing factual information on the whole hydrogen chain, not just applications, and the implications it might have on the lives of citizens it may have is a necessary first step." This conclusion reached by many other scholars has been implemented in current US Hydrogen Policy through the appropriations of hydrogen demonstration and public outreach funding described in section 808 of the Energy Policy Act of 2005. Another concern presented by the Institute for Social, Cultural and Public Policy Research at the University of Salford is that of public distrust of government regulatory committees. In a free-association based survey carried out by Fionnguala Sherry-Brennan of the Manchester Architecture Research Centre (MARC) at the University of Manchester, residents from Unst, Shetland, home of a wind-hydrogen electricity system, were asked to describe words that came to mind when thinking about hydrogen. The survey concluded that details regarding the properties of hydrogen were largely unknown with 9.1 percent of the study population associating the word "hydrogen" with "bomb" and 0.6 percent of the study population associating it with the word "safe." However, there were few safety concerns regarding the use of hydrogen on the island. Concerning a possible hydrogen economy there are safety concerns that need to be addressed, some of which have informed current policy and regulations regarding refueling stations and hydrogen production stations as mentioned earlier in this article.

United States Hydrogen Policy: The principle of a fuel cell was discovered by Christian Friedrich Schönbein in 1838, and the first fuel cell was constructed by Sir William Robert Grove in 1839. The fuel cells made at this time were most similar to today's phosphoric acid fuel cells. Most hydrogen fuel cells today are of the proton exchange membrane (PEM) type. A PEM converts the chemical energy released during the electrochemical reaction of hydrogen and oxygen into electrical energy. The Energy Policy Act of 1992 was the first national legislation that called for large-scale hydrogen research. A five-year program was conducted that investigated the production of hydrogen from renewable energy sources and the feasibility of existing natural gas pipelines to carry hydrogen. It also called for the research into hydrogen storage systems for electric vehicles and the development of fuel cells suitable to power an electric motor vehicle.

Chapter 3
Evaluating Wastes as Low-Cost Substrates for Use in Microbial Citric Acid Production

Bilge Sayın Börekçi
https://orcid.org/0000-0002-1898-0428
Ardahan University, Turkey

ABSTRACT

With the increasing population, developing technology, and industry, the importance given to waste control/effective assessment studies continue with increasing momentum. The use of wastes in the production of biotechnological products is preferred due to its advantages in reducing environmental pollution, preventing nutrient and biomass losses, recycling, and decreasing costs. Citric acid (CA) is an intermediate product formed by the oxidation of carbohydrates to carbon dioxide in the Krebs cycle. This organic acid is used in many industrial areas such as pharmaceuticals and cosmetics. It is also an important organic acid in the food industry and is used as an acidifier, a stabilizer, an antioxidant, a flavor enhancer, and a preservative. Today, CA production is produced by microorganisms through fermentation. In addition, some wastes, such as molasses, glycerol, whey, olive mill wastewater, and various fruit wastes can be evaluated for use in the production of CA. This study reviewed the microbial production of CA using various wastes and some factors affecting the production.

DOI: 10.4018/978-1-7998-9851-1.ch003

INTRODUCTION

Using fermentable substrates as a carbon source is one of the most interesting topics and challenges in biotechnology. For this purpose, it is important to provide low-cost and easily available raw materials (Monrroy et al., 2019). Microbial-based organic acid production, rather than chemical production, can reduce environmental concerns. This progress in production accelerates efficient and environmentally friendly fermentation (Sharma et al., 2021). It is of great interest to make the biotechnological CA production more economical and sustainable through its use of cheap renewable carbon sources such as agricultural industrial wastes and their by-products. These substrates, which are suitable for the production of CA, are rich in carbohydrate and other vital nutrients (Dhillon et al., 2013). The use of food wastes as a substrate in CA production helps to create an effective recycling process to increase the added value (Nayak & Bhushan, 2019). Because, one of the recent problems in this industry has been the inability to make enough profit in CA production due to high raw material and energy costs. Waste generated during food processing accounts for 39% of the total food waste in European Union (EU) countries. However, these wastes can adversely affect human health and the ecosystem by creating significant environmental problems such as greenhouse gas emissions, and water and soil pollution. CA is mainly produced by *Aspergillus niger* using submerged fermentation (SmF) but the solid-state fermentation (SSF) technique attracts considerable attention due to its advantages. Using a similar environment to the microorganism's natural environment, low bacterial contamination risk under low substrate moisture and reduced liquid phase, reduction to the volumes of waste material and liquid effluents, operation in non-sterile conditions, simple fermentation facilities and lower energy consumption of SSF are considered to be advantageous over SmF (Roukas & Kotzekidou, 2020). SSF has gained importance in recent years as an alternative method for CA production to use agro-industrial wastes (Dhillon et al., 2011a). On the other hand, moisture content, substrate particle sizes, nutrient sources, incubation temperature, pH, and inoculum level are very effective in the growth of microorganisms and the formation of the final product (Ali et al., 2016). It is also emphasized that optimizing the medium composition is effective in improving CA production by *A. niger* and *Yarrowia lipolytica* (Hu et al., 2019). Finally, nutrient-rich biomass turns into a suitable substrate for the growth of microorganisms for industrial processes when the suitable conditions are provided. The objective of this study is to compile the wastes used as substrate in the production of CA from the past to present, together with the factors affecting the production, to shed light on future studies.

BACKGROUND

An Overview of CA

CA (2-hydroxypropane-1,2,3-tricarboxylic acid/$C_6H_8O_7.H_2O$) is a weak organic acid, first isolated by Scheele (1784). CA is colorless and easily soluble in water. It is solid at room temperature and its melting point is 153°C. With a molecular weight of 210.14 g/mol, CA involves three functional carboxylic acid groups in its structure and therefore has three different pKa values: 3.13, 4.76 and 6.39 (Behera, 2020). The Krebs cycle is the central oxidative pathway through which carbohydrates, lipids, and proteins are catabolized in aerobic microorganisms and tissues. It is also an important source of intermediates for numerous anabolic pathways leading to the biosynthesis of various biomolecules (Stenesh, 1998). CA, a primary metabolite, is produced in the Krebs cycle by the reaction of pyruvate with acetyl-CoA (Mehyar et al., 2005). CA is the most used and industrially produced organic acid of microbial origin (Cavallo et al., 2017) which demand will never end. Many microorganisms, including bacteria and fungi, are used in CA production (Hesham et al., 2020).

Recently, usage of CA has been researched for new areas such as components in biodegradable packaging, due to its cross-linking capacity, disinfectant agents, extracting agents, fruit preservative components, and environmental remediation. CA is an important organic acid that is low cost, environmentally friendly, and classified as GRAS (Generally Recognized as Safe) status by the FDA (Food and Drug Administration), so it has a wide area and growing rate of use. It should not be forgotten that if it is used as a food additive, it must have a purity of 99.5% (Mores et al., 2020). CA is used mainly in the food, pharmaceutical, and cosmetic industries. Its use in the food industry (70%) is due to some functional properties, such as a pleasant acidic taste and high solubility in water (Nayak & Bhushan, 2019). While it is possible to synthesize CA chemically, chemical synthesis has gradually been abandoned due to the success of microbial fermentation (Soccol et al., 2006).

It has been reported that the largest producer of CA is China. Additionally, CA production was estimated to be around 2 million tons in 2020 (Fickers et al., 2020). It has been thought that the global CA market will be 3.2 billion dollars by 2023, reaching a Compound Annual Growth Rate (CAGR) of 5.1% (Behera et al., 2021). It has also been stated that growth potential of over 5.24% is expected until 2025 (Mores et al., 2020).

Biosynthesis of CA

Yarrowia lipolytica is used in the production of CA with its GRAS status, high productivity, wide range of substrate assimilation, possibility to use renewable substrates or by-products from other processes and high tolerance to metal ions and salt solutions. It has the ability to grow at low and high pH (4 to 8) and over a wide temperature range (18 to 32°C) (da Silva et al., 2020). Although *Y. lipolytica* stands out among the yeasts used in CA production, the biggest disadvantage in production using this strain is considered to be the formation of isocitric acid (ICA), which is a by-product (Fickers et al., 2020). *Y. lipolytica* and several *Candida* species can provide high organic acid production ranging from 180 to 250 g/L. The production of these organic acids from low-cost and renewable carbon sources (such as plant oils, fats, glycerol, ethanol, molasses, alkanes, glucose, and starch hydrolysates, etc.) is a good strategy. Care should be taken to limit growth by limiting the nitrogen source, thiamine, phosphorus, sulfur, and magnesium components while implementing this strategy. Nitrogen restriction in particular triggers CA and ICA secretion. The amount of ICA formed as an undesirable by-product in yeast production varies according to the strain, substrate, and cultivation conditions used (Holz et al., 2009). It has been stated that the cost of the culture medium comprises approximately 50 to 80% of the total production cost. For this reason, it is emphasized that the use of low-cost substrates in production and the utilization of agricultural, industrial, and forestry wastes will provide both sustainable and economic production (Cavallo et al., 2020). For this purpose, SmF and SSF can be used in CA production by yeasts. Figure 1 shows the schematized version of the SSF.

Commercially, CA is only produced by *A. niger* in SmF technique using beet/cane molasses, sucrose, or glucose syrup. However, for maximum production rates generally high initial sugar concentrations (140-220 g/L) are required (Dhillon et al., 2013). The points that need attention in the production of CA by *A. niger* include high substrate concentration, limited nitrogen concentration, a certain amount of trace elements, a high concentration of dissolved oxygen and a low pH level (Sharma et al., 2021). Keeping the pH value at 2 is required for CA production. Adequate adjustment of carbon, nitrogen and phosphorus sources is important for microbial growth. Aeration and oxygen supply have a positive effect by increasing the CA production efficiency and shortening the fermentation time (Nayak & Bhushan, 2019). CA is mainly produced by *A. niger* strains using molasses; however, the disadvantage of this method is that molasses needs to be treated with ferrocyanide to remove trace elements harmful to *A. niger*. But, this method is ecologically problematic, and complex (Kamzolova et al., 2015). Maltose, mannose, galactose, fructose, sucrose, and glucose are among the substrates used for the growth of *A. niger*. Additionally, glucose and sucrose have been noted to give the best results for both growth and CA production when used at concentrations from 10 to 14%. Fungal fermentations can be carried out using surface fermentation (SF), SmF and SSF. Among these, 80% of CA production is obtained by SmF (Amato et al., 2020). Excessive accumulation of CA in certain microorganisms, such as *A. niger*, is due to faulty operations in the

TCA cycle. It is stated that the formation of CA is entirely at the enzymatic level. Therefore, trace elements are part of the cofactors, and their control is important for enzyme activity (Angumeenal & Venkappayya, 2013). Since citrate is one of the inhibitors of glycolysis, the ability of *A. niger* to produce excessive citrate by an active glycolytic pathway has attracted much attention (Papagianni, 2007).

Figure 1. SSF for the production of biotechnological products (Sristava et al., 2019)

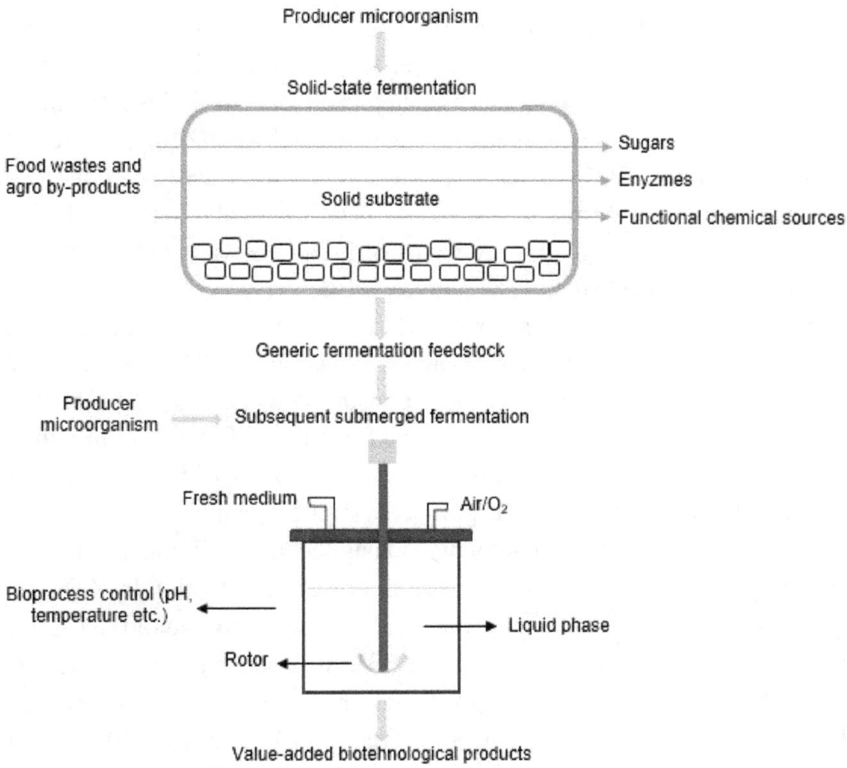

Wastes Used for the Production of CA

Apple Pomace and Apple Pomace Sludge

Apple pomace (AP), which constitutes approximately 25-30% of the total processed biomass, is formed as a solid waste generated after milling and pressing in apple processing plants (Hijosa-Valsero et al., 2017). AP can be used in the production of some value-added products such as some enzymes, organic acids, protein-enriched feeds, edible mushrooms, ethanol, aroma compounds, natural antioxidants, and edible fibers (Vendruscolo et al., 2008). The fact that it contains 12.3% fermentable sugar and is rich in carbohydrates (85%) is important in terms of choosing it as a substrate. The water-soluble components of this waste consist of mono oligosaccharide and polysaccharide whereas water-insoluble components consist of pectic substances, hemicelluloses, and cellulose. Unfortunately, most of the AP cannot be utilized and is dumped on the land. Since AP is not suitable for SmF due to its physical nature, it needs to be diluted, but this process has not been found efficient for CA production. For this reason, research has also been carried out on the production of CA in SSF using AP (Kumar et al., 2010). The apple processing industries generate thousands of tons of AP and AP sludge each year worldwide. AP is rarely used as a feed ingredient or as compost. The apple processing industries suffer from transportation costs for waste treatment and disposal to landfills. Therefore, it is beneficial to produce CA by *A. niger* from rich organic wastes such as AP or AP sludge which can be preferred as a substrate (Dhillon et al., 2011a). Although AP is generally used as cattle feed, only a part of it can be evaluated due to the rapid spoilage of wet pomace (Shalini & Gupta, 2010).

When published literature was examined, it was seen that AP is one of the most studied wastes for CA production. Hang & Woodams (1984) reported that yield was calculated to be 88% based on sugar consumed by *A. niger* NRRL 567 in the presence of 4% methanol in the AP medium after 5 days of fermentation. In another study using the same producer strain, yields varied from 77 to 88%, based on the sugar consumed depending on the pomace varieties (Hang & Woodams, 1986). Differently, a packed-bed bioreactor was designed for SSF and the process was optimized by the Taguchi statistical method. For CA production by *A. niger* BC1, the optimum parameters were determined to be an aeration rate of 0.8 L/min, a bed height of 10 cm, a particle size of 0.6 to 2.33 mm, and a moisture content of 78% (w/w). Under these conditions, 124 g CA was obtained from 1 kg of dry AP with a yield of 80% based on the total sugar (Shojaosadati & Babaeipour, 2002). Similarly, Kumar et al. (2010) recorded that *A. niger* van Tieghem MTCC 281 produced 4.6 g CA per 100 g of AP in the presence of 4% methanol in 5 days at 30°C. Dhillon et al. (2011a) evaluated the AP with the addition of rice husks in the fermentation medium. Production was also optimized using the response surface methodology (RSM) in terms of the moisture content and methanol/ethanol concentration. As a result, the maximum CA productions as 342.41 g/kg and 248.42 g/kg of dry substrate were obtained by *A. niger* NRRL 567. In another study, CA production was realized by *A. niger* NRRL 567 using AP ultrafiltration sludge. Production was optimized with RSM based on the total suspended solids and ethanol/methanol concentrations. They found that the highest CA concentration was 44.9 g/100 g dry substrate optimum conditions (Dhillon et al., 2011b).

Banana Waste

Bananas are one of the most abundant fruits in tropical countries. It is also an organic waste that is rich in carbohydrates and other nutrients (Karthikeyan & Sivakumar, 2010). Banana peel (BP) is around 40% of the whole banana. Generally, BP has been used as a raw material for compost and animal feed, biodegradable film, and fertilizer. Since the BP contains carbon and nitrogen sources necessary for microbial metabolism, its evaluation via fermentation has been shown as an alternative way (Chysirichote, 2020).

Sassi et al. (1991) used banana extract for the CA production by *A. niger* B60. In a study on ethidium bromide treated *A. niger* EB-3 mutant, 67.72 g/L CA was synthesized in 72 h using banana stalks as the support material in SSF (Javed et al., 2010). Karthikeyan & Sivakumar (2010) revealed the optimal conditions for CA production by *A. niger* using BP as a substrate. Kareem & Rahman (2013) reached a CA yield of 82 g/kg dry weight by *A.niger* UABN 210 in the presence of methanol (1%, v/v) and copper ions (10 ppm) in the medium after 96 h. Abbas et al. (2016) studied CA production by *A. niger* and determined the optimum conditions as use of 20% BP, 5% of inoculum level, addition of 0.1% KH_2PO_4 and 0.25% NH_4NO_3 in the medium, and 8 days of fermentation. Sidauruk et al. (2019) increased CA concentration by adding 5% sucrose to the medium which contained 15% BP. Chysirichote (2020) proposed an alternative method to evaluate BP for CA production of *A. niger* ATCC 16888. The optimum initial pH value and aeration rate were determined to be a pH of 5.0 and 1.0 vvm, by applying SSF in a 250 mL flask and a 2 L glass column, respectively. However, due to the rising temperature during fermentation, the production in the packed bed bioreactor decreased compared to the production in the glass column. To solve this problem, a new system was developed and a cooling air jacket built into the bioreactor allowed increased CA production. Under these conditions, CA yield was reported to be 124 mg/g of the dry substrate.

Brewery Waste

It was seen that the use of brewery waste in the production of CA has been investigated in the past, but not in any recent comprehensive studies. In the first study, spent grain liquor (SGL) was supplied from the spent grain recovery process in a brewing plant and lager tank sediment (LTS) was gained after wort fermentation and lagering of beer. SGL and LTS were selected as substrates for the production of CA by *A. niger* ATCC 9142. Eventually, 19 g/L and 11.5 g/L of CA were calculated for SGL and LTS, respectively (Roukas & Kotzekidou, 1986). In their other study in 1987, they investigated the effects of some trace elements, sodium monofluoroacetate, and methanol on CA production by *A. niger* ATCC 9142 using SGL. After 14 days of fermentation, it was found that the addition of 51 ppb Mn^{2+}, 5 ppb Fe^{3+}, 75 ppb Zn^{2+}, and 4% (v/v) methanol in the production medium caused an increase in CA yields to 4.9, 1.9, 10.9, and 16.8% respectively. It was also determined that the addition of sodium monofluoroacetate in concentrations 0.5 to 100 ppm resulted in decreases in CA yield varying between 1 and 70%. In another research using SGL, the effects of additional nitrogen sources, KH_2PO_4, and methanol on CA production by *A. foetidus* were investigated (formerly *A. niger* NRRL 337). Finally, additional nitrogen sources increased mycelial growth and sugar consumption, affecting the CA production negatively. While the addition of KH_2PO_4 had minör effect on the production and mycelial growth, methanol addition (2-4%) markedly increased the production (Hang et al., 1977). The reason for this is due to the shift of production towards cell growth because CA occurs in the exponential growth phase and starts to accumulate after the nitrogen in the environment is depleted (Yalcin et al., 2010). Also, it has been reported that lower alcohols such as methanol, ethanol, n-propanol, isopropanol, or methyl acetate neutralized the negative effect of the metals in CA production when used at about 1 to 5% (Soccol et al., 2006).

Celery by-Product

Only one study was found in which celery by-product (CBP) was evaluated. In that study, to increase the CA production capacity of the domestic *Y. lipolytica* 57, a mutation study was carried out with both UV (ultra violet) and ethyl methane sulfonate. Eventually, it was found that the chemical mutagenesis was more effective. After that, a maximum of 50.1 g/L of CA was synthesized by the selected mutant *Y. lipolytica* K-168 in a synthetic medium. This result provided a yield increase of 57% compared to the production capacity of the domestic strain. Then, the CA production was carried out in another fermentation medium containing CBP. It was indicated that the addition of glucose at different concentrations strongly affected the production, and the highest concentration of CA was 15.78 g/L with the addition of 100 g/L of glucose (Yalcin, 2012).

Coffee Husk

Coffee husk (CH) and coffee pulp (CP) can be used for the production of value-added compounds and the bioremediation of waste materials. When used in SSF, CH acts as solid support for microbial growth. CH and CP contain carbohydrates, lipids, proteins, and polyphenols. The extraction of bioactive compounds such as pigments, production of absorbents, enzymes and ethanol, and biomethanation are among the processes in which CH is used as a raw material. CH and CP contain approximately 50 to 85% total carbohydrates, 14 to 24% reducing sugars, 4 to 12%

protein, 0.5 to 3% lipids, 3 to 10% minerals, 1 to 9% tannins, and 1% caffeine (g/100 g of dry weight). The composition of CHs and CPs varies with processing method, storage conditions, species, and geographic distribution of coffee plants (Kumar et al., 2018). CH causes disposal problems and environmental pollution. The use of CH as fertilizer leads to the loss of valuable biomass. In addition, its use as a fuel is inefficient due to its low calorific value. Therefore, the evaluation of CH for CA production is of interest (Shankaranand & Lonsane, 1994). It was recorded that CH was found to result in higher CA production than wheat bran, rice bran, and de-oiled rice bran (Pandey et al., 2000).

The CH was moistened with a 0.075 M of NaOH solution as a pretreatment, in a research conducted by Shankaranand & Lonsane (1994). The medium was then sterilized and inoculated with *A. niger* CFTRI 30. When iron, copper, and zinc were added to the fermentation medium at 1.0, 0.2, and 0.1 ppm levels, respectively, the CA yield increased by 17%. As a result, 1.5 g CA/10 g of dry CH was produced with a conversion rate of 82% compared to the sugar consumed (in 72 h). Moreover, it was found that adding $(NH_4)_2SO_4$, cellulase, hemicellulase, pectinase, and amylase to the fermentation medium did not increase production. In another study, as a pretreatment, crude CH was washed with deionized water, filtered, dried at 80°C, and then ground into powder. Finally, cellulosic and lignocellulosic residues were sieved at a 0.5 mm pore size to increase the surface area and improve substrate solubility. The addition of methanol (3% v/w) in the medium supplemented with trace elements (Fe^{2+}, Cu^{2+}, and Zn^{2+}) gave the best results for CA production by *A. niger* RCNM 17. Finally, 187.54 g CA/kg dry CH was achieved after 72 h resulting in 78.14% and 83.36% conversion rate depending on the initial sugar concentration present in the CH and sugar consumed during fermentation, respectively (Ramachandra et al., 2013).

Cotton Waste

Kiel et al. (1981) investigated the possible use of CW for CA production by *A. niger* ATCC 9142. It was seen that CA was not produced when CW was used as the sole carbon source. Moreover, in two-stage fermentations, CA production was maximized when mycelium obtained from surface cultures in the CW medium were transferred to sucrose-containing media compared to being transferred directly to sucrose-containing media. As a result, it was revealed that CW can be used to save sucrose and increase production efficiency.

Glycerol-Containing Waste

Biodiesel production can be carried out using various vegetable oils and animal fats. This technology consists of hydrolysis of the oil triglycerides, and methylation with the formation of methylated fatty acids, which are just used only as biodiesel. The main waste from biodiesel production is glycerol, which is generated more than 1 kg per 10 kg of biodiesel (Rymowicz et al., 2010). Biodiesel waste contains approximately 80% glycerol, oil residue, free fatty acids, sodium, and potassium salts, and water (Kamzolova et al., 2011). It is economically more convenient to use crude glycerol as a carbon source without purification. However, depending on the biodiesel production process and the rapeseed used, crude glycerol can contain various impurities in amounts that may lead to the inhibition of microorganisms such as methanol, catalysts, sodium and potassium salts, heavy metals, and soaps (Rywińska et al., 2009). Biotechnological productions such as 1,3-propanediol, hydrogen, propanoic acid, trehalose, single cell oil, n-butanol, glyceric acid, CA, ethanol, polyunsaturated fatty acids, biopolymers offer a very suitable opportunity to evaluate crude glycerol (Sarris et al., 2019).

Rymowicz et al. (2006) obtained 124.5 g/L of CA by acetate-negative mutant *Y. lipolytica* 1.31 using 200 g/L raw glycerol. Under these conditions, yield and productivity were calculated as 0.62 g/g and 0.88 g/L h respectively. Rywińska et al. (2009) compared the possibility of using pure and crude glycerol in the production of CA with acetate-negative mutants. They produced 139 g/L of CA by *Y. lipolytica* Wratislavia AWG7 in a medium containing 200 g/L pure glycerol in fed-batch fermentation after 120 h. However, 131.5 g/L of CA was produced by the same strain from crude glycerol. Rymowicz et al. (2010) studied for CA production by *Y. lipolytica* A-101-1.22 in glycerol-containing waste (GCW) medium and found the fermentation time over 100 h was pointless. To prolong this biosynthesis time, the cell recycles and the repeated batch regimes were applied. Repeated batch culture remained stable for over 1000 h and with 30% feed, every 3 days gave the best results (124.2 g/L of CA with a yield of 0.77 g/g and a productivity of 0.85 g/L h). Kamzolova et al. (2011) selected the mutant *Y. lipolytica* N15, which had the highest acid production in nitrogen-limited medium, among the yeast strains whose acid-formation abilities were examined. Under optimum conditions, 98 g/L of CA was produced in the fermenter which contained pure glycerol while 71 g/L of CA was produced in GCW. One of the important findings was that the volumetric productivity and productivity of the obtained results could compete with the data obtained from the production by *A. niger* with molasses. Thus it was stated that the current process could gain an industrial scale. In addition, technical grade sodium citrate (a substitute for safe phosphate-free detergents) was produced by *Y. lipolytica* VKM Y-2373 from GCW. In the results from the batch experiments, citrate was actively produced for 144 h and then decreased, possibly due to inhibition of the enzymes responsible for citrate formation. They extended the production process up to 500 h with the repeated batch cultivation and produced 79-82 g/L of citrate. Another important step and output of the study was the comparison of the traditional CA production by *A. niger* with the current method. In this sense, it should be noted that pretreatment of molasses with ferrocyanide is an ecologically harmful and longer process, while production by *Y. lipolytica* is shorter and ecologically safe (Kamzolova et al., 2015). Ferreira et al. (2016) studied in a lab-scale stirred tank bioreactor to investigate the effect of oxygen mass transfer rate (kLa) for CA production by *Y. lipolytica* W29 (ATCC 20460) from crude glycerol. kLa was defined as a function of operating conditions (stirring speed and specific airflow rate) and cellular density. It was determined that the increase of initial kLa from 7 h^{-1} to 55 h^{-1} caused an approximately 8-fold increase in the CA concentration while the concentration of ICA decreased. Controlled dissolved oxygen conditions in production were also examined and it was determined that production was positively affected with dissolved oxygen up to 60% of saturation. According to this method, the simplicity of kLa provided an advantage and was an effective parameter in production by *Y. lipolytica* from crude glycerol. It was also proposed an empirical correlation to estimate the kLa value. Surprisingly, it was investigated for the first time to produce organic acids from crude glycerol at pH 3 in the recent study. *Y. lipolytica* AJD pADUTGut1/2 was used as a producer strain. The characteristic of this strain is that it over-expresses glycerol kinase (*GUT1*) and glycerol-3-phosphate dehydrogenase (*GUT2*), which are two genes encoding the enzymes involved in the assimilation of glycerol. As a result, the maximum CA/ICA ratio was determined as 75.9 g/L after 144 h of fermentation. However, ICA reached a fairly high level (42.5 g/L) in the study. For this reason, it was thought that the created process could also be used for ICA production (Rzechonek et al., 2019). On the contrary, in the production of CA by yeast, the pH value is usually greater than 5 (Zhang et al., 2019; Sayın Börekçi, 2020).

Kiwifruit Peel

With the development of the kiwi processing industry, the amount of kiwifruit peels (KP) increased, which causes economic and environmental problems. Since kiwifruit is a rich source of antioxidants including phenols, it is stated that maximum benefit will be obtained if they are preferred to be used in the production of polyphenols (Guo et al., 2018). KP is a waste that represents 10 to 16% of the weight of the original fruit during nectar extraction or slicing from the fruit. In a study using this waste, 100 g of CA/kg of KP was produced by *A. niger* NRRL 567 in a medium containing 2% methanol after 4 days of fermentation (Hang et al., 1987).

Molasses

Although molasses is a by-product of the sugar industry, it can be used as a carbon source in the production of biotechnological products such as enzymes, organic acids, pigments, and polysaccharides (Özdal & Kurbanoğlu, 2019). Molasses generally consist of 30-35% sucrose, 10-25% fructose and glucose, 2-3% non-sugar compounds, minerals and about 45-55% of total fermentable sugars. Molasses is used as an ingredient in the bakery industry also as an animal feed additive and as a fermentation feedstock (Jamir et al., 2021). Evaluation of molasses for CA production is advantageous as it can minimize environmental problems that may arise from other fermentation processes and chemical production (Almakki et al., 2009). However, some studies have proved that beet molasses generally give higher yields than cane molasses. This situation may be due to the relatively high content of trace elements such as magnesium, zinc, manganese, calcium, and iron in cane molasses (Show et al., 2015).

Iqbal et al. (2015) reported that 14.17 g/L of CA was obtained by *A. niger* GCB-117 from sugarcane molasses containing 150 g/L sugar. Jianlong et al. (2000) developed a new method using beet molasses for CA production. For this purpose, an anion-exchange resin-packed-column was connected to a fermenter for the separation of CA from fermentation broth. Consequently, this method increased the CA yield compared to the conventional batch system. Pazouki et al. (2000) investigated the effects of the substrate type and concentration, as well as the addition of methanol on the CA production by *A. niger* NCIM 548 and *C. lipolytica* NCIM 3472. Finally, the best result was 12 g/L for *A. niger* NCIM 548 when 60 g/L molasses was used in the fermentation medium with methanol addition. Additionally, 8.4 g/L of CA was produced by *C. lipolytica* NCIM 3472 when methanol was not added to the medium which contained 100 g/L of glucose. Kim et al. (2015) added 50 mg of invertase (300 to 500 Units/mg) to 1 L of molasses at 55°C, for 2 h. After this step, the sugar content was reduced to 18% (the initial sugar content was 52 to 56%). Then, the yield of CA was maximized up to 91 g/L in the medium prepared (at a C/N ratio of 40/1, a phosphate addition of 1.0 g/L, and a Fe^{2+} ion concentration of less than 50 mg/L). Under the optimum conditions, 18.5 g/L ICA was formed as a by-product. In another study, the pH value of the molasses containing 200 g/L total sugar was adjusted to 5.5 and the solution was heated at 90°C for 15 min. To facilitate the precipitation of heavy metals, the hot liquid treated with 0.1 g/L of $K_4Fe(CN)_6$ was kept at room temperature for 24 h and then centrifuged. To further increase CA production, *A. niger* MO-25 isolate was mutated using ethidium bromide and the resulting mutants were examined for their CA production capacity in the presence of molasses-peptone for 168 h. The mutant EB-12 strain produced 46.3 g/L of CA, while the parent strain produced only 31.08 g/L of CA. After the producer strain selection, an examination of how some supplements might affect CA production was carried out. Consequently, the use of 4 g/L of chicken feather peptone (CFP) together with molasses gave the best results. One of the important results here was that the highest production of CA was achieved with CFP compared to commercial peptones. The addition of KH_2PO_4 (0.15 g/L) to the fermentation medium also supported CA production. CFP was shown to be promising as a source of organic nitrogen and mineral salt, with very good results. CFP are formed as bio-organic waste in poultry processing plants. Millions of tons of CFPs are produced annually in the world, and these feathers cause environmental problems as they are discarded as waste and degrade very slowly. CFPs make up 10% of the total chicken weight, are composed of over 90% protein composed of keratin, a valuable source of amino acids. Keratin wastes must be hydrolyzed with acid or enzymatic hydrolysis to be used by microorganisms (Özdal & Kurbanoğlu, 2019).

Okara (Soy-Residue)

Okara is formed as a by-product during the production of soy milk and tofu. It causes disposal problems; therefore, the use of microbial fermentation to enable recycling is of interest. It has been evaluated in the production of riboflavin, lipase, fructofuranosidase, and single-cell proteins. In a study in which Okara, a cellulosic by-product, was evaluated using a medium prepared with $(NH_4)_2SO_4$ addition, after 11 days of fermentation, the maximum CA was determined to be 5.10 g of CA/100 g of dry solids (Khare et al., 1995).

Olive Mill Wastewater

Olive mill wastewater (OMW) is an important liquid residue produced during the manufacture of olive oil, which is difficult to treat due to its high phenolic content. This waste causes environmental pollution due to the phenolic substances it contains, as well as the high concentrations of sugars, tannins, pigments, and emulsified oil (Papanikolaou et al., 2008). The composition of OMW varies according to the variety of olive used, the technology, and the conditions used to the extraction of the oil. In this waste, biological oxygen demand and chemical oxygen demand values are 200 to 400 times higher than municipal sewage (Sarris et al., 2011). Due to the anticarcinogenic and therapeutic effects of phenolic compounds found in OMW, a recently developed technology allows them to be separated and used in various fields. Besides that, OMW, without removing phenolic compounds, can be used as a substrate by many microorganisms for obtaining value-added substances such as microbial mass, exo-polysaccharides, enzymes, ethanol, CA, and microbial lipids. This waste is returned to the environment without any treatment. Phenolic substances in OMW are also responsible for the toxic effect and dark color (Sarris et al., 2017). OMW contains various carbon sources, organic compounds, and minerals. It is an important point that only a few microorganisms can use this waste as a substrate (Dourou et al., 2016).

Papanikolaou et al. (2008) filtered and centrifuged OMW before use. Glucose was added to the OMW because this waste did not contain sufficient sugar, and additional sugar was required for the production of CA. Production was carried out in shake flasks and the pH value was kept between 5.0 to 6.0 under aseptic conditions. At the end of the fermentation, 28.9 g/L of CA was obtained by *Y. lipolytica* ACA-DC 50109 (initial glucose concentration of OMW was 65 g/L). Similarly, Sarris et al. (2011) grew *Y. lipolytica* ACA-YC 5033 in OMW+ 35 g/L glucose, and 18.9 g/L of CA was produced after 144 h in shake flasks. However, Sarris et al. (2017) reported that 52 g/L of CA was achieved with the same strain by increasing the concentration of added glucose to 80 g/L and the fermentation time to 384 h. Dourou et al. (2016) enriched OMW with 50 g/L of glycerol and found that 30.3 g/L of CA was produced by *Y. lipolytica* LGAM S(7). In another study, Spanish type olive processing wastewater was mixed with white grape pomace (1:1, w/w) for CA production by *A. niger* B60. The process was also optimized in terms of some parameters. After this step, production was scaled-up to a 5 L flask. Under the most favorable conditions, 83 g/L of CA and a yield of 0.54 g/g were calculated after 15 days of fermentation using sugar-enriched wastewater by adding sucrose (100 g/L) and NH_4NO_3 (1.1 g/L) (Papadaki & Mantzouridou, 2019).

Orange Peel

Annual citrus production exceeds 120 million tons in tropical and subtropical regions. Approximately 40-60% of the fruit is thrown away as waste and these wastes cause bad odor and land, air and water pollution. These wastes contain about 75-85% moisture and most of them are dried to be converted into cattle feed pellets, which require high energy consumption. The drying process causes the release of large amounts of volatile organic compounds, sulfur dioxide, nitrogen oxides, methanol, formaldehyde, carbon dioxide, and other potentially hazardous air pollutants from the decomposition of organic compounds (Mahato et al., 2021). Orange peel (OP) consists of soluble sugars, fiber, protein, ashes, and fat. OP contains high concentrations of sugar and it causes mold growth that requires intervention due to the danger of uncontrolled production of greenhouse gases (Koutinas et al., 2015). There are many biotechnological application areas of OP such as producing animal feed, fertilizers, essential oils, pectin, ethanol, methane, single-cell proteins, and macerating enzymes and pectin-lyase (Hamdy, 2013). Along with these applications, this waste is also used in the production of CA.

In the first study, press liquor from the dewatering of OP was selected for CA production of *A. niger* NRRL 599. The best results were obtained with a pH from 5.0 to 6.2 in the presence of 40 mL/kg of methanol in the fermentation medium. Furthermore, thermal pretreatment of the press liquor caused an increase in production

(Aravantinos-Zafiris et al., 1994). In another study, OP was subjected to autohydrolysis as a pretreatment. The positive effect of methanol was also investigated and 9.2 g/L of CA was synthesized by *A. niger* CECT 2090 after 3 days (Rivas et al., 2008). Torrado et al. (2011) examined SmF and SSF fermentation in their study using OP and reported that SSF gave better results and a higher yield because there was no need for additional nutrients or pretreatment except sterilization. Hamdy (2013) reached 640 g/kg OP of CA by *A. niger* van Tieg 1867 after 72 h in shake flasks. It can be stated that the advantages of autohydrolysis are that it limits the corrosion problem, does not form sludge, and is more affordable for capital and operating costs (Hijosa-Valsero et al., 2017).

Pineapple Waste

Pineapple waste (PW) mainly contains cellulose and starch. This waste can be used as a substrate for bromelain, vinegar, wine, food/feed yeast, and alcohol (Femi-Ola et al., 2009). Pineapple peel constitutes 10% (w/w) of the fruit weight and the disposal of this waste is important due to economic and environmental concerns (Kareem et al., 2010). PW contains a high concentration of biodegradable organic material and suspended solids; moreover, it has a high biological oxygen requirement (Abdullah & Mat, 2008). PW is used as animal feed, or unfortunately, sometimes it cannot be used and is thrown into the soil. At the same time, it has been emphasized that this waste may be a suitable substrate for CA production in SSF, as it is rich in carbohydrates and some other nutrients (Imandi et al., 2008). The fact that pineapple is a perennial plant allows it to be used in the production of CA as a sustainable raw material and provides a means to reduce costs (Ayeni et al., 2019).

Tran et al. (1995) showed that PW was a better substrate for CA production by *A. foetidus* ACM 3996 compared to apple pomace, wheat bran, and rice bran. The highest CA was reported as 16.1 g per 100 g dried PW in the presence of 3% methanol in the medium. Tran et al. (1998) achieved a yield of 19.4 g CA per 100 g dry fermented PW by *A. niger* ACM 4992. Optimum conditions for the process have been found as 65% (w/w) initial moisture content, 3% (v/w) methanol, 30°C, pH 3.4, particle size 2 mm, and 5 ppm Fe^{2+}. Another output of the study was the comparison of the production in shake flasks, tray, and rotary drum bioreactors, and the best results were achieved in shake flasks. Kumar et al. (2003) evaluated the potential use of pineapple, mixed fruit, and maosmi waste for the production of CA by *A. niger* DS 1. As an interesting finding in the study, for CA production, a moisture level of 60% was suitable in the absence of methanol whereas a moisture level of 70% was found effective in the presence of methanol. Also, the stimulating effect of methanol on CA production was more effective at high moisture levels. After 8 days of fermentation, the highest CA yields of 54.2%, 46.5%, and 50% were obtained from pineapple, mixed fruit and maosmi residues, respectively. Imandi et al. (2008) carried out an optimization study using Plackett-Burman design for CA production by *Y. lipolytica* NCIM 3589. The optimum conditions were determined to be yeast extract 0.34 (%w/w), moisture content of the substrate 70.71 (%), KH_2PO_4 0.64 (%w/w) and Na_2HPO_4 0.69 (%w/w). Under optimum conditions, CA concentration was 202.35 g/kg ds (g CA produced/kg of dried PW as the substrate). In another SSF study carried out by *A. niger* KS-7, the effects of different concentrations of glucose, sucrose, NH_4NO_3, and $(NH_4)_3PO_4$ on CA production were investigated separately. After 5 days, the best results were found in a medium containing 15% (w/v) sucrose, 0.25% (w/v) NH_4NO_3 and 2% (v/v) methanol. Under these conditions, the yield was determined to be more than 90% based on the sugar consumed (Kareem et al., 2010).

Pomegranate Peel

Although it varies depending on the cultivar, pomegranate peel (PP) constitutes an important part of the fruit weight such as 49-55%. Interestingly, this waste has a stronger biological activity than its pulp, but unfortunately, like many other wastes, it is often used as animal feed and therefore most of it is thrown away and poses an environmental problem due to its moisture content (Magangana et al., 2020). PP is a rich source of phenolic compounds such as tannins and flavonoids. Due to its antioxidant properties, PP extract is preferred in food recipes, animal feedstocks, cosmetics, tinctures, and certain therapeutic formulas. This waste is used to obtain industrially important value-added products such as dietary fibers, color pigments, dyes, medicinal components, and bioactive compounds. The modified form is used as solvent extraction and heavy metal absorber, while the dried form serves as a substrate for the production of certain enzymes, biogas, and single-celled proteins (Saleem et al., 2020).

Though PP is used in many fields, only two studies were found in which this waste was evaluated for its suitability for CA production. In one of them, CA production was realized by *A. niger* B60 with dried and non-dried PPs under non-aseptic conditions. The maximum concentration of CA was calculated after 8 days of fermentation using undried samples was 306.8 g/kg dry peel. In this production, temperature was adjusted to 25°C, the moisture content to 75%, and the initial pH value to 8.0. Moreover, the addition of 3% (w/w) methanol to the medium containing dried and non-dried PP wastes significantly increased the concentration of CA in both cases (Roukas & Kotzekidou, 2020). When these results were evaluated, it was thought that PP is a very promising waste for CA production and studies can continue.

Ram Horn

Kurbanoğlu & Kurbanoğlu (2003) hydrolyzed ram horn (RH) by treatment with H_2SO_4 and created the RH hydrolysate. Then, it was reported that the most suitable concentration of RH hydrolysate for SmF with *A. niger* NRRL 330 was 8% and this process increased the production approximately 26% compared to the control medium. In another SmF study conducted using the same strain, the same pretreatments on RH were carried out. Compared to the control sample, RH addition to the fermentation medium at a rate of 4% increased the yield. Also, the residual sugar concentration was reduced and mycelial growth was stimulated. It has been accordingly suggested that RH can be considered a suitable mineral and amino acid supplement for CA production. Precautions should be taken for the disposal of ram horn (RH), known as fibrous protein, which occurs in slaughterhouses (in amounts of up to tons) every year. This waste can be also converted into biomass, protein concentrate, or amino acids using proteases produced from some microorganisms (Kurbanoğlu, 2004).

Some Other Lignocellulosic Wastes

Lignocellulosic biomass such as rice straw, corn stover, and wheat straw are preferred for the production of biofuels and some biochemical products due to its abundance and availability. Studies on the use of these agricultural wastes for the production of CA especially by *A. niger* have been continuing (Hu et al., 2019). Cellulose is the most abundant carbohydrate derived from the photosynthesis of plants and one of the most common renewable biological resources. Today, significant amounts of cellulose are left in agricultural lands to be burned and cause environmental concerns. It is known that *Y. lipolytica* is unable to utilize cellulosic materials due to the deficiency of cellulase activity (Liu et al., 2015a). Corn steep liquor (CSL), which is a by-product of the corn steeping process, is a substance rich in amino acids, vitamins, and polypeptides. CSL contains approximately 1 mg/kg of biotin. In the production of CA, the use of CSL replaces peptone or yeast extract as an alternative organic nitrogen and vitamin source, adding an extra economic dimension to the production. It has been reported to date that this substance can be used in the production of various organic acids or other products (Liu et al., 2015b). Corn cobs (CC)/corn husks (CHs) formed as a result of processing sweet corn is a waste that can be used as animal feed and can also be sent back to the field for land application (Hang & Woodams, 2000; 2001).

Hang & Woodams (1998) produced CA (250 g/kg of dry matter CC) by *A. niger* NRRL 2001 with more than 50% overall yield. In another study, CHs was evaluated for CA production by the same strain. As a pretreatment, CHs were diluted with 0.5 mol/L of NaOH. After this step, the mixture was acidified to pH 5.0 with concentrated HCl and autoclaved. Furthermore, after inoculation, a commercial apple juice processing enzyme preparation and methanol were added to the fermentation medium. After 120 h of fermentation, the CA yield was calculated to be 259 g/kg of dry matter of CHs (Hang & Woodams, 2000). In the study by the same researchers in 2001 with CC, the most appropriate substrate concentration was determined to be 5 g/100 mL, the fermentation time was 72 h, and the enzyme concentration was 1 mL/100mL. The highest CA yield was calculated to be 603.5 g/kg of dry matter (Hang & Woodams, 2001).

Sugarcane and Cassava Bagasse

Sugarcane bagasse (SB) is a renewable material, one of the largest cellulosic agro-industrial by-products. Electricity generation, pulp, and paper production and some products such as alcohol, alkaloids, mushrooms, protein-enriched animal feed (single-cell protein), enzymes, and biofuel are among the production processes that use bagasse (Parameswaran, 2009). SB contains approximately 46 to 52% w/w water and 43 to 52% fiber (including cellulose 50%, hemicellulose 25% and lignin 25%). Since thousands of tons of bagasses are discarded every year, this pollutes the environment. This waste is also used in the production of biotechnological products with high added value. SB benefits as substrate and/or carrier in SSF processes (Khosravi-Darani & Zoghi, 2008).

The first report on the potential use of sugarcane-pressmud in CA production was presented by Shankaranand & Lonsane (1993). They explored that $K_4Fe(CN)_6$ treatment enhanced the conversion yield and shortened the fermentation time by 24 h. Maximum CA yield was calculated by *A. niger* CFTRI 30 after 120 h in a ferrocyanide-treated medium which initially contained 12.5% (w/w) effective sugar and 0.1% (w/w) NH_4NO_3. On the other side, cassava bagasse (CB) was found to be a better substrate than a vegetable sponge and SB (Kolicheski et al., 1997). Vandenberghe et al. (2000) defined the best substrate as CB, when compared to SB and coffee husk. At the end of the production, 88 g/kg dry matter of CA was obtained. Khosravi-Darani & Zoghi (2008) compared acid, alkali, and urea treatments to determine the most suitable pretreatment for SB used in CA production. As a result, it was concluded that the most effective method was urea pretreatment. Accordingly, 137.6 g/kg dry SB and CA yield of 96% based on sugar consumed was obtained. Yadegary et al. (2012) compared the CA production with untreated or pre-treated SB. It was found that the highest results were calculated from the samples treated with alkaline followed by acid, compared to the control sample. Gasiorek (2018) studied sugar beet pulp in a horizontal rotating drum bioreactor for CA production. Approximately 150 g per kg substrate dry matter CA yield was found under the determined optimum conditions. Moreover, the optimum conditions did not stimulate the formation of biomass.

Waste Cooking Oil

A significant problem for catering services is the issue of waste cooking oil (WCO). The main constituent of this waste is triacylglycerols that are involved in microbial metabolism by the action of lipases, which are extracellular lipolytic enzymes. Lipases hydrolyze triacylglycerols to form glycerol and related fatty acids, and the hydrolyzate can be used to synthesize CA and free fatty acid in mitochondria and cytoplasm of *Y. lipolytica* in turn (Liu et al., 2015c). WCOs are no longer reutilized because the fatty acids have already oxidized. It has been reported that the 500 million tons of waste produced every year can be used for the production of erythritol and CA by *Y. lipolytica* (Liu et al., 2018).

In the first study, *Y. lipolytica* SWJ-1b synthesized 30.3 g/L of CA after 336 h in a fermenter containing 80 g/L of WCO in the fermentation medium. The ICA, and biomass concentrations were calculated to be 6.9, and 6.1 g/L, respectively. When the WCO concentration was increased from 120 to 140 g/L, a decrease in the concentration of CA and an increase in the concentration of ICA were detected. In this study, the effects on CA production of some substances added to the medium in different amounts were examined separately. As a conclusion, the highest CA concentrations were achieved when 0.2 g/L of $(NH_4)_2SO_4$ and 1.0 g/L magnesium were added to the fermentation medium and no KH_2PO_4 was added (Liu et al., 2015c). Liu et al. (2018) evaluated WCO again for CA production about how initial NaCl concentration affected the production. They found that the highest CA concentration was obtained when 20 g/L of NaCl was used. Moreover, the synergistic effect of osmotic pressure and pH level was also investigated, and the best results were detected as 0.75 osmol/L and a pH of 6.0.

Wastewater Generated by CA Production

Since the wastewater generated by the production of CA is also a problem, a process has been developed in one study to reuse and evaluate this wastewater in the production of CA. In that study, environments with tap water and fermentation process water were compared in terms of CA production. The CA wastewater was first treated using anaerobic digestion. Anaerobic digestion effluent was then subjected to ultrafiltration and nanofiltration processes to eliminate the inhibitory effects of Na^+ and Mg^{2+} concentrations. Finally, it was proven that the anaerobic digestion effluent could compare with the tap water. The creation of this process has benefited the submerged fermentation industry for wastewater management, recycling, and ultimately a cleaner production process (Zhang et al., 2017).

Whey

Whey, the liquid resulting from the precipitation and removal of milk casein during cheese making, is the most abundant pollutant in dairy wastewater, representing 85 to 95% of milk volume (Zotta et al., 2020). Whey is a waste that is generated in large quantities as a by-product of cheese production (Yalcin et al., 2009). Lactose is important as an usable carbohydrate, accounting for 4.9% of the weight of whey (Mehyar et al., 2005). Additionally, disposal of whey causes serious environmental problems (El-Samragy et al., 1996).

Abou-Zeid et al. (1983) researched the usage of date seed hydrolysate and whey combination for CA production by *C. lipolytica*. Then, 25 g/L of glucose and 2 g/L of $(NH_4)_2SO_4$ were added to the medium and CA concentration was increased to 45 g/L. Also, El-Holi & Al-Delaimy (2003) added in fermentation medium to glucose, fructose, galactose, and sucrose to enhance CA production. Finally, they produced 106.5 g/L of CA in whey containing 15% sucrose. Similarly, Roobini & Gnanavel (2018) also found that the highest CA concentration by *A.niger* ATCC 9642 after 20 days of fermentation was determined in whey contained 15% sucrose. In addition, Mehyar et al. (2005) created a medium which contained 20% molasses in cheese whey and obtained 32.4 g/L of CA by *A. niger* ATCC 6275. The addition of methanol and tricalcium phosphate in the same medium caused a significant increase in CA production. In this case, the CA concentrations were 38.4 and 42.4 g/L, respectively. In conclusion, it was thought that this study was important as it simultaneously evaluated the use of two wastes for CA production. Yalcin et al. (2009) did not apply any pretreatment to whey. On the other hand, only glucose (100 g/L) and fructose (150 g/L) were added to stimulate CA production. The pH value of whey was adjusted to 5.2 with $CaCO_3$, then sterilized at 121°C for 15 minutes. After sterilization, *Y. lipolytica* 57 (domestic strain) and *Y. lipolytica* NBRC 1658 were inoculated into production medium. The highest CA concentrations obtained by domestic strain with the addition of fructose and glucose were 49.23 g/L and 42.31 g/L respectively. Since high nitrogen concentrations may adversely affect CA production, Aslan et al. (2016) heated the whey at 90°C for 15 minutes and removed the whey proteins by filtration and centrifugation, thus reducing the nitrogen concentration and obtaining a fermentation medium with a high C:N ratio. Whey prepared in this way contained 43.6 g/L of total sugar and 3.1 g/L of protein. *Y. lipolytica* (B9) (a domestic strain that can assimilate lactose) was selected as a producer microorganism and sodium alginate was used as the entrapping agent for cell immobilization. Optimum conditions were determined to be temperature of 20°C, a pH of 5.5, 20 g/L of additional lactose, 2%

of sodium alginate, 150 beads/100 mL, and incubation time of 120 h. Under these conditions, the CA concentration and CA:ICA ratio were calculated to be 33.3 g/L and 6.79 g/L, respectively, in non-sterile partly deproteinized cheese whey. Interestingly, immobilized cells were reused in three continuous reaction cycles without any reduction in the maximum CA concentration. Additionally, supplemental lactose increased CA production up to 20 g/L, but additional nitrogen and phosphorus sources caused a sustained decrease. It was also stated that the combination of low pH and temperature was a good barrier factor in preventing contamination, and it was important to use the cold-adapted lactose positive *Y. lipolytica* strain for the first time in non-sterile conditions.

FUTURE RESEARCH DIRECTIONS

It is a great idea to obtain a sustainable process with the use of agro-industrial residues to minimize environmental problems. With the need to meet the increasing CA demands and overcome the challenges posed by current pollution, it is thought that industrialists and researchers need to evaluate wastes as a substrate and optimize the process variables in CA production. In addition, it is important that increasing the production capacity of the strains to be used in CA production by genetic manipulations (mutant strains, gene transfer, etc.) is also very effective. Finally, it was concluded that implementation of one or more of these is strategic steps to be taken to make production more efficient and economical.

CONCLUSION

One of the biggest advantages of biotechnological studies to the industry is considered to be waste recycling. Agricultural wastes are abundant, inexpensive, and renewable resources for the sustainable production of value-added biochemicals such as bioethanol, organic acids, food additives, and enzymes. Similarly, as can be seen from the literature review in the present study, high yields can be obtained for CA production by *A. niger* and *Y. lipolytica*, when some agricultural and industrial wastes are evaluated as substrates. CA is an important organic acid used in many industrial fields, especially pharmaceuticals, cosmetics, and food, with an increasing market share. Most of the CA production in the world is produced by microbial fermentation and the importance given to the studies where wastes are evaluated as potential substrates continue.

ACKNOWLEDGMENT

This research received no specific grant from any funding agency in the public, commercial, or not-for-profit sectors.

REFERENCES

Abbas, N., Safdar, W., Ali, S., Choudhry, S., & Ilahi, S. (2016). Citric acid production from *Aspergillus niger* using banana peel. *International Journal of Scientific and Engineering Research*, *7*(1), 1580–1583.

Abdullah, A., & Mat, H. (2008). Characterisation of solid and liquid pineapple waste. *Reaktor*, *12*(1), 48–52. doi:10.14710/reaktor.12.1.48-52

Abou-Zeid, A. Z. A., Baghlaf, A. O., Khan, J. A., & Makhashin, S. S. (1983). Utilization of date seeds and cheese whey in production of citric acid by *Candida lipolytica*. *Agricultural Wastes*, *8*(3), 131–142. doi:10.1016/0141-4607(83)90112-9

Ali, S. R., Anwar, Z., Irshad, M., Mukhtar, S., & Warraich, N. T. (2016). Biosynthesis of citric acid from single and co-culture-based fermentation technology using agro-wastes. *Journal of Radiation Research and Applied Sciences*, *9*(1), 57–62. doi:10.1016/j.jrras.2015.09.003

Almakki, A., Mirghani, M. E., & Kabbashi, N. A. (2019). Production of citric acid from sugarcane molasses by *Aspergillus niger* using submerged fermentation. *Biological and Natural Resources Engineering Journal*, *2*(1), 47–55.

Amato, A., Becci, A., & Beolchini, F. (2020). Citric acid bioproduction: The technological innovation change. *Critical Reviews in Biotechnology*, *40*(2), 199–212. doi:10.1080/07388551.2019.1709799 PMID:31903797

Angumeenal, A. R., & Venkappayya, D. (2013). An overview of citric acid production. *Lebensmittel-Wissenschaft + Technologie*, *50*(2), 367–370. doi:10.1016/j.lwt.2012.05.016

Aravantinos-Zafiris, G., Tzia, C., Oreopoulou, V., & Thomopoulos, C. D. (1994). Fermentation of orange processing wastes for citric acid production. *Journal of the Science of Food and Agriculture*, *65*(1), 117–120. doi:10.1002/jsfa.2740650117

Arslan, N. P., Aydoğan, M. N., & Taşkın, M. (2016). Citric acid production from partly deproteinized whey under non-sterile culture conditions using immobilized cells of lactose-positive and cold-adapted *Yarrowia lipolytica* B9. *Journal of Biotechnology*, *231*, 32–39. doi:10.1016/j.jbiotec.2016.05.033 PMID:27234881

Ayeni, A., Daramola, M. O., Taiwo, O., Olanrewaju, O. I., Oyekunle, D. T., Sekoai, P. T., & Elehinafe, F. B. (2019). Production of citric acid from the fermentation of pineapple waste by *Aspergillus niger. The Open Chemical Engineering Journal, 13*(1), 88–96. doi:10.2174/1874123101913010088

Behera, B. C. (2020). Citric acid from *Aspergillus niger*: A comprehensive overview. *Critical Reviews in Microbiology, 46*(6), 727–749. doi:10.1080/1040841X.2020.1828815 PMID:33044884

Behera, B. C., Mishra, R., & Mohapatra, S. (2021). Microbial citric acid: Production, properties, application, and future perspectives. *Food Frontiers, 2*(1), 62–76. doi:10.1002/fft2.66

Cavallo, E., Charreau, H., Cerrutti, P., & Foresti, M. L. (2017). *Yarrowia lipolytica*: A model yeast for citric acid production. *FEMS Yeast Research, 17*(8), fox084. doi:10.1093/femsyr/fox084 PMID:29106510

Cavallo, E., Nobile, M., Cerrutti, P., & Foresti, M. L. (2020). Exploring the production of citric acid with *Yarrowia lipolytica* using corn wet milling products as alternative low-cost fermentation media. *Biochemical Engineering Journal, 155*, 107463. doi:10.1016/j.bej.2019.107463

Chysirichote, T. (2020). Valorization of banana peel for citric acid production under solid state fermentation with *Aspergillus niger. Chemical and Biochemical Engineering Quarterly, 34*(1), 49–57. doi:10.15255/CABEQ.2019.1740

da Silva, L. V., Coelho, M. A. Z., da Silva, M. R. S., & Amaral, P. F. F. (2020). Investigation of mitochondrial protein expression profiles of *Yarrowia lipolytica* in response to citric acid production. *Bioprocess and Biosystems Engineering, 43*(9), 1703–1715. doi:10.100700449-020-02363-z PMID:32363419

Dhillon, G. S., Brar, S. K., Kaur, S., & Verma, M. (2013). Screening of agro-industrial wastes for citric acid bioproduction by *Aspergillus niger* NRRL 2001 through solid state fermentation. *Journal of the Science of Food and Agriculture, 93*(7), 1560–1567. doi:10.1002/jsfa.5920 PMID:23108761

Dhillon, G. S., Brar, S. K., Verma, M., & Tyagi, R. D. (2011a). Enhanced solid-state citric acid bio-production using apple pomace waste through surface response methodology. *Journal of Applied Microbiology, 110*(4), 1045–1055. doi:10.1111/j.1365-2672.2011.04962.x PMID:21294819

Dhillon, G. S., Brar, S. K., Verma, M., & Tyagi, R. D. (2011b). Apple pomace ultrafiltration sludge–A novel substrate for fungal bioproduction of citric acid: Optimisation studies. *Food Chemistry*, *128*(4), 864–871. doi:10.1016/j.foodchem.2011.03.107

Dourou, M., Kancelista, A., Juszczyk, P., Sarris, D., Bellou, S., Triantaphyllidou, I. E., Rywinska, A., Papanikolaou, S., & Aggelis, G. (2016). Bioconversion of olive mill wastewater into high-added value products. *Journal of Cleaner Production*, *139*, 957–969. doi:10.1016/j.jclepro.2016.08.133

El-Holi, M. A., & Al-Delaimy, S. (2003). Citric acid production from whey with sugars and additives by *Aspergillus niger. African Journal of Biotechnology*, *2*(10), 356–359. doi:10.5897/AJB2003.000-1073

El-Samragy, Y. A., Khorshid, M. A., Foda, M. I., & Shehata, A. E. (1996). Effect of fermentation conditions on the production of citric acid from cheese whey by *Aspergillus niger. International Journal of Food Microbiology*, *29*(2-3), 411–416. doi:10.1016/0168-1605(95)00072-0 PMID:8796442

Femi-Ola, T. O., Oluyege, J. O., & Gbadebo, A. O. (2009). Citric acid production from pineapple waste. *Continental Journal of Microbiology*, *3*, 1–5.

Ferreira, P., Lopes, M., Mota, M., & Belo, I. (2016). Oxygen mass transfer impact on citric acid production by *Yarrowia lipolytica* from crude glycerol. *Biochemical Engineering Journal*, *110*, 35–42. doi:10.1016/j.bej.2016.02.001

Fickers, P., Cheng, H., & Sze Ki Lin, C. (2020). Sugar alcohols and organic acids synthesis in *Yarrowia lipolytica*: Where are we? *Microorganisms*, *8*(4), 574. doi:10.3390/microorganisms8040574 PMID:32326622

Gasiorek, E. (2008). Effect of operating conditions on biomass growth during citric acid production by solid-state fermentation. *Chemical Papers*, *62*(2), 141–146. doi:10.247811696-008-0003-5

Guo, C., Qiao, J., Zhang, S., Ren, X., & Li, M. (2018). Purification of polyphenols from kiwi fruit peel extracts using macroporous resins and high-performance liquid chromatography analysis. *International Journal of Food Science & Technology*, *53*(6), 1486–1493. doi:10.1111/ijfs.13729

Hamdy, H. S. (2013). Citric acid production by *Aspergillus niger* grown on orange peel medium fortified with cane molasses. *Annals of Microbiology*, *63*(1), 267–278. doi:10.100713213-012-0470-3

Hang, Y. D., Luh, B. S., & Woodams, E. E. (1987). Microbial production of citric acid by solid state fermentation of kiwifruit peel. *Journal of Food Science*, *52*(1), 226–227. doi:10.1111/j.1365-2621.1987.tb14014.x

Hang, Y. D., Splittstoesser, D. F., Woodams, E. E., & Sherman, R. M. (1977). Citric acid fermentation of brewery waste. *Journal of Food Science*, *42*(2), 383–384. doi:10.1111/j.1365-2621.1977.tb01504.x

Hang, Y. D., & Woodams, E. E. (1984). Apple pomace: A potential substrate for citric acid production by *Aspergillus niger*. *Biotechnology Letters*, *6*(11), 763–764. doi:10.1007/BF00133071

Hang, Y. D., & Woodams, E. E. (1986). Solid state fermentation of apple pomace for citric acid production. *MIRCEN Journal of Applied Microbiology and Biotechnology*, *2*(2), 283–287. doi:10.1007/BF00933494

Hang, Y. D., & Woodams, E. E. (1998). Production of citric acid from corncobs by *Aspergillus niger*. *Bioresource Technology*, *65*(3), 251–253. doi:10.1016/S0960-8524(98)00015-7

Hang, Y. D., & Woodams, E. E. (2000). Corn husks: A potential substrate for production of citric acid by *Aspergillus niger*. *Lebensmittel-Wissenschaft + Technologie*, *33*(7), 520–521. doi:10.1006/fstl.2000.0711

Hang, Y. D., & Woodams, E. E. (2001). Enzymatic enhancement of citric acid production by *Aspergillus niger* from corn cobs. *Lebensmittel-Wissenschaft + Technologie*, *34*(7), 484–486. doi:10.1006/fstl.2001.0784

Hesham, A. E. L., Mostafa, Y. S., & AlSharqi, L. E. O. (2020). Optimization of citric acid production by immobilized cells of novel yeast isolates. *Mycobiology*, *48*(2), 122–132. doi:10.1080/12298093.2020.1726854 PMID:32363040

Hijosa-Valsero, M., Paniagua-García, A. I., & Díez-Antolínez, R. (2017). Biobutanol production from apple pomace: The importance of pretreatment methods on the fermentability of lignocellulosic agro-food wastes. *Applied Microbiology and Biotechnology*, *101*(21), 8041–8052. doi:10.100700253-017-8522-z PMID:28932978

Holz, M., Förster, A., Mauersberger, S., & Barth, G. (2009). Aconitase overexpression changes the product ratio of citric acid production by *Yarrowia lipolytica*. *Applied Microbiology and Biotechnology*, *81*(6), 1087–1096. doi:10.100700253-008-1725-6 PMID:18850095

Hu, W., Li, W. J., Yang, H. Q., & Chen, J. H. (2019). Current strategies and future prospects for enhancing microbial production of citric acid. *Applied Microbiology and Biotechnology, 103*(1), 201–209. doi:10.100700253-018-9491-6 PMID:30421107

Imandi, S. B., Bandaru, V. V. R., Somalanka, S. R., Bandaru, S. R., & Garapati, H. R. (2008). Application of statistical experimental designs for the optimization of medium constituents for the production of citric acid from pineapple waste. *Bioresource Technology, 99*(10), 4445–4450. doi:10.1016/j.biortech.2007.08.071 PMID:17936623

Iqbal, J., & Utara, U. (2015). Isolation of *Aspergillus niger* strains from soil and their screening and optimization for enhanced citric acid production using cane molasses as carbon source. *Journal of Applied Environmental and Biological Sciences, 5*(4), 128–137.

Jamir, L., Kumar, V., Kaur, J., Kumar, S., & Singh, H. (2021). Composition, valorization and therapeutical potential of molasses: A critical review. *Environmental Technology Reviews, 10*(1), 131–142. doi:10.1080/21622515.2021.1892203

Javed, S., Asgher, M., Sheikh, M. A., & Nawaz, H. (2010). Strain improvement through UV and chemical mutagenesis for enhanced citric acid production in molasses-based solid state fermentation. *Food Biotechnology, 24*(2), 165–179. doi:10.1080/08905436.2010.482014

Jianlong, W., Xianghua, W., & Ding, Z. (2000). Production of citric acid from molasses integrated with in-situ product separation by ion-exchange resin adsorption. *Bioresource Technology, 75*(3), 231–234. doi:10.1016/S0960-8524(00)00067-5

Kamzolova, S. V., Fatykhova, A. R., Dedyukhina, E. G., Anastassiadis, S. G., Golovchenko, N. P., & Morgunov, I. G. (2011). Citric acid production by yeast grown on glycerol-containing waste from biodiesel industry. *Food Technology and Biotechnology, 49*(1), 65–74.

Kamzolova, S. V., Vinokurova, N. G., Lunina, J. N., Zelenkova, N. F., & Morgunov, I. G. (2015). Production of technical-grade sodium citrate from glycerol-containing biodiesel waste by *Yarrowia lipolytica. Bioresource Technology, 193*, 250–255. doi:10.1016/j.biortech.2015.06.092 PMID:26141285

Kareem, S. O., Akpan, I., & Alebiowu, O. O. (2010). Production of citric acid by *Aspergillus niger* using pineapple waste. *Malaysian Journal of Microbiology, 6*(2), 161–165.

Kareem, S. O., & Rahman, R. A. (2013). Utilization of banana peels for citric acid production by *Aspergillus niger. Agriculture and Biology Journal of North America, 4*(4), 384–387. doi:10.5251/abjna.2013.4.4.384.387

Karthikeyan, A., & Sivakumar, N. (2010). Citric acid production by Koji fermentation using banana peel as a novel substrate. *Bioresource Technology, 101*(14), 5552–5556. doi:10.1016/j.biortech.2010.02.063 PMID:20219361

Khare, S. K., Jha, K., & Gandhi, A. P. (1995). Citric acid production from okara (soy-residue) by solid-state fermentation. *Bioresource Technology, 54*(3), 323–325. doi:10.1016/0960-8524(95)00155-7

Khosravi-Darani, K., & Zoghi, A. (2008). Comparison of pretreatment strategies of sugarcane baggase: Experimental design for citric acid production. *Bioresource Technology, 99*(15), 6986–6993. doi:10.1016/j.biortech.2008.01.024 PMID:18334291

Kiel, H., Guvrin, R., & Henis, Y. (1981). Citric acid fermentation by *Aspergillus niger* on low sugar concentrations and cotton waste. *Applied and Environmental Microbiology, 42*(1), 1–4. doi:10.1128/aem.42.1.1-4.1981 PMID:16345802

Kim, K. H., Lee, H. Y., & Lee, C. Y. (2015). Pretreatment of sugarcane molasses and citric acid production by *Candida zeylanoides. Microbiology and Biotechnology Letters, 43*(2), 164–168. doi:10.4014/mbl.1503.03006

Kolicheski, M. B., Soccol, C. R., Martin, B., Medeiros, T., & Rainbault, M. (1997). Citric acid production on three cellulosic supports in solid state fermentation. In S. Roussos, B. K. Lonsane, M. Raimbault, & G. Viniegra-Gonzalez (Eds.), *Advances in solid state fermentation* (pp. 449–462). Kluwer Academic Publishers. doi:10.1007/978-94-017-0661-2_37

Koutinas, M., Patsalou, M., Stavrinou, S., & Vyrides, I. (2015). High temperature alcoholic fermentation of orange peel by the newly isolated thermotolerant *Pichia kudriavzevii* KVMP 10. *Letters in Applied Microbiology, 62*(1), 75–83. doi:10.1111/lam.12514 PMID:26510181

Kumar, D., Jain, V. K., Shanker, G., & Srivastava, A. (2003). Utilisation of fruits waste for citric acid production by solid state fermentation. *Process Biochemistry, 38*(12), 1725–1729. doi:10.1016/S0032-9592(02)00253-4

Kumar, D., Verma, R., & Bhalla, T. C. (2010). Citric acid production by *Aspergillus niger* van. Tieghem MTCC 281 using waste apple pomace as a substrate. *Journal of Food Science and Technology, 47*(4), 458–460. doi:10.100713197-010-0077-2 PMID:23572671

Kumar, S. S., Swapna, T. S., & Sabu, A. (2018). Coffee husk: A potential agro-industrial residue for bioprocess. In R. Singhania, R. Agarwal, R. Kumar, & R. Sukumaran (Eds.), *Waste to wealth. Energy, environment, and sustainability* (pp. 97–109). Springer. doi:10.1007/978-981-10-7431-8_6

Kurbanoglu, E. B. (2004). Enhancement of citric acid production with ram horn hydrolysate by *Aspergillus niger. Bioresource Technology*, *92*(1), 97–101. doi:10.1016/j.biortech.2003.08.002 PMID:14643992

Kurbanoglu, E. B., & Kurbanoglu, N. I. (2003). Production of citric acid from ram horn hydrolysate by *Aspergillus niger. Process Biochemistry*, *38*(10), 1421–1424. doi:10.1016/S0032-9592(03)00029-3

Liu, X., Lv, J., Xu, J., Xia, J., He, A., Zhang, T., Xu, X., & Xu, J. (2018). Effects of osmotic pressure and pH on citric acid and erythritol production from waste cooking oil by *Yarrowia lipolytica. Engineering in Life Sciences*, *18*(6), 344–352. doi:10.1002/elsc.201700114 PMID:32624914

Liu, X., Lv, J., Xu, J., Zhang, T., Deng, Y., & He, J. (2015c). Citric acid production in *Yarrowia lipolytica* SWJ-1b yeast when grown on waste cooking oil. *Applied Biochemistry and Biotechnology*, *175*(5), 2347–2356. doi:10.100712010-014-1430-0 PMID:25488499

Liu, X., Lv, J., Zhang, T., & Deng, Y. (2015a). Citric acid production from hydrolysate of pretreated straw cellulose by *Yarrowia lipolytica* SWJ-1b using batch and fed-batch cultivation. *Preparative Biochemistry & Biotechnology*, *45*(8), 825–835. doi:10.1080/10826068.2014.979203 PMID:25356914

Liu, X., Wang, X., Xu, J., Xia, J., Lv, J., Zhang, T., Wu, Z., Deng, Y., & He, J. (2015b). Citric acid production by *Yarrowia lipolytica* SWJ-1b using corn steep liquor as a source of organic nitrogen and vitamins. *Industrial Crops and Products*, *78*, 154–160. doi:10.1016/j.indcrop.2015.10.029

Magangana, T. P., Makunga, N. P., Fawole, O. A., & Opara, U. L. (2020). Processing factors affecting the phytochemical and nutritional properties of pomegranate (*Punica granatum L.*) peel waste: A review. *Molecules (Basel, Switzerland)*, *25*(20), 4690. doi:10.3390/molecules25204690 PMID:33066412

Mahato, N., Sharma, K., Sinha, M., Dhyani, A., Pathak, B., Jang, H., Park, S., Pashikanti, S., & Cho, S. (2021). Biotransformation of citrus waste-I: Production of biofuel and valuable compounds by fermentation. *Processes (Basel, Switzerland)*, *9*(2), 220. doi:10.3390/pr9020220

Mehyar, G. F., Delaimy, K. S., & Ibrahim, S. A. (2005). Citric acid production by *Aspergillus niger* using date-based medium fortified with whey and additives. *Food Biotechnology, 19*(2), 137–144. doi:10.1081/FBT-200063458

Monrroy, M., Rueda, L., Aparicio, A. L., & García, J. R. (2019). Fermentation of *Musa paradisiaca* peels to produce citric acid. *Journal of Chemistry, 2019*, 1–8. doi:10.1155/2019/8356712

Mores, S., de Souza Vandenberghe, L. P., Júnior, A. I. M., de Carvalho, J. C., de Mello, A. F. M., Pandey, A., & Soccol, C. R. (2020). Citric acid bioproduction and downstream processing: Status, opportunities, and challenges. *Bioresource Technology*, 124426. PMID:33249260

Nayak, A., & Bhushan, B. (2019). An overview of the recent trends on the waste valorization techniques for food wastes. *Journal of Environmental Management, 233*, 352–370. doi:10.1016/j.jenvman.2018.12.041 PMID:30590265

Özdal, M., & Kurbanoglu, E. B. (2019). Citric acid production by *Aspergillus niger* from agro-industrial by-products: Molasses and chicken feather peptone. *Waste and Biomass Valorization, 10*(3), 631–640. doi:10.100712649-018-0240-y

Pandey, A., Soccol, C. R., Nigam, P., Brand, D., Mohan, R., & Roussos, S. (2000). Biotechnological potential of coffee pulp and coffee husk for bioprocesses. *Biochemical Engineering Journal, 6*(2), 153–162. doi:10.1016/S1369-703X(00)00084-X PMID:10959086

Papadaki, E., & Mantzouridou, F. T. (2019). Citric acid production from the integration of Spanish-style green olive processing wastewaters with white grape pomace by *Aspergillus niger. Bioresource Technology, 280*, 59–69. doi:10.1016/j.biortech.2019.01.139 PMID:30754006

Papagianni, M. (2007). Advances in citric acid fermentation by *Aspergillus niger*: Biochemical aspects, membrane transport and modeling. *Biotechnology Advances, 25*(3), 244–263. doi:10.1016/j.biotechadv.2007.01.002 PMID:17337335

Papanikolaou, S., Galiotou-Panayotou, M., Fakas, S., Komaitis, M., & Aggelis, G. (2008). Citric acid production by *Yarrowia lipolytica* cultivated on olive-mill wastewater-based media. *Bioresource Technology, 99*(7), 2419–2428. doi:10.1016/j.biortech.2007.05.005 PMID:17604163

Parameswaran, B. (2009). Sugarcane bagasse. *Biotechnology for Agro-Industrial Residues Utilisation*, 239-252.

Pazouki, M., Felse, P. A., Sinha, J., & Panda, T. (2000). Comparative studies on citric acid production by *Aspergillus niger* and *Candida lipolytica* using molasses and glucose. *Bioprocess Engineering*, *22*(4), 353–361. doi:10.1007/PL00009115

Ramachandra, Y. L., Narayanamurthy, G., Jois, S., Chavan, A., & Satwadi, P. R. (2013). Production of citric acid in basal coffee husk medium by *Aspergillus niger* under solid state fermentation. *Advances in Biological Research (Faisalabad)*, *7*(6), 234–240.

Rivas, B., Torrado, A., Torre, P., Converti, A., & Domínguez, J. M. (2008). Submerged citric acid fermentation on orange peel autohydrolysate. *Journal of Agricultural and Food Chemistry*, *56*(7), 2380–2387. doi:10.1021/jf073388r PMID:18321055

Roobini, U., & Gnanavel, D. G. (2018). Citric acid Production from whey by *Aspergillus niger. IOSR Journal of Engineering*, *8*(9), 65–71.

Roukas, T., & Kotzekidou, P. (1986). Production of citric acid from brewery wastes by surface fermentation using *Aspergillus niger. Journal of Food Science*, *51*(1), 225–228. doi:10.1111/j.1365-2621.1986.tb10876.x

Roukas, T., & Kotzekidou, P. (1987). Influence of some trace metals and stimulants on citric acid production from brewery wastes by *Aspergillus niger. Enzyme and Microbial Technology*, *9*(5), 291–294. doi:10.1016/0141-0229(87)90006-8

Roukas, T., & Kotzekidou, P. (2020). Pomegranate peel waste: A new substrate for citric acid production by *Aspergillus niger* in solid-state fermentation under non-aseptic conditions. *Environmental Science and Pollution Research International*, *27*(12), 13105–13113. doi:10.100711356-020-07928-9 PMID:32016865

Rymowicz, W., Fatykhova, A. R., Kamzolova, S. V., Rywińska, A., & Morgunov, I. G. (2010). Citric acid production from glycerol-containing waste of biodiesel industry by *Yarrowia lipolytica* in batch, repeated batch, and cell recycle regimes. *Applied Microbiology and Biotechnology*, *87*(3), 971–979. doi:10.100700253-010-2561-z PMID:20376633

Rymowicz, W., Rywińska, A., Żarowska, B., & Juszczyk, P. (2006). Citric acid production from raw glycerol by acetate mutants of *Yarrowia lipolytica. Chemical Papers*, *60*(5), 391–394. doi:10.247811696-006-0071-3

Rywińska, A., Rymowicz, W., Zarowska, B., & Wojtatowicz, M. (2009). Biosynthesis of citric acid from glycerol by acetate mutants of *Yarrowia lipolytica* in fed-batch fermentation. *Food Technology and Biotechnology*, *47*(1), 1–6.

Rzechonek, D. A., Dobrowolski, A., Rymowicz, W., & Mirończuk, A. M. (2019). Aseptic production of citric and isocitric acid from crude glycerol by genetically modified *Yarrowia lipolytica*. *Bioresource Technology*, *271*, 340–344. doi:10.1016/j.biortech.2018.09.118 PMID:30292133

Saleem, A., Hussain, A., Chaudhary, A., Iqtedar, M., Javid, A., & Akram, A. M. (2020). Acid hydrolysis optimization of pomegranate peels waste using response surface methodology for ethanol production. *Biomass Conversion and Biorefinery*, 1–12.

Sarris, D., Galiotou-Panayotou, M., Koutinas, A. A., Komaitis, M., & Papanikolaou, S. (2011). Citric acid, biomass and cellular lipid production by *Yarrowia lipolytica* strains cultivated on olive mill wastewater-based media. *Journal of Chemical Technology and Biotechnology (Oxford, Oxfordshire)*, *86*(11), 1439–1448. doi:10.1002/jctb.2658

Sarris, D., Rapti, A., Papafotis, N., Koutinas, A. A., & Papanikolaou, S. (2019). Production of added-value chemical compounds through bioconversions of olive-mill wastewaters blended with crude glycerol by a *Yarrowia lipolytica* strain. *Molecules (Basel, Switzerland)*, *24*(2), 222. doi:10.3390/molecules24020222 PMID:30634450

Sarris, D., Stoforos, N. G., Mallouchos, A., Kookos, I. K., Koutinas, A. A., Aggelis, G., & Papanikolaou, S. (2017). Production of added-value metabolites by *Yarrowia lipolytica* growing in olive mill wastewater-based media under aseptic and non-aseptic conditions. *Engineering in Life Sciences*, *17*(6), 695–709. doi:10.1002/elsc.201600225 PMID:32624815

Sassi, G., Ruggeri, B., Specchia, V., & Gianetto, A. (1991). Citric acid production by *A. niger* with banana extract. *Bioresource Technology*, *37*(3), 259–269. doi:10.1016/0960-8524(91)90193-N

Sayın Börekçi, B., Kaban, G., & Kaya, M. (2020). Citric acid production of yeasts: An overview. *The Eurobiotech Journal*, *5*(2), 79–91. doi:10.2478/ebtj-2021-0012

Shalini, R., & Gupta, D. K. (2010). Utilization of pomace from apple processing industries: A review. *Journal of Food Science and Technology*, *47*(4), 365–371. doi:10.100713197-010-0061-x PMID:23572655

Shankaranand, V. S., & Lonsane, B. K. (1994). Coffee husk: An inexpensive substrate for production of citric acid by *Aspergillus niger* in a solid-state fermentation system. *World Journal of Microbiology & Biotechnology*, *10*(2), 165–168. doi:10.1007/BF00360879 PMID:24420939

Sharma, N., Sahota, P. P., & Singh, M. P. (2021). Organic acid production from agricultural waste. In B. K. Kashyap, M. K. Solanki, D. V. Kamboj, & A. K. Pandey (Eds.), *Waste to Energy: Prospects and Applications* (pp. 415–437). Springer.

Shojaosadati, S. A., & Babaeipour, V. (2002). Citric acid production from apple pomace in multi-layer packed bed solid-state bioreactor. *Process Biochemistry*, *37*(8), 909–914. doi:10.1016/S0032-9592(01)00294-1

Show, P. L., Oladele, K. O., Siew, Q. Y., Aziz Zakry, F. A., Lan, J. C. W., & Ling, T. C. (2015). Overview of citric acid production from *Aspergillus niger. Frontiers in Life Science*, *8*(3), 271–283. doi:10.1080/21553769.2015.1033653

Sidauruk, M. G. E., Hutauruk, K. S. N., Martgrita, M. M., & Manurung, A. (2019). Citric acid production from toba banana peel (*Musa acuminata* colla) through submerged fermentation using *Aspergillus niger. Microbiology Indonesia*, *13*(4), 2–2.

Soccol, C. R., Vandenberghe, L. P., Rodrigues, C., & Pandey, A. (2006). New perspectives for citric acid production and application. *Food Technology and Biotechnology*, *44*(2).

Srivastava, N., Srivastava, M., Ramteke, P., & Mishra, P. (2019). Solid-state fermentation strategy for microbial metabolites production: An overview. In V. K. Gupta & A. Pandey (Eds.), *New and future developments in microbial biotechnology and bioengineering* (pp. 345–354). Elsevier. doi:10.1016/B978-0-444-63504-4.00023-2

Srivastava, N., Srivastava, M., Ramteke, P. W., & Mishra, P. K. (2019). Solid-state fermentation strategy for microbial metabolites production: An overview. *New and Future Developments in Microbial Biotechnology and Bioengineering*, 345-354.

Stenesh, J. (1998). The citric acid cycle. In *Biochemistry* (pp. 273–291). Springer. doi:10.1007/978-1-4757-9427-4_11

Torrado, A. M., Cortés, S., Salgado, J. M., Max, B., Rodríguez, N., Bibbins, B. P., Converti, A., & Domínguez, J. M. (2011). Citric acid production from orange peel wastes by solid-state fermentation. *Brazilian Journal of Microbiology*, *42*(1), 394–409. doi:10.1590/S1517-83822011000100049 PMID:24031646

Tran, C. T., & Mitchell, D. A. (1995). Pineapple waste-a novel substrate for citric acid production by solid-state fermentation. *Biotechnology Letters*, *17*(10), 1107–1110. doi:10.1007/BF00143111

Tran, C. T., Sly, L. I., & Mitchell, D. A. (1998). Selection of a strain of *Aspergillus* for the production of citric acid from pineapple waste in solid-state fermentation. *World Journal of Microbiology & Biotechnology*, *14*(3), 399–404. doi:10.1023/A:1008821413239

Vandenberghe, L. P., Soccol, C. R., Pandey, A., & Lebeault, J. M. (2000). Solid-state fermentation for the synthesis of citric acid by *Aspergillus niger*. *Bioresource Technology*, *74*(2), 175–178. doi:10.1016/S0960-8524(99)00107-8

Vendruscolo, F., Albuquerque, P. M., Streit, F., Esposito, E., & Ninow, J. L. (2008). Apple pomace: A versatile substrate for biotechnological applications. *Critical Reviews in Biotechnology*, *28*(1), 1–12. doi:10.1080/07388550801913840 PMID:18322854

Yadegary, M., Hamidi, A., Alavi, S. A., Khodaverdi, E., Yahaghi, H., Sattari, S., Bagherpour, G., & Yahaghi, E. (2013). Citric acid production from sugarcane bagasse through solid state fermentation method using *Aspergillus niger* mold and optimization of citric acid production by Taguchi method. *Jundishapur Journal of Microbiology*, *6*(9). Advance online publication. doi:10.5812/jjm.7625

Yalcin, S. K. (2012). Enhancing citric acid production of *Yarrowia lipolytica* by mutagenesis and using natural media containing carrot juice and celery byproducts. *Food Science and Biotechnology*, *21*(3), 867–874. doi:10.100710068-012-0112-3

Yalcin, S. K., Bozdemir, M. T., & Ozbas, Z. Y. (2009). Utilization of whey and grape must for citric acid production by two *Yarrowia lipolytica* strains. *Food Biotechnology*, *23*(3), 266–283. doi:10.1080/08905430903106860

Yalcin, S. K., Bozdemir, M. T., & Ozbas, Z. Y. (2010). Citric acid production by yeasts: Fermentation conditions, process optimization and strain improvement. *Current Research. Technology and Education Topics in Applied Microbiology and Microbial Biotechnology*, *9*, 1374–1382.

Zhang, H., Xu, J., Su, X., Bao, J., Wang, K., & Mao, Z. (2017). Citric acid production by recycling its wastewater treated with anaerobic digestion and nanofiltration. *Process Biochemistry*, *58*, 245–251. doi:10.1016/j.procbio.2017.04.022

Zhang, S., Jagtap, S. S., Deewan, A., & Rao, C. V. (2019). pH selectively regulates citric acid and lipid production in *Yarrowia lipolytica* W29 during nitrogen-limited growth on glucose. *Journal of Biotechnology*, *290*, 10–15. doi:10.1016/j.jbiotec.2018.10.012 PMID:30496777

Zotta, T., Solieri, L., Iacumin, L., Picozzi, C., & Gullo, M. (2020). Valorization of cheese whey using microbial fermentations. *Applied Microbiology and Biotechnology*, *104*(7), 2749–2764. doi:10.100700253-020-10408-2 PMID:32009200

KEY TERMS AND DEFINITIONS

Autohydrolysis: A technique that involves the use of hot water or steam as a reagent to break down a lignocellulosic biomass.

Biomass: The total quantity or weight of living organisms.

Bioremediation: Waste management technique that to remove toxic waste from polluted environment by certain microorganisms.

Compound Annual Growth Rate: Annual growth of investments over a specific period of time.

FDA: The Food and Drug Administration that is responsible for protecting the public health.

Tofu: Product prepared by coagulating soy milk.

Volumetric Productivity: Product produced per unit volume of fermentation medium per unit time.

Chapter 4
Green Polymer Chemistry and Bio-Based Plastic

Ankur Khapre
University of Hyderabad, India

Payal Panwar
BITS Pilani, India

ABSTRACT

The overview briefly surveys the practice of green chemistry in polymer science. The conservation and sustainable utilization of polymers is achieved through documentation of indigenous knowledge through the ideology for syntheses of green polymers. The economic importance of polymers includes the study of the relationship between people and polymers. The chapter offers a systematic approach to study the variety of materials. It intersects many fields including the need for green polymer chemistry, environmentally friendly methods for syntheses of green polymers, biodegradable polymers, recycling of polymers, and many more topics. This study aims to provide an updated survey of the green synthesised polymers in order to obtain recyclable materials for various industrial and indoor applications. A number of barriers that hinder the adoption of newer technology have also been discussed. All of these areas are experiencing an increase in research activity with the development of new tools and technologies. Examples are given of recent developments in green polymer chemistry.

DOI: 10.4018/978-1-7998-9851-1.ch004

INTRODUCTION

Green chemistry is the development of chemical products and processes that decrease or eliminate the usage of hazardous chemicals. The term "sustainability" refers to growth that fulfills current demands without jeopardizing future generations' ability to satisfy their own. These ideas have gained attraction in recent years and have become major study subjects. Several books and review papers on toxicity reduction in polymers have been published in recent years (Mat lack *et al*, 2021), and there is growing interest in green chemistry. Many recent symposia on this issue have been held at national ACS meetings, demonstrating this.

Polymer is the only substance on the planet that is both highly appreciated for its use and widely reviled. We have conflicting, contradictory, and vacillating sentiments regarding polymers, and we've never determined whether they're good, terrible, or ugly. Their newness is possibly one of the reasons behind their ambivalence. The fast expansion of polymer manufacture was a twentieth-century phenomenon, and anything less than a century old is considered new in historical terms. Polymers are newbies among materials, and we haven't had enough time to form opinions about them. It would be unintelligent not to give polymers due attention because of their obvious use. They are frequently not only less expensive than other materials, but they often offer better characteristics. Their cheap cost has life-saving implications, such as in drought-prone parts of Africa, where lightweight plastic water pails have replaced clay and stone containers, allowing people to bring water in even from remote wells during acute water scarcity. Polymers are also well-suited to the modern information-age applications of cell phones, bank cards, and laptops. Even when it comes to basic comfort, no one can disagree that polymers are remarkable performers (Stevens *et al,* 2002).

Modern polymer technology has green routes. Polymers have an important part in both natural and man-made technologies as structural and multifunctional macromolecular materials that are extremely adaptable and diverse. Polymers ensure a good quality of life and act as pacemakers for contemporary technology, thus modern life would be impossible without them.

In a green economy, reduced demand for resources and energy, as well as waste minimization, prevention of pollution and dangers, reduction of greenhouse gas emissions, optimization of industrial processes, and effective waste recycling are all critical. These components are essential to sustainable chemistry, often known as green chemistry, which was coined in the 1990s. Biomaterials and biotechnology are not synonymous with "green" in this sense. In fact, many existing polymers and polymerization processes meet the demands of green chemistry. Polyolefins are well-known examples of effective sustainable materials like polyethylene and polypropylene, which account for almost half of worldwide polymer output. Polymers have become key components of nearly any consumer product that satisfies human society's diverse demands. (Andrade *et al*, 2003)

Most polymers have unique property combinations that no other class of materials can match.

- Lightweight
- Excellent corrosion resistance
- Reasonable cost
- High flexibility in terms of characteristics and applications

During the twentieth century, a wide range of synthetic polymers became commercially available. Petrochemistry, which first appeared in the 1950s, has a strong technological foundation. The use of oil and gas as fossil raw resources in the chemical industry and polymer synthesis has considerably increased cost-effectiveness and simplified macromolecular material manufacturing. Since then, the polymer has been defined by its appealing mix of low cost, ease of production, and innovation.

The decline in the use of less competitive natural polymers, which account for less than 1% of today's polymer production of 300 million tons per year, was attributed to the rise of low-cost synthetic polymers with far superior properties, produced in high energy and resource-efficient polymerization processes. We are witnessing a renaissance of renewable polymers and a strong push toward the creation of bio-based macromolecular materials at the dawn of the twenty-first century.

Is polymer's future going to be green? There are a number of causes for this paradigm change and the anticipated transition from petrochemistry to bio-economy. In theory, there are three main renewable polymer techniques that can be used in a green economy. Bio-refining of biomass and chemical conversion of carbon dioxide are used in the approach to create synthetic crude oil and green monomers for extremely resource and energy-efficient polymer manufacturing processes without imperiling existing recycling technologies.

The following are some of the most important green polymer production principles: (Umlaut *et al,* 2012)

- Production method that is clean and lean.
- High levels of safety.
- No auxiliary chemicals are used.
- By reducing toxicity.
- Materials manufacture and applications have high energy efficiency.

As a result, polymer wastes became an important source of raw materials and energy. Polymers are designed to be stiff, soft, rubbery, conducting or insulating, optically transparent or opaque, permeable or impermeable, stable or (bio)degradable in today's highly efficient industrial polymerization processes. Food and medical packaging materials are two prominent examples of polymer uses and lightweight engineering. (Stevens *et al,* 2002)

WHY NEED GREEN POLYMERIZATION METHODS?

Polymer materials have almost permeated every area of the national economy and people's lives since the synthesis of the first synthetic polymer in the 1970s, due to their wide variety, great performance, reasonable price, and a range of other useful uses. The utilization of polymer materials has grown to the point that worldwide output is anticipated to surpass 500 million metric tons by 2050 (Sardon and Dove *et al,* 2018). Because of their stability and long durability, discarded synthetic polymeric materials such as common plastic materials tend to collect in the environment rather than disintegrate.

Figure 1. White pollution presentation a) entire planet, b) plastic wastes on mountain range and forest, c) wider use of agro - based mulch and polymer waste, d) ocean life, plastic trapping, (Westminster City Council | Commercial Waste Services, 2020)

Dumping in landfills and incineration are the primary methods of solid waste disposal for plastic products (Duan *et al,* 2019). These techniques, however, are not environmentally sustainable since landfills pollute groundwater and waste materials are burned, releasing harmful gases into the atmosphere (Khatoon *et al,* 2019). Furthermore, following their end-of-life, recovery, and recycling of plastic materials has not kept pace with current use, resulting in massive buildup over time, compounding the disposal problem and contributing to the environmental burden. Only thirteen percent of all plastic is presently recycled, according to reports twelve percent by mechanical means and one percent by monomer recycling. The great bulk of forty percent plastic trash ends up in landfills, twenty-five percent is burned, and the remaining twenty-two percent ends up in unorganized dumps or is rejected in recycling (Degnan *et al,* 2019).

As a result, one of the hotspots in academia should be creating a large number of practical civil biodegradable polymer materials, as well as engineering novel catalysts or polymerization techniques to lower the application cost of biodegradable polymeric materials. Consumers, after all, are rarely ready to pay extra for a product that is both natural and biodegradable. Richard A. Gross addressed it. A biodegradable polymer is one with a high molecular weight that can be broken down by micro-organisms and enzymes to lower molecular weight components. Polyester Poly-caprolactone and Poly-lactase acid, amide-containing polymer, polyurethane, and almost every natural macromolecule have hetero atoms in the polymer backbone that are vulnerable to hydrolytic cleavage of ester groups (-COO-), amide bonds (-CONH-), and ether bonds (-O-). Aliphatic polyesters and related copolymers are important biodegradable materials. These polymers have the potential to be useful as a biodegradable packaging material as well as in other fields such as medicine, pharmacology, and agriculture (Sun *et al,* 2018). In the following part, we'll go through how polyester is used. It has been observed that Aliphatic Polyesters may be hydrolyzed to produce an oligomer and then a monomer (Sriyapai and Sashiwa *et al,* 2017). The crystallinity, porosity, and pH of the solution all influence the rate of deterioration. Low crystallinity and excessive porosity both speed up the rate of deterioration.

Natural macromolecules, as the name indicates, are macromolecules that may be taken directly from nature and generated by living organisms. The most researched natural materials include natural rubber, lignin, and polysaccharides. Cellulose, Chitin, Chitosan, Amylose, Sodium Alginate, and other polysaccharides are examples of natural macromolecules. Others In addition to the above important polyesters and polysaccharides, there are other types of biodegradable polymers, such as biodegradable polyurethane (Peng *et al,* 2014 & Howard *et al* 2002), amide-containing polymer polypeptide (Hayashi *et al,* 2002), proteins (Tosta *et al* 2019) and so on.

To date, poly (vinyl alcohol) is the only biodegradable polymer containing solely carbon atoms in the polymer backbone (Scott *et al,* 2002), and it is now utilized in the textiles, paper, and packaging sectors as paper coatings, adhesives, and films. However, as a managed end-of-life cycle alternative in regulated industrial compost facilities, the feature of biodegradability may be beneficially employed for managing plastic waste coming notably from certain industries such as food packaging.

Packaging is one of the most important industries for both consuming plastics and producing trash (Agarwal *et al,* 2020).

BIODEGRADABLE POLYMERS PRESENT OPPORTUNITY AND CHALLENGES:

Polyolefins (polyethylene and polypropylene) and polyethylene-terephthalate are the most common polymers found as Micro-Plastics in the environment (Li *et al,* 2019). These high-molecular-mass polymers with an extremely stable carbon-carbon or C-heteroatom backbone survive in the environment for tens of thousands of years. Therefore, biodegradable polymers are very often discussed as one of the solutions to the present plastic pollution issues by substituting Micro-Plastics creating polyolefins in general or for a few specific applications of polymers intended for use in the natural environment.

Based on recent and previous studies, the use of biodegradable polymers looks to be extremely promising; nevertheless, numerous factors of environmental sustainability, structure-property interactions, and biodegradation in the complex natural environment must be considered. To understand the function of biodegradable polymers in the context of Micro-Plastics, the current situation, and future directions, it is necessary to first revisit some of the fundamentals of biodegradation (Eskander *et al,* 2017 & Tokiwa *et al,* 2009).

The first stage is fragmentation, which involves breaking down a high molar mass macromolecular chain into oligomers with polar functional chain ends and monomers without losing particular polymer characteristics like molar mass and strength. This step can take place due to hydrolysis, oxidation, or any other means, depending on the chemical structure of the polymer backbone and the environment in which the polymer is being disposed and used.

In the second step, oligomers with polar chain ends and monomers are mineralized by micro-organisms forming ultimately carbon dioxide, methane, water, and biomass. The product changes depending on the amount of oxygen available. Several review studies exist that describe the biodegradation process in detail (Chandra *et al,* 1998 & Lucas *et al,* 2008). One type of biodegradable polymer is aliphatic polyesters having readily hydrolyzable ester units in the chain. A copolymer of terephthalic acid, butanediol, and adipic acid (poly-butylenes adipate-co-butylenes terephthalate) is an example of latter copolymers. Semi-crystalline are poly-caprolactone, poly-lactide, poly-glycolide.

Their copolymers, and particular polyesters generated by bacteria: poly-hydroxy-alkanoates are some examples of aliphatic biodegradable polyesters described in the literature (Shah *et al,* 1998). A variety of processes can contribute to the initial phase of biodegradation in the natural surroundings, speeding up the total decomposition.

ENVIRONMENT FRIENDLY METHOD'S FOR POLYMERS SYNTHESIS:

1. Polymer from carbon dioxide:

Carbon dioxide is a greenhouse gas that exists in large quantities in the air. Before it is dissolved into the oceans, the typical lifespan of a CO2 molecule is estimated to be on the order of decades (Revelle *et al,* 1957). Carbon dioxide is a desirable feedstock for the chemical industry since it is a naturally occurring, plentiful, low-cost, nonflammable, and renewable chemical component (Darensbourg *et al,* 2007 & Eberhardt *et al,* 2002). Carbon dioxide may be recovered in huge quantities via power plant carbon capture, fossil fuel combustion, and steam reforming to create hydrogen from water and coal. Carbon dioxide must be activated with energy before it can be used in industry. Carbon dioxide is reacted with energy-rich strained rings, such as oxidants, to generate cyclic carbonates in industrially feasible methods. The present technology, as well as future approaches, have been briefly outlined.

Here are some important methods to synthesize polymers by CO2:

a. **Copolymerization of CO2 and Epoxides:**

Inoue and coworkers used organ metallic catalysts to effectively copolymerize CO2 with epoxides in 1969 (Kindermann *et al,* 2017). By alternating copolymerization of CO2 with propylene oxide, they were able to make high molecular weight polycarbonates.

Figure 2. Synthesis of poly-cyclohexene carbonate from cyclohexene oxide (Darensbourg, 1995)

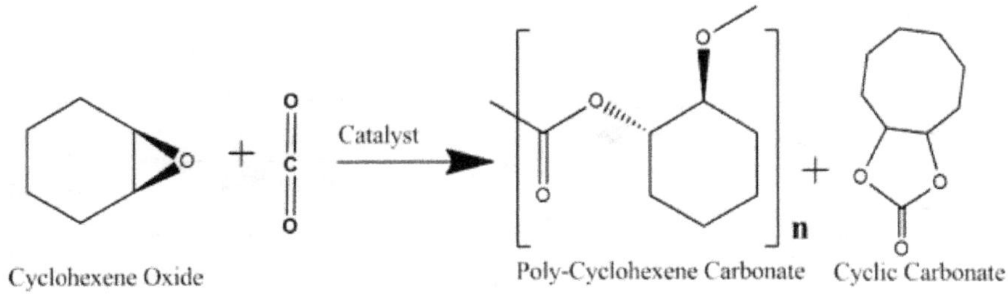

b. **Copolymerization of CO2 and Alkynes:**

The carbon-carbon triple bonds are used as the key components in the copolymerization of CO2 with alkynes. This is a one-of-a-kind method for producing poly-alkanoates with a high molecular weight average (MW 31,400). The reactions were carried out at a relatively low temperature (80°C for 12 hours). An industrially feasible process would be both efficient and resilient, resulting in a high yield of the end product. By continually adding alkyl-di-halide in the presence of catalyst and CO2, the produced telechelic polymer may be utilized to manufacture high molecular weight functional polymers. (Wang *et al.* 2020)

Figure 3. Polymerization reaction of dienes, alkyl halide and CO2 in presence of Ag2WO4 and Cs2CO3 (Wang. 2020)

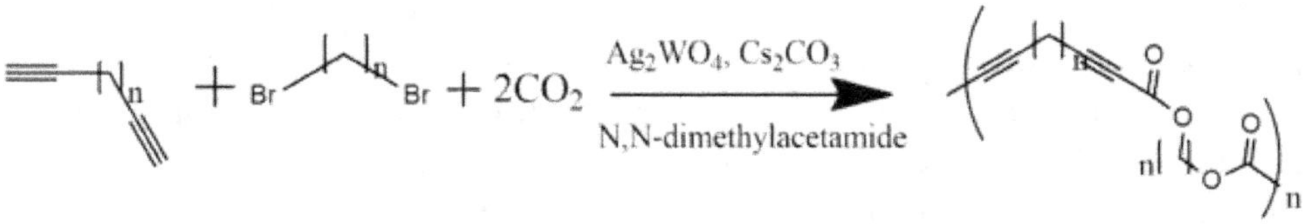

c. **Copolymerization of CO2 and Olefins:**

The copolymerization of CO2 with olefins has piqued attention; nevertheless, it is e endothermic propagation phase, and homo-polymerization of olefins has occurred in many instances owing to reaction failure. Furthermore, the synthesized lactones were polymerized in the presence of 1,1-azobis cyclohexane-1-carbonitrile and zinc chloride, resulting in the formation of terpolymers (CO2, butadiene or isoprene) and (CO2, butadiene or 1,3-pentadiene) with molecular weights 5,500 and 16,000, respectively, and yields of the product are 46 and 35 percent. (Nakano *et al.* 2014)

Figure 4. Copolymerization of CO2 with butadiene by radical polymerization (Tamura, 2016)

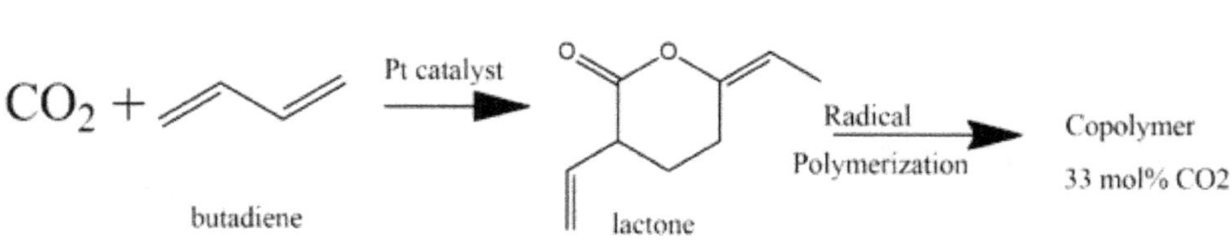

d. Copolymerization of CO2 and Diols:

This method involves direct copolymerization. By taking into account traditional polycarbonate manufacturing techniques which typically involve significant costs or dangerous chemicals. In the presence of condensing agents, CO2 was reacted with p- and m-xylene glycols to produce poly-xylem carbonate (Tamura *et al.* 2016).

e. Poly-Methacrylates from CO2:

Under moderate circumstances, it is feasible to make five-member cyclic carbonates by reacting CO2 with epoxides. However, the many steps necessary to convert CO2 make these reactions inefficient. Using CO2 is generally made in a two-step reaction that necessitates an extra purification step as well as a long reaction time.

Figure 5. Synthesis of poly-gycidyl methacrylate from CO2 (Ochiai, 2008)

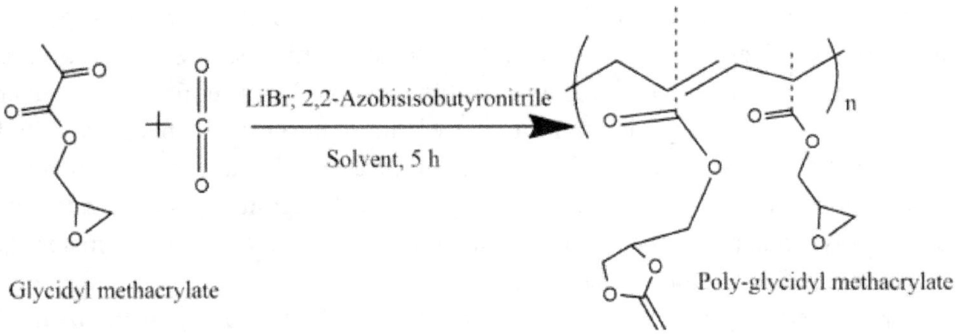

f. Bioderived Epoxides from CO2:

Although another limonene oxide, which is structurally similar to cyclohexane oxide and has a cheap cost, is common. Under moderate circumstances, Byrne et al. reported the alternating copolymerization of limonene oxide and CO2 to produce polycarbonates (Byrne *et al,* 2004). Di-iminate zinc acetate was used as a catalyst for copolymerization. (Saini *et al.* 2017)

Figure 6. Poly(limonene) dicarbonate from cis/trans limonene oxide (Saini, 2017)

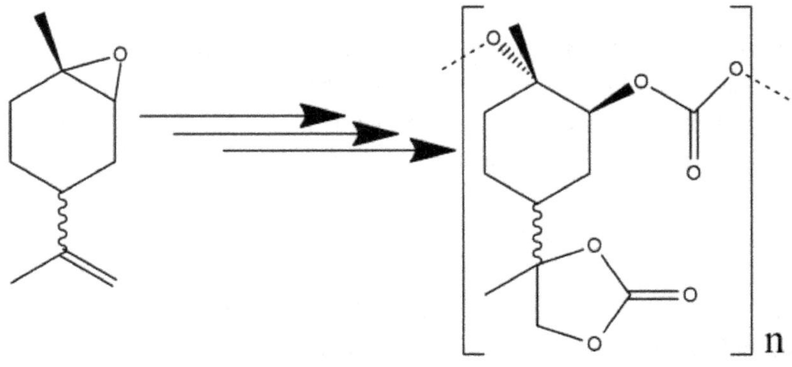

2. Microwave-Assisted Single-Step Synthesis:

The microwave-assisted synthesis of organic molecules has piqued curiosity since its discovery. When compared to typical synthetic techniques, direct activation of the reagents saves a significant amount of time and produces significantly greater yields. In the realm of organic synthesis, these benefits are currently being exploited more often (Da Silva *et al,* 2007 & Geresh *et al,* 2004). Because it may increase the functional characteristics of natural polysaccharides, graft copolymerization of natural polysaccharides is becoming a valuable resource for producing new materials (Singh *et al,* 2007 & Masuhiro *et al,* 2005).

The grafted polymers are often made by the traditional redox grafting method (Nayak & Tripathi *et al,* 2001), microwave irradiation (Masuhiro *et al,* 2005 & Singh *et al,* 2004), gamma-ray irradiation (Wang *et al,* 2008), or electron beam irradiation (Vahdat *et al,* 2007). Furthermore, the percentage grafting in microwave irradiation-based synthesis is dependent on the reaction parameters, namely monomer concentration, power, and irradiation duration, the last two are electronically regulated, assuring excellent repeatability. The main benefit is that it allows for virtually immediate uniform heating of materials in a controlled manner (Deshayes *et al,* 1999 & Wiesbrock *et al,* 2004). Concerns about the environment and a sustainable eco-friendly approach to the creation of materials for high technology have led to the development of improved flocculants for wastewater treatment based on modified polysaccharides (Poly-acrylamide) based flocculants, on the other hand, are extremely efficient at very low dosages, fragile, and cheap, and are commonly employed in wastewater and industrial effluent treatment. Brown algae cell walls are used to extract it. It's widely used in the food industry as an emulsifier and to improve viscosity. It is also used in the production of dental impressions and indigestion pills. Reactive dye printing, which is utilized in the textile sector, is an important use of sodium alginate.

Synthesis

The graft copolymer is synthesized via a "microwave initiated" technique (Sodium Alginate-g-Poly-acrylamide) Microwave irradiation was utilized to create free-radical sites on the polysaccharide backbone in this technique. The reaction vessel and its contents were eventually cooled and left undisturbed for 24 hours to finish the grafting reactions once the microwave irradiation for the desired amount of time was completed. The addition of a saturated HQ solution brought the reaction to a halt. The gel-like material that had remained in the reaction vessel had now been dumped into an excess of acetone. The grafted polymer precipitate was collected and dried in a hot air oven. It was crushed and sieved after that. The suggested mechanism for graft copolymer production utilizing microwave irradiation is shown in [Scheme 6].

Figure 7. Schematic representation for the synthesis of SAG-g-PAM from SAG using microwave irradiation (Sen, 2010)

Grafting of acrylamide onto the backbone of sodium alginate

Initiation:

$$\text{SAGOH} + \text{M} \xrightarrow{\text{MW}} \text{SAGO}^{*} + \text{M}^{*}$$

Propagation:

$$\text{SAGO}^{*} + \text{M} \longrightarrow \text{SAGOM}^{*}$$

$$\text{SAGO M}^{*} + \text{M} \longrightarrow \text{SAGOMM}^{*}$$

$$\text{SAGOMM}^{*}_{n-1} + \text{M} \longrightarrow \text{SAGOM}^{*}_{n}$$

Termination:

$$\text{SAGOM}^{*}_{n} + \text{SAGOM}^{*}_{n} \longrightarrow \text{GRAFT COPOLYMER}$$

SAGOH - Sodium Alginate
M- Monomer(acrylamide)
MW- Microwave photon

According to the results of the aforementioned experiments, various grades of Sodium Alginate-g-Polyacrylamide may be produced utilizing just microwave irradiation. A number of physicochemical characterization approaches have been used to confirm that Poly-acrylamide has been grafted onto the Sodium Alginate backbone in the synthesized graft copolymers, indicating that Poly-acrylamide has been grafted onto the Sodium Alginate backbone. Furthermore, the created material has the potential to be used as a flocculent.

3. Green Polymers Using Nature's Catalyst:

Enzymes are nature's catalysts that are designed to accelerate specific reactions taking place in the cell and its immediate surroundings. For example, lipases catalyze the hydrolysis of triglycerides into fatty acids and glycerol's in vivo (Scheme 7). Enzymes, on the other hand, are known to catalyses a variety of synthetic transformations in organic solvents, in addition to their native, aqueous reaction media (Klibanov *et al,* 2001 & Wong *et al,* 1994).

Figure 8. Lipase Catalysis (Klibanov, 2001)

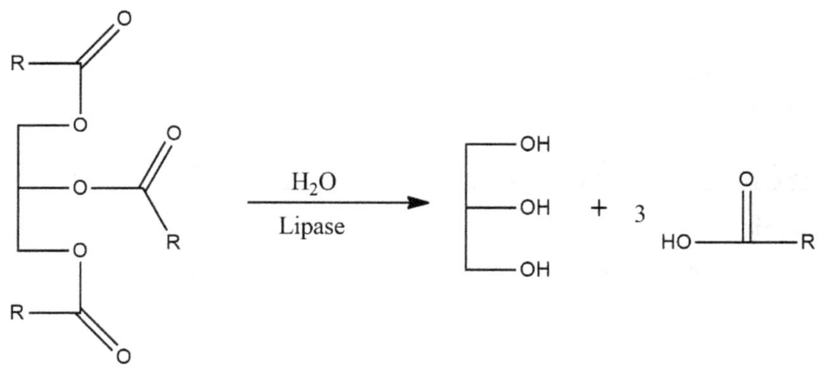

In chemical processes such as trans-esterification, esterification, aminolysis, and Michael addition, lipases are the most commonly used enzymes (Santaniello *et al,* 1993 & Fernandez *et al,* 2006). Because of its high stability and reactivity, Candida antarcticalipase B (Fig. 2) is an excellent catalyst for these conversions (Anderson *et al,* 1998). Although trans-esterification processes are reversible, when enolate esters are employed, they produce unstable enols as by-products, which tautomerize to yield the corresponding aldehydes or ketones. Because vinyl esters have less steric hindrance and hence greater reaction rates than isopropenyl esters, they are preferred (Faber *et al,* 1992).

4. Enzymes in polymer chemistry:

The use of enzymes in polymer synthesis was thoroughly examined (Kobayashi *et al,* 2016 & Uyama *et al,* 2006). Poly-condensation, oxidative polymerization, and ring-opening polymerization are the most common examples. The following are some examples. Poly-condensation Polyesters can be made by poly-condensing hydroxyl-acids or their esters (O'Hagan *et al,* 1994) (A-B type condensation) and Di-carboxylic acids or their analogs with glycols using a lipase catalyst (Chaudry *et al,* 1995).

a. **Oxidative Polymerization:**

Some oxidoreductases, such as peroxides and lactase, have been used to produce poly aromatic chemicals via oxidative polymerization of phenols and aniline derivatives (Feber *et al,* 2004). Horseradish peroxides are the most commonly utilized oxidoreductase in oxidative polymerizations. The enzyme's active site is an iron-protoporphyrin complex (Scheme 8). The process includes the oxidation of Fe3 to an iron-oxo derivative at the cost of H2O2, followed by the extraction of an electron from an electron-rich substrate, such as phenol, to a substrate radical, a phenoxy radical (Scheme 8).

b. **Ring Opening Polymerization:**

The production of polyesters via lipase-catalyzed ring-opening polymerization of lactones is frequent use of enzymes. for example, used lipase from Pseudomonas fluorescens to demonstrate ROP of a racemic combination of a-methyl-b-propiolactone. The enzyme was stereoselective for the (S)-enantiomer, resulting in an (S)-enriched polymer (Uyama et al, 2006) (AA-BB type condensation) (Scheme 8)

5. Production of Biodegradable Polymers from Agricultural Waste:

Biopolymers, biodegradable polymers, and polymers are made from agricultural waste such as fruit seeds, fruit peels, coconut shells, potato peels, orange tree pruning, wheat straw, corn starch, and rice husks, all of which are renewable sources. The goal is to determine if biopolymers, biodegradable polymers, and polymers derived from agricultural waste were totally biodegradable or only recyclable. Given that more than 400 million tonnes of trashes are created each year, the present rate of global plastic manufacturing is unsustainable. Furthermore, by 2050, the rate is predicted to grow fourfold, with a corresponding increase in agricultural plastic waste. Shade nets, mulching materials, and pesticide containers all contribute to agricultural plastic trash. The lack of simple, scalable, and ecologically friendly recycling technologies had an impact on worldwide plastics waste recycling—only 9% of plastics are recycled. Plastics pose a hazard to the environment that goes beyond the absence of appropriate recycling technologies; inadequate synthetic processes have hampered the development of environmentally acceptable polymer composites. Beyond the circular business model typology, the manufacturing of biodegradable polymers from agricultural waste can help to improve sustainability.

Figure 9. Lipase-catalyzed polycondensation between bis(2,2,2-trichloroethyl) adipate and 1,4-butanediol (Linko, 1995)

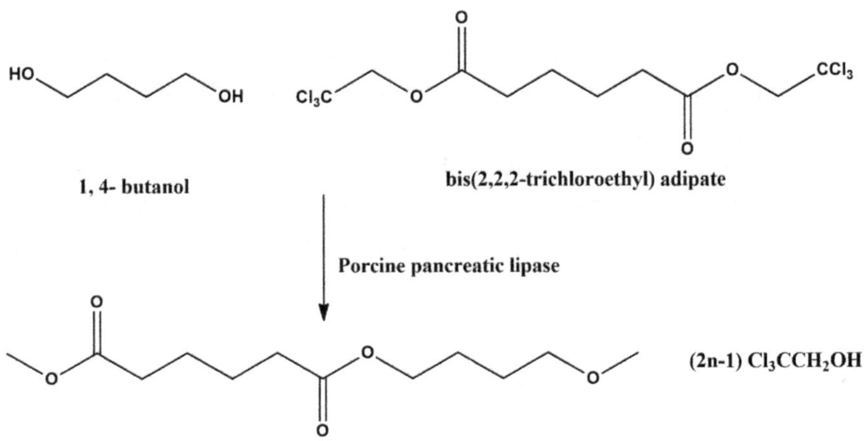

a. **Production of Biodegradable Polymers:**

As illustrated in Figure 3, biodegradable polymers are a special type of polymer that is environmentally good (biocompatible and biodegradable). Depending on the intended products and accessible materials, the production method for these biopolymers is divided into four categories. Chemical synthesis techniques, bacterial synthesis methods, biopolymer blends, and renewable sources are the categories Bio-based polymers can be made from a variety of renewable sources. Plant-based precursors comprising lignocelluloses fibers have been used to make bio-based polymers. The recycling of agricultural waste through composting and the manufacturing of fertilizers boost global carbon emissions, according to data collected in Italy. Olive waste-based compost and anaerobic digester-based compost (AD), of instance, yielded 64 and 67 kg of CO2 equivalent per milligram, respectively.

b. **Thermoplastic Starch-Based Polymers:**

Starch is a polysaccharide that can be found in agro-waste from tubers, legumes, and cereals, and it's an excellent carbon source for bio-based polymers (Nayak *et al,* 2016). Because of their efficient reinforcing characteristics, abundance, and adjustable properties, thermoplastic starch-based polymers are viable alternatives to petroleum-based polymers (Syafri *et al,* 2009). The inclusion of L-lactate and a catalyst is the initial step in the synthesis of starch-based polymers from Agro-wastes. Because they are flexible and light, these materials are important for the future of sustainable food packaging, according to current research (Suffo *et al,* 2020).

c. **Pineapple Peels and Tomato Pomace are used to Make Bio-Based Plastics:**

The extraction of biopolymers from agricultural waste is used in the manufacturing of bio-based polymers from pineapple peel, which follows a conventional procedure. The first steps entail analyzing the chemical composition, particularly the C/N and C/P ratios, which are used to estimate polymer yields. Except for the melting poly-condensation stage, the manufacture of bio-based polymers from tomato pomade is comparable to that of bio-based polymers from pineapple peels (Vega-castro *et al,* 2016).

d. **Production of Lactic Acid, PLA and PHA from Agro-Wastes:**

Specific bacteria strains for fermentation and hydrolysis, as well as the availability of agro-waste as starting materials, impact the synthesis of lactic acid and poly-lactic acid (Djukic-vukovic *et al,* 2019). Rhizopus, Pediococcus, and Streptococcus are among the fungus and bacteria strains used in industrial applications (Djukic-vukovic *et al,* 2019). Because of the differences in the fermentation processes that lead to the formation of fermentable sugars like starch and cellulose, the availability of a diverse range of bacteria and fungus species has an influence on the material characteristics of the final product. Lactic acid is produced during the fermentation process, which is polymerized to generate PLA.

DEGRADATION

Biodegradable polymers are described as "polymers susceptible to breakdown by biological activity, with degradation accompanied by a decrease in mass" by the International Union of Pure and Applied Chemistry (IUPAC) (McNaught *et al,* 1997). Other definitions call for a biodegradable substance to be mineralized during biodegradation into carbon dioxide, water, and biomass (standard CEN/TR 15351:2006). So, first and foremost, we must define how "biodegradability" should be defined. Do we simply need to convert polymers into water-soluble chemicals to tackle the plastic waste of accumulating plastics in the environment, or do we require complete mineralization? Furthermore, the biodegradation time range is critical. According to the European Union's newly released plastics policy, consistent criteria for identifying and labeling compostable and biodegradable plastics will be proposed to allow for correct sorting and avoid misleading environmental claims (European Union, 2018). Biodegradation of polymers may be broken down into four stages: i) biodeterioration, ii) depolymerization, iii) bioassimilation, and iv) mineralization (Lucas *et al,* 2008).

The development of a microbial biofilm leads to surface degradation, in which the polymeric substance is broken into smaller particles in the first phase. Polymer degradation is influenced by both abiotic and biotic causes in general. Plastics breakdown in the environment through four different mechanisms: B. Sethi works at K.L. Mehta D.N. College in Faridabad and studies photo-degradation, thermo-oxidative degradation, hydrolytic degradation, and biodegradation. Natural plastic deterioration is understood to begin with photo-degradation because UV sunlight supplies the energy of activation needed to start the incorporation of oxygen atoms in the polymer. The activation energy necessary to start the incorporation of oxygen atoms into the plasma is activated by UV sunlight, leading to thermo-oxidative deterioration (Ojida *et al,* 2012011 & Tocháček *et al,* 2014). This causes the plastic to split into smaller and smaller parts until the polymer chains reach low molecular weight sufficiently to be metabolized by microbes (Ojeda *et al,* 2011). Either the carbon of the polymer chains is converted into carbon dioxide, or incorporated into bio-molecules (Zhang *et al,* 2014). This whole process, however, is exceedingly sluggish to break down plastics entirely for 50 or more years (Müller *et al,* 2001). Polymeric products are destroyed by an enhanced bond cleavage process, dissemination of radicals, intermediate oxidation, and dispersion of end products through the depth of penetration. Heat, light, moisture mechanical loads contribute mainly to synergetic actions. Another place is the inner process of structural alterations (White *et al,* 1994 & Searle *et al,* 2010). The level of cross-linking, the molecular size and tangling, the presence of the inorganic filler and sample history are key parameters that affect the diffusion via oxygen penetration and diffusion through the spread of degrading products (Raquez *et al,* 2011).

RECYCLING

The environmental impact of plastic wastes has been recently put on the top of the agenda of each government's politics. There are several approaches to tackle the problems associated with plastic trash. These are innovative process recycling design, which ensures even sophisticated packaging is properly recycled; development of new packaging materials that are more efficient in protecting products but still easy to recycle; and development of biologically degradable packing materials that satisfy all requirements of conventional materials; improvement of the collection process (Ragheb and Miller *et al,* 2018). Based on their optimal qualities, plastics are considerably higher than aluminum or other materials, such as corrosion resistance, low density, high strength, and easy-to-use design. For instance, density is a characteristic extremely significant as it provides information about the building's inherent strength, as in the case of flax reinforcement, when PP and LDPE are the best (due to their low density) since they want to make as light as possible a composite. Glass transition temperature (Tg) is another essential feature for researching mechanical polymer characteristics since when they are refrozen below Tg, the flexibility of amorphous polymers is greatly diminished. There are no dimension changes or segmental movement in the polymer at these temperatures. In thermoplastic polymers, the mechanical characteristics of greatest tensile strength and tensile module are also significant (Krevelen *et al,* 2009, Park et al, 2002 & Nichols *et al,* 2001).

In the case of thermoplastic polymers, recycling and combustion are the common elements of recovery strategies. Some issues occur during incineration, including the generation of hazardous fumes and the residual ash including plum and cadmium. Recycling offers benefits such as environmental concerns reduction and material and energy saving (Mohammadzadeh *et al,* 2009, Favis *et al,* 2017 & Francis *et al,* 2001).

1. **Primary Recycling:**

Due to their simplicity and cheap costs, the most prevalent technique is primary recycling. The reuse in its original structure is referred to in this procedure. The drawback of this technique is that the number of cycles for each material is limited (Hopewell *et al,* 2009 & Francis *et al,* 2016).

2. **Secondary Recycling or Mechanical Recycling:**

Only thermoplastic polymers may be utilized in this procedure as they can be recast into finished goods. During the procedure, mechanical recycling does not entail a modification of the polymer. This process is represented by a physical technique in which the plastic waste is sliced, shredded, or washed into granulates, flakes, or pellets of suitable production quality and then melt to produce the new product by extrusion.

3. **Feedstock or Chemical Recycling:**

This technique can be used as a supplement to mechanical recycling. The process in which polymers are chemically transformed into monomers or partly polymers is specified for chemical recycling. The monomers produced can be utilized to recreate the original or a similar polymeric product for further polymerizations. This technique can convert plastic materials, starting with monomers, oligomers or combinations, into smaller molecules that may be utilized as feedstock material (Francis *et al,* 2017 & Olah *et al,* 2008).

4. **Energy Recovery or Quaternary Recycling:**

This means the recovery of the energy component of the plastic. Incineration is the most effective means of reducing the amount of organic material that incorporates energy recovery. This technology is an excellent option since it produces substantial energy from polymers, but because of the health risk of hazardous chemicals, like the recycling of dioxins, is not ecologically acceptable.

CONCLUSION

The chapter ideology & methodology of Green polymer Chemistry review major research work based on ideology of green chemistry. In order to address the challenge of sustainability, the progress in green polymer chemistry during the previous decade must increase at an accelerated pace. The revolution of one day is supposed to become the next's new orthodoxy. If the principles of Green Polymer Chemistry concept and technique are merely adopted as part of daily chemistry, then focus, emphasis, and monikers of green chemistry are no longer required. And when that day arrived, it is impossible to conceive the problems that polymer chemistry will face.

FUTURE ENHANCEMENT

Future problems for polymer chemistry are as varied as the scientific imagination and address the most fundamental questions of sustainability. It should come as no surprise that a number of these problems are being tackled for reasons ranging from economic to scientific. Although numerous novel materials and synthesis techniques are announced each year, green polymer chemistry accounts for only a modest percentage of the total. The research obstacles in attaining green synthesis chemistry principles are plenty, and a thorough explanation of each would be impossible.

In the subject of polymer chemistry, the concept and practice of green polymer chemistry can be presented to the students at all levels on the basis of education. To effectively integrate green chemistry in their teachings and research, the educators require proper tools, knowledge, and skills.

The above discussion illustrates the information on the ideology & methodology of Green polymer Chemistry shows that further studies on the screening of polymer chemistry are essential to minimize polymer risks for the environment. Yet, if we strive to replace all commodity polymers with biodegradable alternatives in order to achieve a more sustainable future, we need to change the way we deal with polymers in general. Or quite simply put in *"We can't hope to achieve polymer's promise for the 21st century if we stick with wasteful 20th-century habits of polymer production and consumption.*

REFERENCES

A European Strategy for Plastics in a Circular Economy. (2018). *European Union.*

Agarwal, S. (2020). Biodegradable Polymers: Present Opportunities and Challenges in Providing a Microplastic-Free Environment. *Macromolecular Chemistry and Physics, 221*(6), 2000017. doi:10.1002/macp.202000017

Anderson, E. M., Larsson, K. M., & Kirk, O. (1998). One Biocatalyst–Many Applications: The Use of candida Antarctica B-Lipase in Organic Synthesis. *Biocatalysis and Biotransformation, 16*(3), 181–204. doi:10.3109/10242429809003198

Andrady, A. L. (2003). *Plastics and the Environment* (1st ed.). Wiley-Interscience. doi:10.1002/0471721557

Andrady, A. L. (2011). Microplastics in the marine environment. *Marine Pollution Bulletin, 62*(8), 1596–1605. doi:10.1016/j.marpolbul.2011.05.030 PMID:21742351

Busto, E., Gotor-Fernández, V., & Gotor, V. (2006). Enantioselective Synthesis of 4-(Dimethylamino)pyridines through a Chemical Oxidation-Enzymatic Reduction Sequence. Application in Asymmetric Catalysis. *Advanced Synthesis & Catalysis, 348*(18), 2626–2632. doi:10.1002/adsc.200600274

Byrne, C. M., Allen, S. D., Lobkovsky, E. B., & Coates, G. W. (2004). Alternating Copolymerization of Limonene Oxide and Carbon Dioxide. *Journal of the American Chemical Society, 126*(37), 11404–11405. doi:10.1021/ja0472580 PMID:15366863

Chandra, R. (1998). Biodegradable polymers. *Progress in Polymer Science, 23*(7), 1273–1335. doi:10.1016/S0079-6700(97)00039-7

Chaudhary, A., Beckman, E. J., & Russell, A. J. (1995). Rational Control of Polymer Molecular Weight and Dispersity during Enzyme-Catalyzed Polyester Synthesis in Supercritical Fluids. *Journal of the American Chemical Society, 117*(13), 3728–3733. doi:10.1021/ja00118a009

da Silva, D. A., de Paula, R. C., & Feitosa, J. P. (2007). Graft copolymerisation of acrylamide onto cashew gum. *European Polymer Journal, 43*(6), 2620–2629. doi:10.1016/j.eurpolymj.2007.03.041

Darensbourg, D. J. (2007). Making Plastics from Carbon Dioxide: Salen Metal Complexes as Catalysts for the Production of Polycarbonates from Epoxides and CO2. *Chemical Reviews, 107*(6), 2388–2410. doi:10.1021/cr068363q PMID:17447821

Degnan, T., & Shinde, S. L. (2019). Waste-plastic processing provides global challenges and opportunities. *MRS Bulletin, 44*(06), 436–437. doi:10.1557/mrs.2019.133

Deshayes, S., Liagre, M., Loupy, A., Luche, J. L., & Petit, A. (1999). Microwave activation in phase transfer catalysis. *Tetrahedron, 55*(36), 10851–10870. doi:10.1016/S0040-4020(99)00601-8

Djukić-Vuković, A., Mladenović, D., Ivanović, J., Pejin, J., & Mojović, L. (2019). Towards sustainability of lactic acid and poly-lactic acid polymers production. *Renewable & Sustainable Energy Reviews, 108*, 238–252. doi:10.1016/j.rser.2019.03.050

Duan, H., Song, G., Qu, S., Dong, X., & Xu, M. (2019). Post-consumer packaging waste from express delivery in China. *Resources, Conservation and Recycling, 144*, 137–143. doi:10.1016/j.resconrec.2019.01.037

Eberhardt, R., Allmendinger, M., & Rieger, B. (2003). DMAP/Cr(III) Catalyst Ratio: The Decisive Factor for Poly(propylene carbonate) Formation in the Coupling of CO2 and Propylene Oxide. *Macromolecular Rapid Communications, 24*(2), 194–196. doi:10.1002/marc.200390022

Eskander, S. (2017). *EI-Din H, Saleh M. Environmental Science and Engineering* (Vol. 8). Studium Press.

Faber, K. (2004). *Biotransformations in Organic Chemistry* (5th ed.). Springer-Verlag. doi:10.1007/978-3-642-18537-3

Favis, B. D., & Corroller, P. (2017). *Polymeric material and process for recycling plastic blends*. U.S. Patent 9670344.

Francis, R. (2016). *Recycling of Polymers: Methods, Characterization and Applications*. John Wiley & Sons. doi:10.1002/9783527689002

Geresh, S., Gdalevsky, G. Y., Gilboa, I., Voorspoels, J., Remon, J. P., & Kost, J. (2004). Bioadhesive grafted starch copolymers as platforms for peroral drug delivery: A study of theophylline release. *Journal of Controlled Release*, *94*(2–3), 391–399. doi:10.1016/j.jconrel.2003.10.019 PMID:14744489

Hayashi, T., Kanai, H., Yodoya, S., Oka, M., & Hayashi, T. (2002). Biodegradation of random co-polypeptide hydrogels consisting of N-hydroxypropyl l-glutamine as one component. *European Polymer Journal*, *38*(1), 139–146. doi:10.1016/S0014-3057(01)00137-9

Hebeish, A., Abd El-Thalouth, I., El-Kashouti, M. A., & Abdel-Fattah, S. H. (1979). Graft copolymerization of acrylonitrile onto starch using potassium permanganate as initiator. *Angewandte Makromolekulare Chemie*, *78*(1), 101–108. doi:10.1002/apmc.1979.050780107

Hopewell, J., Dvorak, R., & Kosior, E. (2009). Plastics recycling: Challenges and opportunities. *Philosophical Transactions of the Royal Society of London. Series B, Biological Sciences*, *364*(1526), 2115–2126. doi:10.1098/rstb.2008.0311 PMID:19528059

Howard, G. T. (2002). Biodegradation of polyurethane: A review. *International Biodeterioration & Biodegradation*, *49*(4), 245–252. doi:10.1016/S0964-8305(02)00051-3

Ilyas, R., Sapuan, S., Ibrahim, R., Abral, H., Ishak, M., Zainudin, E., Atikah, M., Mohd Nurazzi, N., Atiqah, A., Ansari, M., Syafri, E., Asrofi, M., Sari, N. H., & Jumaidin, R. (2019). Effect of sugar palm nanofibrillated cellulose concentrations on morphological, mechanical and physical properties of biodegradable films based on agro-waste sugar palm (Arenga pinnata (Wurmb.) Merr) starch. *Journal of Materials Research and Technology*, *8*(5), 4819–4830. doi:10.1016/j.jmrt.2019.08.028

Kalia, S., Sharma, A., & Kaith, B. S. (2007). Ab initio study of gas phase and water-assisted tautomerization of maleimide and formamide. *Journal of Chemical Sciences*, *119*(6), 617–624. doi:10.100712039-007-0077-6

Kang, H. M., Cai, Y. L., & Liu, P. S. (2006). Synthesis, characterization and thermal sensitivity of chitosan-based graft copolymers. *Carbohydrate Research*, *341*(17), 2851–2857. doi:10.1016/j.carres.2006.08.001 PMID:17027942

Khatoon, N., Jamal, A., & Ali, M. I. (2018). Lignin peroxidase isoenzyme: A novel approach to biodegrade the toxic synthetic polymer waste. *Environmental Technology*, *40*(11), 1366–1375. doi:10.1080/09593330.2017.1422550 PMID:29291675

Kindermann, N., Cristòfol, L., & Kleij, A. W. (2017). Access to Biorenewable Polycarbonates with Unusual Glass-Transition Temperature (Tg) Modulation. *ACS Catalysis*, *7*(6), 3860–3863. doi:10.1021/acscatal.7b00770

Klibanov, A. M. (2001). Improving enzymes by using them in organic solvents. *Nature*, *409*(6817), 241–246. doi:10.1038/35051719 PMID:11196652

Laycock, B., Nikolić, M., Colwell, J. M., Gauthier, E., Halley, P., Bottle, S., & George, G. (2017). Lifetime prediction of biodegradable polymers. *Progress in Polymer Science*, *71*, 144–189. doi:10.1016/j.progpolymsci.2017.02.004

Li, Y., Wang, Z., Sun, L., Liu, L., Xu, C., & Kuang, H. (2019). Nanoparticle-based sensors for food contaminants. *Trends in Analytical Chemistry*, *113*, 74–83. doi:10.1016/j.trac.2019.01.012

Linko, Y. Y., Wang, Z. L., & Seppälä, J. (1995). Lipase-catalyzed synthesis of poly(1,4-butyl sebacate) from sebacic acid or its derivatives with 1,4-butanediol. *Journal of Biotechnology*, *40*(2), 133–138. doi:10.1016/0168-1656(95)00039-S

Lucas, N., Bienaime, C., Belloy, C., Queneudec, M., Silvestre, F., & Nava-Saucedo, J. E. (2008). Polymer biodegradation: Mechanisms and estimation techniques – A review. *Chemosphere*, *73*(4), 429–442. doi:10.1016/j.chemosphere.2008.06.064 PMID:18723204

Lucas, N., Bienaime, C., Belloy, C., Queneudec, M., Silvestre, F., & Nava-Saucedo, J. E. (2008). Polymer biodegradation: Mechanisms and estimation techniques – A review. *Chemosphere*, *73*(4), 429–442. doi:10.1016/j.chemosphere.2008.06.064 PMID:18723204

Masuhiro, T., Shafiul, I., Takayuki, A., Alessandra, B., Giuliano, F., & … . (2005).. . *AUTEX Research Journal*, *5*, 40.

Matlack, A. (2021). *Introduction to Green Chemistry* (2nd ed.). T&F India.

McNaught, A. D., & Wilkinson, A. (1997). *Compendium of Chemical Terminology (IUPAC Chemical Data)* (2nd ed.). Wiley.

Mohammadzadeh, M. (2009). *Characterization of Recycled Thermoplastic Polymers* [Master's Thesis]. University of Borås, Borås, Sweden.

Mülhaupt, R. (2012). Green Polymer Chemistry and Bio-based Plastics: Dreams and Reality. *Macromolecular Chemistry and Physics, 214*(2), 159–174. doi:10.1002/macp.201200439

Müller, R. J., Kleeberg, I., & Deckwer, W. D. (2001). Biodegradation of polyesters containing aromatic constituents. *Journal of Biotechnology, 86*(2), 87–95. doi:10.1016/S0168-1656(00)00407-7 PMID:11245897

Nakano, R., Ito, S., & Nozaki, K. (2014). Copolymerization of carbon dioxide and butadiene via a lactone intermediate. *Nature Chemistry, 6*(4), 325–331. doi:10.1038/nchem.1882 PMID:24651200

Nayak, B., & Singh, R. (2001). Synthesis and characterization of grafted hydroxypropyl guar gum by ceric ion induced initiation. *European Polymer Journal, 37*(8), 1655–1666. doi:10.1016/S0014-3057(01)00035-0

Nayak, P. L. (1999). Biodegradable Polymers: Opportunities and Challenges. *Journal of Macromolecular Science, Part C: Polymer Reviews, 39*(3), 481–505. doi:10.1081/MC-100101425

Nichols, C., Moore, T., & Griffith, S. (2001). *Thermoplastic Polymers with Improved Infrared Reheat Properties*. U.S. Patent 09973436.

O'Hagan, D., & Zaidi, N. A. (1994). Enzyme-catalysed condensation polymerization of 11-hydroxyundecanoic acid with lipase from Candida cylindracea. *Polymer, 35*(16), 3576–3578. doi:10.1016/0032-3861(94)90928-8

Ochiai, B., Hatano, Y., & Endo, T. (2008). Fixing Carbon Dioxide Concurrently with Radical Polymerization for Utilizing Carbon Dioxide by Low-Energy Cost. *Macromolecules, 41*(24), 9937–9939. doi:10.1021/ma801960q

Ojeda, T., Freitas, A., Birck, K., Dalmolin, E., Jacques, R., Bento, F., & Camargo, F. (2011). Degradability of linear polyolefins under natural weathering. *Polymer Degradation & Stability, 96*(4), 703–707. doi:10.1016/j.polymdegradstab.2010.12.004

Olah, G. A., Goeppert, A., & Prakash, G. K. S. (2008). Chemical Recycling of Carbon Dioxide to Methanol and Dimethyl Ether: From Greenhouse Gas to Renewable, Environmentally Carbon Neutral Fuels and Synthetic Hydrocarbons. *The Journal of Organic Chemistry, 74*(2), 487–498. doi:10.1021/jo801260f PMID:19063591

Park, H. M., Li, X., Jin, C. Z., Park, C. Y., Cho, W. J., & Ha, C.-S. (2002). Preparation and properties of biodegradable thermoplastic starch/clay hybrids. *Macromolecular Materials and Engineering*, *287*(8), 553–558. doi:10.1002/1439-2054(20020801)287:8<553::AID-MAME553>3.0.CO;2-3

PD CEN/TR 15351. (2006). BSI.

Peng, Y. H., Shih, Y. H., Lai, Y. C., Liu, Y. Z., Liu, Y. T., & Lin, N. C. (2014). Degradation of polyurethane by bacterium isolated from soil and assessment of polyurethanolytic activity of a Pseudomonas putida strain. *Environmental Science and Pollution Research International*, *21*(16), 9529–9537. doi:10.100711356-014-2647-8 PMID:24633845

Properties of Polymers: Their Correlation with Chemical Structure; Their Numerical Estimation and Prediction from Additive Group Contributions. (2014). Elsevier Science & Technology.

Raquez, J. M., Bourgeois, A., Jacobs, H., Degée, P., Alexandre, M., & Dubois, P. (2011). Oxidative degradations of oxodegradable LDPE enhanced with thermoplastic pea starch: Thermo-mechanical properties, morphology, and UV-ageing studies. *Journal of Applied Polymer Science*, *122*(1), 489–496. doi:10.1002/app.34190

Reihmann, M., & Ritter, H. (2006). Synthesis of Phenol Polymers Using Peroxidases. *Enzyme-Catalyzed Synthesis of Polymers*, 1–49.

Revelle, R., & Suess, H. E. (1957). Carbon Dioxide Exchange Between Atmosphere and Ocean and the Question of an Increase of Atmospheric CO2 during the Past Decades. *Tellus*, *9*(1), 18–27. doi:10.3402/tellusa.v9i1.9075

Saini, P. K., Fiorani, G., Mathers, R. T., & Williams, C. K. (2017). Zinc versus Magnesium: Orthogonal Catalyst Reactivity in Selective Polymerizations of Epoxides, Bio-derived Anhydrides and Carbon Dioxide. *Chemistry (Weinheim an der Bergstrasse, Germany)*, *23*(18), 4260–4265. doi:10.1002/chem.201605690 PMID:28295663

Santaniello, E., Ferraboschi, P., & Grisenti, P. (1993). Lipase-catalyzed transesterification in organic solvents: Applications to the preparation of enantiomerically pure compounds. *Enzyme and Microbial Technology*, *15*(5), 367–382. doi:10.1016/0141-0229(93)90123-J

Sardon, H., & Dove, A. P. (2018). Plastics recycling with a difference. *Science*, *360*(6387), 380–381. doi:10.1126cience.aat4997 PMID:29700253

Sashiwa, H., Fukuda, R., Okura, T., Sato, S., & Nakayama, A. (2018). Microbial Degradation Behavior in Seawater of Polyester Blends Containing Poly(3-hydroxybutyrate-co-3-hydroxyhexanoate) (PHBHHx). *Marine Drugs*, *16*(1), 34. doi:10.3390/md16010034 PMID:29342118

Satyanarayana, K. G., Arizaga, G. G., & Wypych, F. (2009). Biodegradable composites based on lignocellulosic fibers—An overview. *Progress in Polymer Science*, *34*(9), 982–1021. doi:10.1016/j.progpolymsci.2008.12.002

Scott, G. (2002). Why Degradable Polymers? *Degradable Polymers*, 1–15.

Searle, N. D., McGreer, M., & Zielnik, A. (2010). *Weathering of Polymeric Materials. Encyclopedia of Polymer Science and Technology*. Published.

Sen, G., Singh, R. P., & Pal, S. (2010). Microwave-initiated synthesis of polyacrylamide grafted sodium alginate: Synthesis and characterization. *Journal of Applied Polymer Science*, *115*(1), 63–71. doi:10.1002/app.30596

Shah, T. V., & Vasava, D. V. (2019a). A glimpse of biodegradable polymers and their biomedical applications. *E-Polymers*, *19*(1), 385–410. doi:10.1515/epoly-2019-0041

Singh, N., Hui, D., Singh, R., Ahuja, I., Feo, L., & Fraternali, F. (2017). Recycling of plastic solid waste: A state of art review and future applications. *Composites. Part B, Engineering*, *115*, 409–422. doi:10.1016/j.compositesb.2016.09.013

Singh, V., Tiwari, A., Pandey, S., & Singh, S. K. (2007). Peroxydisulfate initiated synthesis of potato starch-graft-poly(acrylonitrile) under microwave irradiation. *Express Polymer Letters*, *1*(1), 51–58. doi:10.3144/expresspolymlett.2007.10

Sriyapai, P., Chansiri, K., & Sriyapai, T. (2018). Isolation and Characterization of Polyester-Based Plastics-Degrading Bacteria from Compost Soils. *Microbiology*, *87*(2), 290–300. doi:10.1134/S0026261718020157

Stevens, E. S. (2002). *Green Plastics* (1st ed., Vol. 1). Amsterdam University Press. doi:10.1515/9780691214177

Suffo, M., Mata, M. D. L., & Molina, S. (2020). A sugar-beet waste based thermoplastic agro-composite as substitute for raw materials. *Journal of Cleaner Production*, *257*, 120382. doi:10.1016/j.jclepro.2020.120382

Sun, T., Li, G., Ning, T. Y., Zhang, Z. M., Mi, Q. H., & Lal, R. (2018). Suitability of mulching with biodegradable film to moderate soil temperature and moisture and to increase photosynthesis and yield in peanut. *Agricultural Water Management*, *208*, 214–223. doi:10.1016/j.agwat.2018.06.027

Tamura, M., Ito, K., Honda, M., Nakagawa, Y., Sugimoto, H., & Tomishige, K. (2016). Direct Copolymerization of CO2 and Diols. *Scientific Reports*, *6*(1), 24038. doi:10.1038rep24038 PMID:27075987

Tocháček, J., & Vrátníčková, Z. (2014). Polymer life-time prediction: The role of temperature in UV accelerated ageing of polypropylene and its copolymers. *Polymer Testing*, *36*, 82–87. doi:10.1016/j.polymertesting.2014.03.019

Tokiwa, Y., Calabia, B., Ugwu, C., & Aiba, S. (2009). Biodegradability of Plastics. *International Journal of Molecular Sciences*, *10*(9), 3722–3742. doi:10.3390/ijms10093722 PMID:19865515

Tosta, M. R., Prates, L. L., Christensen, D. A., & Yu, P. (2019). Biodegradation kinetics by microorganisms, enzymatic biodigestion, and fractionation of protein in seeds of cool-climate-adapted oats: Comparison among oat varieties, between milling-type and feed-type oats, and with barley grain. *Journal of Cereal Science*, *89*, 102814. doi:10.1016/j.jcs.2019.102814

Tripathy, T., & Singh, R. P. (2001). Characterization of polyacrylamide-grafted sodium alginate: A novel polymeric flocculant. *Journal of Applied Polymer Science*, *81*(13), 3296–3308. doi:10.1002/app.1786

Uyama, H., & Kobayashi, S. (2006). Enzymatic Synthesis of Polyesters via Polycondensation. *Advances in Polymer Science*, *194*, 133–158. doi:10.1007/12_031

Uyama, H., & Kobayashi, S. (2006). Enzymatic Synthesis of Polyesters via Polycondensation. *Advances in Polymer Science*, *194*, 133–158. doi:10.1007/12_031

Vahdat, A., Bahrami, H., Ansari, N., & Ziaie, F. (2007). Radiation grafting of styrene onto polypropylene fibres by a 10MeV electron beam. *Radiation Physics and Chemistry*, *76*(5), 787–793. doi:10.1016/j.radphyschem.2006.05.009

Vega-Castro, O., Contreras-Calderon, J., León, E., Segura, A., Arias, M., Pérez, L., & Sobral, P. J. (2016). Characterization of a polyhydroxyalkanoate obtained from pineapple peel waste using Ralsthonia eutropha. *Journal of Biotechnology*, *231*, 232–238. doi:10.1016/j.jbiotec.2016.06.018 PMID:27316828

Wang, J., Qin, A., & Tang, B. Z. (2020). Multicomponent Polymerizations Involving Green Monomers. *Macromolecular Rapid Communications*, *42*(6), 2000547. doi:10.1002/marc.202000547 PMID:33314433

Wang, J. P., Chen, Y. Z., Zhang, S. J., & Yu, H. Q. (2008). A chitosan-based flocculant prepared with gamma-irradiation-induced grafting. *Bioresource Technology*, *99*(9), 3397–3402. doi:10.1016/j.biortech.2007.08.014 PMID:17888656

Westminster City Council | Commercial Waste Services. (2020, July 6). *Commercial Waste*. https://cleanstreets.westminster.gov.uk/

White, J. R., & Turnbull, A. (1994). Weathering of polymers: Mechanisms of degradation and stabilization, testing strategies and modelling. *Journal of Materials Science*, *29*(3), 584–613. doi:10.1007/BF00445969

Wiesbrock, F., Hoogenboom, R., & Schubert, U. S. (2004). Microwave-Assisted Polymer Synthesis: State-of-the-Art and Future Perspectives. *Macromolecular Rapid Communications*, *25*(20), 1739–1764. doi:10.1002/marc.200400313

Wong, C. H., & Whitesides, G. M. (1994). *Enzymes in Synthetic Organic Chemistry*. Pergamon.

Yamada-Onodera, K., Mukumoto, H., Katsuyaya, Y., Saiganji, A., & Tani, Y. (2001). Degradation of polyethylene by a fungus, Penicillium simplicissimum YK. *Polymer Degradation & Stability*, *72*(2), 323–327. doi:10.1016/S0141-3910(01)00027-1

Zhang, F., Zheng, B., Zhang, J., Huang, X., Liu, H., Guo, S., & Zhang, J. (2010). Horseradish Peroxidase Immobilized on Graphene Oxide: Physical Properties and Applications in Phenolic Compound Removal. *The Journal of Physical Chemistry C*, *114*(18), 8469–8473. doi:10.1021/jp101073b

Zhang, X., Zhou, L., Pi, H., Guo, S., & Fu, J. (2014). Performance of layered double hydroxides intercalated by a UV stabilizer in accelerated weathering and thermal stabilization of PVC. *Polymer Degradation & Stability*, *102*, 204–211. doi:10.1016/j.polymdegradstab.2014.01.003

Chapter 5
Green Solvents for Sustainable Organic Synthesis:
An Overview

Anuradha Singh
https://orcid.org/0000-0002-8145-1369
Department of Chemistry, S. S. Khanna Girls Degree College, University of Allahabad, Prayagraj, India

Kavya Singh
S. S. Khanna Girls Degree College, University of Allahabad, Prayagraj, India

Swastika Singh
S. S. Khanna Girls Degree College, University of Allahabad, Prayagraj, India

ABSTRACT

Green or sustainable chemistry is now recognised as an important scientific subject, and research in this area has led to the creation of cleaner and safer chemical processes, with many new advances occurring each year. Researchers are very interested in unconventional or alternative solvents for chemical synthesis. Chemical solvents make up around 80% of all chemical compounds utilised in a variety of essential chemical processes, notably in the chemical industry. Unfortunately, many of these solvents are volatile organic compounds (VOCs) produced from petrochemicals, which pose a variety of health and environmental hazards. In this chapter, different environmentally benign solvent alternatives have been discussed based on their green matrix in chemical processes.

DOI: 10.4018/978-1-7998-9851-1.ch005

BACKGROUND

Green chemistry is one of the most tempting environmentally-conscious approaches in chemistry for sustainability, which entails 12 Principles as shown in Figure 1, for the design, synthesis, and applications of chemical products. Green Chemistry necessitates a paradigm shift from the conventional chemical process which mainly targets high product yield, to one that targets to minimize the toxic and/or hazardous substances during the development and manufacturing of the chemical products (Anastas & Warner, 1998).

The 12 Principles of Green Chemistry, expressly claim the use of "safer solvents and auxiliaries" for sustainable chemical processes. And who could argue with that? The replacement of conventional toxic and volatile solvents by sustainable solvents could improve chemical processes, reduce the economical cost and decrease the processing steps with a lower environmental impact, including their regeneration and recycle, process intensification, atom efficiency. Solvents have attracted a lot of attention in the context of sustainable or green chemistry. As a result, a plethora of scientific research articles dealing with "green solvents" or alternative solvents have been published (Anastas & Eghbali, 2010; Dolzhenko, 2021; Gómez-López et al., 2020; Häckl & Kunz, 2018; Jahangirian et al., 2017; Pacheco-Fernández & Pino, 2019; Savitskaya et al., 2021; Zimmerman et al., 2020). The present chapter highlights different environmentally benign solvent alternatives based on their green matrix in chemical processes.

INTRODUCTION

For any chemical reaction, the reaction vessel contains three important components viz reagents/catalysts, solvent systems and energy inputs. Each component of the reaction vessel, major or minor, affects the final yield of the reaction product. Though solvents do not take part directly in the reaction vessel but play a crucial role by providing a reaction medium to enable chemical reactions to occur more rapidly. The component of reaction & role of solvents in the chemical processes has been shown in Figure 2. Solvents play a role in the process by facilitating energy and mass transformations, modifying reactivity, forming a homogenous dispersion, and allowing solid deposition. It also helps in product separation and its purification. A good solvent also allows for thermodynamic and kinetic control of a desired chemical process (Lomba et al., 2011).

Figure 1. Principles of green chemistry

Further solvents are the largest auxiliary waste in any reaction and estimated about 80% of the total mass of chemicals involved in chemical industries comprises solvents, with recovery efficiencies between 50–80%. The extravagant use of organic solvents in the chemical industries has emerged in an uncurbed rise in issues related to environmental pollution, and waste treatment. Every year, about 20 million tonnes of volatile organic solvents are discharged into the environment, the vast majority of which are used in chemical synthesis. Because of the high need for solvents in the manufacturing of complex chemicals and the conversion of raw materials into active medicinal components, the chemical and pharmaceutical industries generate substantial amounts of chemical waste roughly 85% solvent by mass (Wagare et al., 2021).

Figure 2. Component of reaction and role of solvents (shown in box) in the chemical process

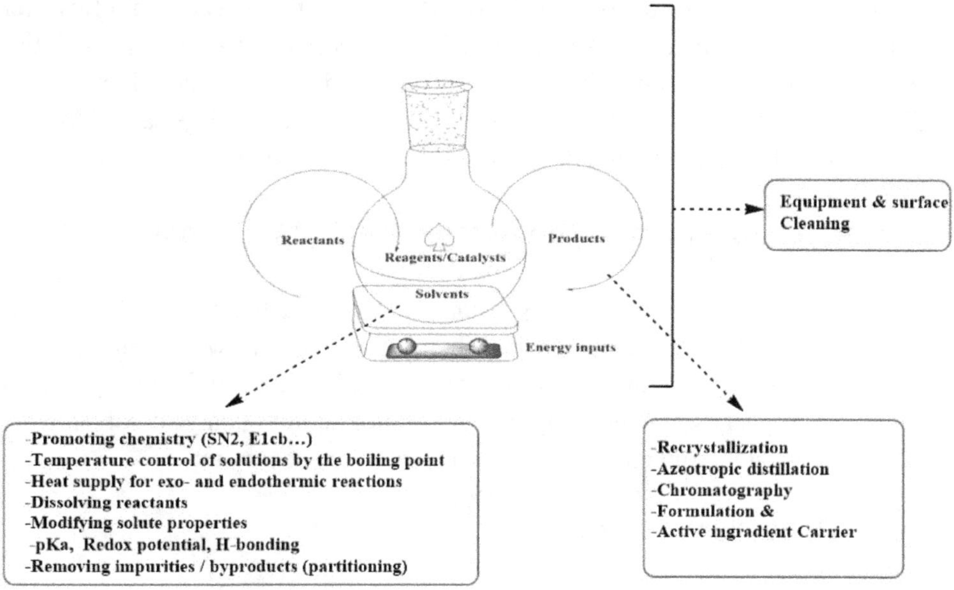

Organic solvents had been associated with a multitude of environmental and health issues in past decades. The use of toxic, flammable, or environmentally damaging solvents appears unnecessary because these properties do not affect the function or progress of the system to which the solvent is applied. However, the negative consequences of solvent use are frequently linked to the beneficial properties of the solvent required for the application. The volatility of solvents allows for solvent recovery and purification via distillation, but it also produces unwanted air emissions and poses a risk of worker exposure (Uzma et al., 2008). The use of hazardous solvents is deleterious to organs, e.g. carbon tetrachloride and chloroform are hepatotoxic. The kidney failure occurs while glycol ethers and chlorinated solvents are used (Gupta, 2020; Hume & Ho, 2019). Likewise, amide solvents have the high polarity required to dissolve a wide range of substrates and speed up reactions (Buhler & Reed, 1990), but this functionality frequently implies reproductive toxicity. Hydrocarbon solvents, on the other end of the polarity scale, have the ability to dissolve oils in extractions and perform separations (Virot et al., 2008), but they are also highly combustible, and their low water solubility (high logP) has been linked to bioaccumulation and aquatic toxicity (Tebby et al., 2011). Many conventional organic solvents are flammable, toxic, form smog and released hazardous waste (Yaseen et al., 2021).

The use of hazardous solvents such as benzene, dichloromethane, and chloroform, which were recognised as carcinogens as well as hepatotoxic, was usually restricted by the World Health Organization (WHO) and the International Agency for Research on Cancer (IARC) evaluation (Wagare et al., 2021). Because of their ozone-depleting effects, chlorofluorocarbons (CFCs) were also banned by the Montreal Protocol in 1987 (Young et al., 2021). Since then, a considerable amount of research has been directed toward alternatives to chlorinated solvents in general.

Approaches to Minimize the Release/Use of Solvents

To reduce the use of harmful organic solvents and search for better substitute to them is the prime aim of green chemistry. The chemical industry views solvent selection as a key component in the overall sustainability profile of a manufacturing process. Following approaches are generally used to minimize the release of solvents and contaminated water from chemical industries into the ecosystem:

Solvent Reduction or Recycling

In chemical industries, by invoked "closed-loop systems" remarkable progress has been made in the direction of improve the ability to recycle the solvents. Solvent recovery is the operation of extracting beneficial materials from waste generated during the chemical manufacturing process. fractionation, azeotropic distillation, and extractive distillation are some methods of solvents recovery (Aboagye et al., 2021).

Reactions in a Dry Medium

Another attractive area in sustainable green technology could be shifting to solvent-free processes in chemical synthesis (Avila-Ortiz & Juaristi, 2020). Neat reaction, atom economy, good yields, short reaction time and easy purification process are some characteristics of reactions in solvent-free conditions. The main attraction of these reactions is the minimization of by-products/waste by bypass the use of conventional volatile organic solvents. These reactions can be carried out by using a microwave, ultrasonic, ultraviolet, visible, infrared irradiation, grinding, and milling technology. Solvent-free systems include aldol condensations, sequential aldol and Michael additions, Stobbe condensations, *O*-silylation of alcohols with silyl chlorides, and clay-catalyzed synthesis of trans chalcones (Cintas et al., 2020).

Limitations

- The isolation of products and the subsequent purification procedures often require column chromatography or at least filtration followed by recrystallization, involving necessarily the use of solvent.
- The possibility of producing "hotspots," which can lead to runaway reactions and consequently the increased likelihood of unwanted side reactions. As a result, measuring the heat of reaction in solvent-free systems is critical, as is effective heat disposal. If extremely exothermic reactions are identified, that are otherwise suited to solventless conditions, the problem might be solved by improved reactor design (Loupy, 2004).

Alternative or Green Solvents

This chapter's major focus is on the use of green or alternative solvents in chemical processes. over the last 15 years, novel solvent alternatives have picked up steam and have gradually started to replace traditional solvents. Water, fluorous solvents, ionic liquids, organic carbonates, carbon dioxide, and biosolvents are among them. In reality, these many solvents do not compete with one another, but rather complement one another, each with its own set of advantages and disadvantages (Cvjetko Bubalo et al., 2015; Pena-Pereira & Tobiszewski, 2017; Talele & Shahare, 2021).

How to Choose Sustainable Solvents?

Selection of an appropriate and sustainable solvent for a particular chemical reaction is crucial to attain the sustainability in process. Before selecting a solvent for a reaction, several factors should be considered such as

- The products and reagents should be chemically compatible.
- In conventional reactions, the solvent should dissolve the reactants in order to form a homogeneous media. In this case, the polarity or nonpolarity of solutes and solvents should be taken into account.
- The reaction temperature, as well as the melting and boiling points of solvents and reagents, should be taken into account (Jimenez-Gonzalez, 2019; Sheldon, 2019).
- Several efforts have been made to establish guidelines for the selection of sustainable solvents. To assist chemists in selecting more sustainable solvents, a few pharmaceutical companies (such as GSK, AstraZeneca, Pfizer, and Sanofi) and institutions (such as the ACS Green Chemistry Institute Pharmaceutical Roundtable (GCI-PR) and Innovative Medicines Initiative

(IMI)), have devised data-rich solvent selection guides with visual cues ("traffic light" system) for quick shortlisting. There is no universal approach to select a particular solvent for a chemical process, but recognizing solvent properties along with environmental, safety, and health (ESH) characteristics have been used to identify sustainable solvent and their ranking (Capello et al., 2007; Clarke et al., 2018). For the sustainability of solvents, the twelve principles of green chemistry may be applied on solvent properties which represented in Figure 3. Furthermore, selection criterion for sustainability of solvent has been shown in Figure 4.

Figure 3 Solvent properties & the twelve principles of green chemistry

In general, the use of safer solvents is the 'low hanging branch' for incorporating Green Chemistry Principles into conventional synthetic procedure because of the accessibility of several solvent screening manuals that have been published within the pharmaceutical sector and have been used as a basis for guiding the selection of safer solvent choices for chemical reactions, extractions, and purification process. Some important manuals are as follow:

- Pfizer solvent selection tool

This tool was developed with worker safety (acute and chronic toxicity) in mind, as well as process safety (flammability and reactivity) and environmental and regulatory consciousness. Solvents were classified as "preferred," "useful," or "undesirable" using this method (Byrne et al., 2016).

- GlaxoSmithKline (GSK) Solvent Selection Guide

The guide is accompanied by a table that identifies the issues of concern in the following categories: waste, impact on the environment, health impacts, flammability and explosion hazard, reactivity and stability, life-cycle score, and a legislation flag that includes the information to any regulatory restrictions. Solvents are classified into three types: "few concerns," "some issues," and "serious issues" (Byrne et al., 2016).

- The Sanofi solvent selection guide

The inclusion of statutory categories makes the Sanofi solvent choice guide industrially applicable, with necessity taking precedence over any personal idea of what a green solvent is. The overall score and description of other issues make the tool useful for users in exploratory chemistry laboratories who are not immediately confronted with the regulatory limitations of solvent usage. Each category is evaluated on a scale of 1 to 10, with 10 being the most serious issue. The Unified version of general solvent selection guides for chemists has been summarized in Table 1(American Chemical Society Green Chemistry Institute., n.d.; Byrne et al., 2016). Overall, most solvent selection guides come to similar conclusions, although there are minor deviations.

- Innovative Medicines Initiative (IMI)-CHEM21 solvent selection guide

The Innovative Medicines Initiative (IMI)-CHEM21 public-private partnership is a European consortium comprises six pharmaceutical companies from the European Federation of Pharmaceutical Industries and Associations (EFPIA), ten universities and five small to medium enterprises. Based on a review of publicly accessible solvent selection guidelines, the CHEM21 solvent selection guide was developed. A set of Safety, Health, and Environment criteria, linked with the Global Harmonized System (GHS) and European standards, is proposed to rate less traditional solvents. A technique based on a basic combination of these factors provides a preliminary overall ranking of any solvent. This allows for a more straightforward assessment of the greenness of bio-derived solvents. A methodology was developed that allows for hazard-driven assessment of any solvent's Safety, Health, and Environment, as well as an overall ranking into three categories by choice (recommended, problematic or hazardous). Because this model aligned well with the classical solvents, it may be used to perform a preliminary greenness evaluation of novel solvents (Clarke et al., 2018).

Figure 4. Selection criterion for sustainability of solvent

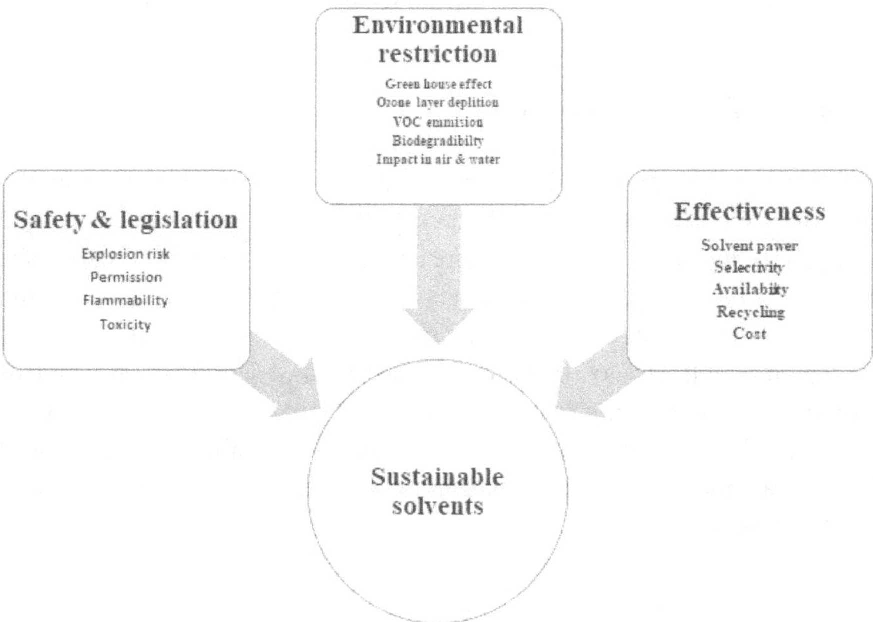

Table 1. The Unified version of general solvent selection guides for chemists (Byrne et al., 2016)

Class	Solvent	Conclusion (Pfizer)	Conclusion (GSK)	Conclusion (Sanofi)
Alcohols	Methanol	Preferred	Some issues	Recommended
	Ethanol	Preferred	Some issues	Recommended
	1-Propanol	Preferred	Some issues	Recommended
	i-Propanol	Preferred	Some issues	Recommended
	1-Butanol	Preferred	Few issues	Recommended
	2-Butanol		Few issues	Recommended
	t-Butanol	Preferred	Some issues	Substitution advisable
	Ethylene glycol	Usable		Substitution advisable
	2-Methoxyethanol		Major issues	Substitution requested
Hydro-carbons	*n*-Pentane	Undesirable		Banned
	Hexane(s)	Undesirable	Major issues	Substitution requested
	Cyclohexane	Usable	Some issues	Substitution advisable
	Methylcyclohexane	Usable		Substitution advisable**
	Heptane	Usable	Some issues	Substitution advisable
	Isooctane	Usable	Some issues	
	Benzene	Undesirable	Major issues	
	Toluene	Usable	Some issues	Substitution advisable
	Xylene(s)	Usable	Some issues	Substitution advisable
Dipolar aprotic	DMSO	Usable	Some issues	Substitution advisable
	Acetonitrile	Usable	Major issues	Recommended
	DMF	Undesirable	Major issues	Substitution requested
	DMAc*	Undesirable	Major issues	Substitution requested
	NMP	Undesirable	Major issues	Substitution requested
Esters	Methyl acetate		Some issues	Substitution advisable
	Ethyl acetate	Preferred	Some issues	Recommended
	n-Propyl acetate		Few issues	Recommended
	i-Propyl acetate	Preferred	Few issues	Recommended
Ethers	THF	Usable	Major issues	Substitution advisable
	2-MeTHF	Usable	Some issues	Recommended
	TBME	Usable	Some issues	Substitution advisable
	CPME		Some issues	Substitution requested
	Diethyl ether	Undesirable	Major issues	
	Di-*i*-propyl ether	Undesirable	Major issues	Substitution advisable
	1,2-DME	Undesirable	Major issues	Substitution requested
	1,4-Dioxane	Undesirable	Major issues	Substitution requested
Ketones	Acetone	Preferred	Some issues	Recommended
	Methylethyl ketone	Preferred	Major issues	Recommended
	MIBK		Some issues	Recommended
Halo-genated	Dichloromethane	Undesirable	Major issues	Substitution advisable
	1,2-Dichloroethane	Undesirable	Major issues	
	Chloroform	Undesirable	Major issues	
	CCl_4	Undesirable	Major issues	
Miscellan-eous	Water	Preferred	Few issues	Recommended
	Acetic acid	Usable		Substitution advisable
	Pyridine	Undesirable		Substitution advisable

GREEN OR SUSTAINABLE SOLVENTS

According to green chemistry, the use of auxiliary materials such as solvents should be avoided whenever possible and it should be non hazardous. In terms of raw materials, many solvent resources are derived from fossil reserves such as petroleum, natural gas, or coal. The main issue with fossil resources is their non-renewability and rapidly diminishing availability (Lomba et al., 2011).

Green solvents are those that are made from renewable resources such as water, bio-solvents, and supercritical carbon dioxide. They can be used as alternatives to harmful traditional solvents. Ionic liquids, organic carbonates, and fluorous solvents are also classified as green solvents due to their unique properties, which are discussed further below. It should be noted that solvents has its own set of advantages and disadvantages, and they do not compete with one another, but rather complement one another.

Water

Water as a sustainable green solvent in organic synthesis has attracted attention in recent decades due to its innocuous nature, ample availability, low cost and tunable physicochemical properties. It also shows some alluring properties related with reactivity like peculiar selectivities, influences of hydrogen-bond network on reaction behaviour, adjustable pH values, application of biphasic reaction systems and the usage of salts for the salting-in or salting-out impact. The use of water as a solvent for organic reactions is one of the finest solutions to the problem of solvent toxicity and disposal. The chemistry in natural systems (biochemical reactions) is based on water. The use of water as a solvent for synthetic chemistry holds great promise for the future in terms of the cheaper and less hazardous production of chemicals (Filly et al., 2016; Hartonen & Riekkola, 2017).

Water as a solvent has several advantages, including ease of operation and high efficiency in many organic processes using water-soluble substrates, reagents, and renewable resources, such as carbohydrates, without the need for derivatizations. Many reactions that were previously thought to occur solely in organic solvents have been engineered to run in water over the last two decades. Since the seminal work on aqueous Diels-Alder reactions by Breslow (Breslow, 1991; Otto et al., 1996), there has been profound research activities in the development of organic reactions in aqueous medium such as catalytic hydrogenation and hydroformylation, metal-mediated carbon-carbon bond formation, water-tolerant Lewis acid catalysis including solid acids, and transition metal-catalyzed carbon-carbon bond formations. These reactions enable the direct modification of water-soluble and renewable compounds in water without the need for the protection-deprotection processes usually associated with traditional chemistry (Denmark & Wynn, 2001; Okuhara, 2002).

To simplify catalyst recycling and product isolation, development of high atom efficiency catalytic processes is an another potential opportunity for greener aqueous synthesis. The reactant and product should have no or very little water solubility in the ideal situation. As a consequence, simple phase separations may be used to isolate the product, and the catalyst can be easily recycled. As a result, the catalyst-bearing aqueous solution may be recycled without discharge or regenerate it. Thus poor atom-efficiency activities that lead to the build up of undesirable materials in the solution can be avoided. Catalytic hydrogenations, hydroformylations, the Wacker oxidation, and certain polymerization processes are all successful examples in this category (Bényei & Joó, 1990; Cortes-Clerget et al., 2021).

Hence, it is certainly a valuable approach, but it should be kept in mind that water has also some disadvantages: its temperature-restricted range of liquid state and especially its significant heat capacity, which makes distillation processes are extremely energy-consuming. The same is true for pervaporation and reverse osmosis. All these techniques may be necessary when water must be separated from a non-volatile reaction product or impurities. In these cases, the high energy consumption may lead to the choice of another solvent (Hauk et al., 2021).

Ionic Liquids (ILs)

The design and process of chemical synthesis have evolved away from traditional reaction conditions and toward more environmentally friendly alternatives. The use of ionic liquids (ILs) in organic synthesis is a smart step in the right direction Ionic liquids (ILs) are salts composed of discrete anions and cations with a melting point lower than the boiling point of water. Ionic liquids are also referred as neoteric solvents, designer solvents, ionic fluids, and molten salts (Malolan et al., 2021).

These ILs are made up of two ions that interact with one other via from a combination of Coulombic forces, π–π interactions, hydrogen bonds, and dispersion forces as shown in Figure 5 along with some frequently utilised cations and anions. They may dissolve both inorganic compounds due to their large organic cationic part and smaller inorganic anionic part. Changing anion–cation combinations of ionic liquids also changes their physicochemical properties, although this cannot be achieved *in situ* as with other tunable solvents. The suitable combination of cations and anions can produce desirable ILs, in which solubility, hydrophobicity, viscosity, density, as well as other characteristics can be fine-tuned to meet a specific purpose Liquid salts and inorganic salts can be combined to form multicomponent ILs or task specific ionic liquids or functionalized ionic liquids as shown in Figure 6. The term functionalized ionic liquids really refers to an endeavour to capitalise on ionic liquids' potential "design" capability and transform them into true working systems rather than merely reaction medium. ILs are effectively non-volatile due to extremely high enthalpies of vaporization (ΔHvap). Columbic forces hold the ions in ILs together, resulting in a vapour pressure approaching zero above the liquid surface. As a result, during their transport, handling, and usage, ILs do not produce potentially harmful VOCs. Because ILs do not oxidise, they are non-flammable and nonexplosive. ILs can be kept for a long period without degradation. The ionic nature of ILs, increases the rate of the reaction (Ratti, 2014).

Figure 5. Task specific ionic liquid and their physicochemical properties

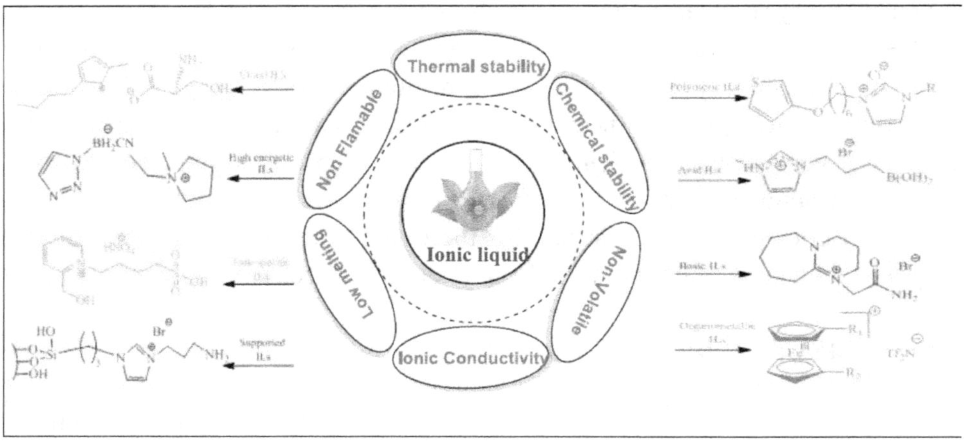

Application of ILs in Organic Synthesis

ILs have been widely used as solvents and catalysts in organic chemistry. These compounds often perform better compared with that of the conventional solvents and catalysts. The appropriate physiochemical characteristics aided in the usage of ILs in the inorganic, organic, and inorganic as well as biocatalytic reactions. The ILs functioned as alternative reaction media for the majority of reactions mainly associated with transition metal catalysis leading to better organic transformations. In figure 6 some prominent application has been shown.

Davis Jr. and Forrester (1999) (Xing et al., 2005), coining the term "task specific ionic liquid" for ILs in which a functional group is incorporated as part of the cation and/or anion structure as shown in Figure 5.

The covalent bonding of a functional group to a cation/anion or both of an ordinary ionic liquid in "task specific ionic liquid" gives it the ability to serve not only as a solvent but also as a reagent and/or catalyst in chemical processes. For example

- Bronsted acidic ionic liquids containing sulphonic acid groups were utilised as solvents and/or catalysts in esterification and other acid-catalyzed processes (Xing et al., 2005).
- Ionic liquids with attached amines have the ability to extract carbon dioxide from gas streams (Bates et al., 2002).
- Ionic liquids with large aromatic head groups exhibit increased activity for aromatic extraction in aqueous biphasic solutions.
- In the production of ethoxybenzene, ionic liquids containing a tethered hydroxyl group (OH) were employed as a phase transfer catalyst (Feng et al., 2007).
- Ionic liquids with metal ligating groups are used to remove metal ions from aqueous solutions (Ratti, 2014).
- Ionic liquids with carboxylate groups attached have been utilised as supports for "IL-phase" synthesis, a flexible extension of the solid phase synthesis idea (Ratti, 2014).

Limitation

Ionic liquids (ILs) are still getting popular due to their numerous favourable physicochemical characteristics. ILs have frequently been regarded as environmentally friendly solvents. However, environmental, health, and safety evaluations of ILs have revealed some concerns regarding their safety, and their greenness status is still unknown. Bystrzanowska, M et al conducted a detailed examination of more than 300 commercially available ILs to clarify the situation regarding their greenness using multicriteria decision analysis, a method that allow to rate multiple choices based on important criteria. They included toxicity to various creatures, biodegradability, danger statements, and preventive procedures while handling them (Bystrzanowska et al., 2019).

Biosolvents

Bio-based solvents are generated from agricultural residues, as the name implies. The agricultural source of the biomass used to produce these solvents may be divided into four groups: (a) lignocellulosic; (b) sugar and starch; (c) protein and oil derived; and (d) other forestry and food residues. The solvents obtained from these categories can be further classified based on their functional groups as shown in Figure 8.

Figure 6. Structure of ILs and their applications in organic synthesis

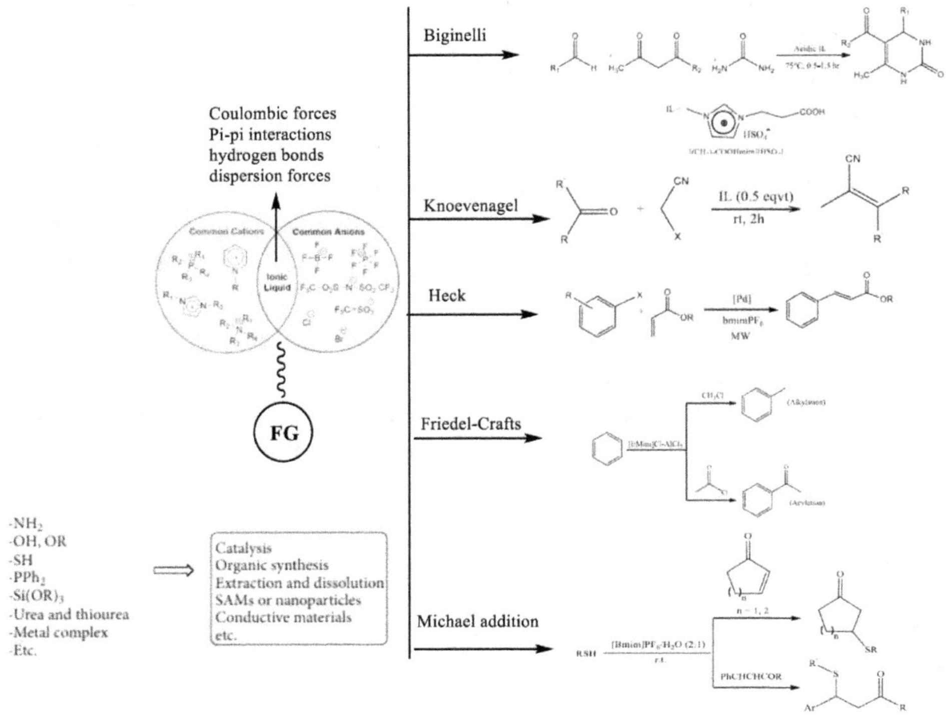

Figure 7. Biobased solvents

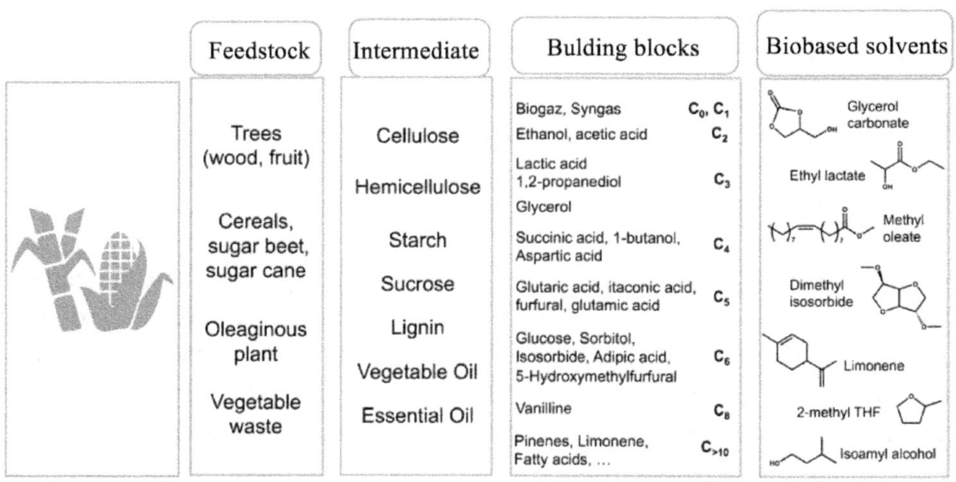

Some important bio-solvents are as follows:

2-Methyl Tetrahydrofuran (2-MeTHF)

2-MeTHF was initially designed as a biofuel, but it is now exploited as a renewable alternative to THF. 2-MeTHF has a lower water solubility, greater stability, and lower volatility than THF. It can be synthesized from xylose and glucose, both of which are obtained from biomass via various feedstock intermediates such as levulinic acid and furfural. Despite being derived from renewable sources, the CHEM21 selection criteria classify 2-MeTHF problematic as its high flammability (Alcantara & de Maria, 2018).

γ-Valerolactone (GVL)

Many outstanding characteristics of GVL make it a great candidate as a sustainable alternative to hazardous dipolar aprotic solvents such as acetonitrile, dimethylformamide (DMF), or N,N-dimethylacetamide (DMA). The polarity of GVL and other common polar solvents is quite comparable; for example, the observed GVL dielectric constant at 25°C is 36.47, whereas the dielectric values of DMF, NMP, CH_3CN, and DMA are 36.7, 32.0, 37.5, and 37.8 respectively. It is made from cellulose by converting hydroxymethylfurfural to levulinic acid, or from hemicellulose by converting furfural to levulinic acid. GVL is a non-toxic liquid that is stable at room temperature. At the moment, GVL has just a few industrial uses. Due of high production costs, GVL has limited industrial uses at present (Oklu et al., 2019).

Cyrene

Cyrene (dihydrolevoglucosenone) is a bicyclic, chiral, seven-membered heterocyclic cycloalkanone that is a bio-based and completely biodegradable dipolar aprotic solvent. Cyrene is extraced from plantation radiata pine. It is utilized as a green solvent in many chemical processes. Cyrene is nontoxic, harmless, and a greener alternative than a toxic organic solvent such as Dimethlyformamide (DMF) and *N*-methyl-2-pyrrolidone (NMP) (Camp, 2018).

Glycerol

Glycerol, a natural polyol is the main co-product of the vegetable oil industry. The use of glycerol as a solvent or as a precursor for the synthesis of biomass-based solvents has recently emerged as a feasible and promising approach. Specially, intrinsic properties of glycerol, such as low toxicity, good biodegradability, low vapor pressure, large availability and easy handling and storage, offer complementary advantages to the usual reported green solvents. The environmentally benign properties of glycerol, low price and the beneficial effect of glycerol on reaction rates and selectivity, it is not an exaggeration that glycerol is, no doubt, a qualified green solvent. The poor solubility of glycerol in common volatile organic solvents makes it an excellent solvent for extracting compounds from the glycerol phase. As with ionic liquids, the poor solubility of glycerol in supercritical CO2 (scCO2) provides new instruments for (1) performing catalysis in a low-cost, long-term medium and (2) selectively and cleanly recovering reaction products from glycerol. Remarkably, glycerol have combine advantages of water (low price, non toxicity) and ionic liquid (high boiling point, low solubility in $scCO_2$), as solvent has resulted in few cases in an improved reactivity and/or selectivity, and also in an easier product separation and an improved catalyst recycling (Gu & Jérôme, 2010). Because glycerol is very viscous, it can cause problems in a many chemical processes.

Polyethylene Glycol Polymers

Polyethylene glycol polymers (PEGs) have also developed as a new class of environmentally friendly solvents. PEG and PEG-supported catalysis are alternate reaction media in green chemistry that are offering current innovation in the use of bio-based compounds. PEG and its aqueous solutions are regarded as outstanding solvent systems for a variety of other popular systems, including ionic liquids, glycerol, supercritical carbon dioxide, and micellar systems. PEG is a hydrophilic polymer that dissolves easily in water and various organic solvents such as toluene, dichloromethane, alcohol, and acetone, but not in aliphatic hydrocarbons such as hexane, cyclohexane, or diethyl ether. PEGs have been used. The main feature allows the recovery of PEGs by precipitation and filtration, which is extremely important in soluble polymer-supported chemistry. Polyethylene glycol polymers (PEGs) have been proved as eco-friendly reaction media as well as new route for the synthesis of important bioactive heterocyclic compound (Chen et al., 2005).

DEEP EUTECTIC SOLVENTS

Deep eutectic solvents (DES) are obtained by combining Lewis acids and bases. Because the melting temperatures of DES are lower than the melting points of the individual components, they are frequently produced by combining two solid reagents to generate a liquid. Different types of DES have been described, and most contain hydrogen-bond acceptors (HBAs) such as choline chloride and hydrogen-bond donors (HBDs) such as urea. These materials have the potential to be used as low-cost "designer" solvents with a variety of customizable physicochemical features. DESs meet a variety of requirements to serve as a green medium, including availability, non-toxicity, biodegradability, recyclability, flammability, and low cost, among others DESs provide significant potential as charge transfer and reaction media for electrochemical energy storage and conversion, as well as a myriad of other applications (EESC). DES are excellent solvents for a variety of chemical processes, for example, extractions, catalysis, organic synthesis, and electro deposition.

DES may be produced from renewable components, including plant metabolites (such as Ammonium salts, sugars, and organic acids, are known as NADES (natural deep eutectic solvents). DES is often made by simply heating and stirring, combining two or more components without further purification steps. In comparison to ionic liquids, which require long, air-sensitive preparations, DES has been proposed as a preferable green solvent.

DES, like ionic liquids, has been proposed as an alternate solvent for improving the sustainability of various chemical processes. The extensive spectrum of bioderived components present in DES, as well as the widespread use of NADES, have greatly boosted the reputation of deep eutectics as environmental benign solvents. Furthermore, NADES' biocompatibility has facilitated the development of several biotechnology and bioengineering applications. Studies on the use of DES in common organic processes such as aldol, cycloaddition, and catalysts for reactions have been reported.

The majority of the research indicated that DES and NADES had strong antibacterial and antifungal activities, which can be beneficial for some chemical processes, particularly in the biotechnology industry. The toxicity might be utilised as a deterrent advantage in comprehending biological metabolic processes as a possible anticancer agent. These solvents have a low toxicity, vapour pressures, excellent thermal stabilities, and a wide range of applications, making them particularly appealing as substitutes for conventional organic solvents However, there are certain issues for pollution of soil and water at the site of disposal (Hansen et al., 2020; Wu et al., 2021).

Supercritical Liquid

Supercritical fluids which have low tuneable densities varied between those of gas and liquid phases in critical conditions have been extensively studied for the past two decades to study their fundamental properties as solvents.

By changing the density of the fluid, their solvation power can be tailored. These SCFs are comparatively chemically inert, eco-friendly, low-cost and have critical point under ambient conditions.

Supercritical water (scH_2O) has been used as an industry-attractive approach for catalytic reactions in aqueous biphasic systems. Similarly, in the sense of green chemistry and catalytic applications in various mono and biphasic systems, supercritical carbon dioxide ($scCO_2$) is an important reaction medium.

Supercritical liquids such as liquid CO_2, which fulfils various characteristics of green mediums proved as a better substitute to organic solvents as shown in Figure 8. Liquid CO_2 offers a wide range of applications in the synthetic chemical process by inculcating both the properties of gases and liquids.

Supercritical fluids (SCFs) are approved green solvents in both chemical industry and academic research. SCFs inculcating both the properties of gases and liquids and exhibit liquid-like properties such as high density and simultaneously low viscosities and high diffusivities similar to the gas state. High diffusivities provides increased reaction rates, and the high solubility of gases makes them suitable media for reactions involving gaseous substrates.

Figure 8. Phase diagram of carbon dioxide and its advantages at supercritical conditions (Soh & Lee, 2019)

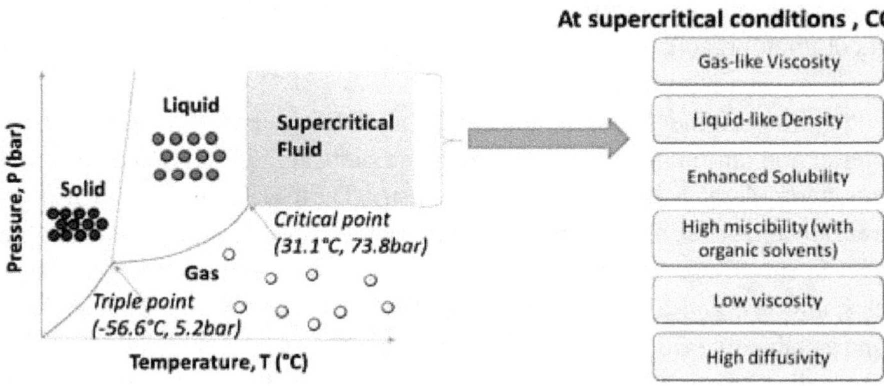

Supercritical liquids can progress the reaction, which is difficult or nearly impossible to achieve by organic solvents. Physical properties of supercritical CO_2 are density:1 g/cm^3, diffusion: below 10-5, the more praise property of supercritical CO_2 is very low viscosity (10^{-2} g cm S) and negligible surface tension, it has more dissolving power toward many non-polar compounds, and it can dissolve into condensed phase drastically by reducing the viscosity and surface tension of the condense phase making processing highly viscous material easily. Moreover, it does not cause harm to ozone layer like chlorofluorocarbons which were used in synthesis of polymer. Supercritical CO_2 is a right choice as a solvent for the processing of pharmaceuticals, biomolecule separation and heat-sensitive phenomenon because of its easily recoverable, low value of critical temperature. Various organic compounds soluble in supercritical CO_2 such as amoxicillin, ketoprofen, piroxicam and nimesulide, isoniazid polynuclear aromatic hydrocarbons, cholesterol, some fat-soluble vitamins A, D, E, K phenols and pyrocatechols (Clarke et al., 2018; Deepthi & Sathi, 2021).

Limitation

The limited solvating power can efficiently be overcome using cosolvents and additives (and is often beneficial for solvent removal), although this may negate some of the aforementioned advantages. Nevertheless, the required high operating pressures resulting in potential operational safety risks, which also result in a relatively high energy demand of scCO2-based processes, pose a challenge to the feasibility for commercial operation of scCO2-based processes.

CONCLUSION

While environmentally conscious solvent design (making something renewable, biodegradable, or less toxic) is admirable, it generally necessitates costly extractions from feedstocks (creating waste) and, in many cases, a more complicated synthesis. If efficiency factors are not taken into account, this might result in a less sustainable solvent or process than the petrochemical option. The economic effect is typically a considerably more revealing indicator of solvent performance (at least from an LCA standpoint), and even if there is a gain in the environmental sustainability of the solvent once created, the solvent's provenance must also be addressed.

REFERENCES

Aboagye, E. A., Chea, J. D., & Yenkie, K. M. (2021). Systems Level Roadmap for Solvent Recovery and Reuse in Industries. *iScience*, *24*(10), 103114. doi:10.1016/j.isci.2021.103114 PMID:34622166

Alcantara, A. R., & de Maria, P. D. (2018). Recent advances on the use of 2-methyltetrahydrofuran (2-MeTHF) in biotransformations. *Current Green Chemistry*, *5*(2), 86–103. doi:10.2174/2213346105666180727100924

American Chemical Society Green Chemistry Institute. (n.d.). *Pharmaceutical Roundtable Solvent Selection Guide*. https://www.acs.org/content/acs/en/greenchemistry/industry-business/pharmaceutical.html

Anastas, P., & Eghbali, N. (2010). Green chemistry: Principles and practice. *Chemical Society Reviews*, *39*(1), 301–312. doi:10.1039/B918763B PMID:20023854

Anastas, P. T., & Warner, J. C. (1998). Green chemistry. *Frontiers*, *640*, 1998.

Avila-Ortiz, C. G., & Juaristi, E. (2020). Novel methodologies for chemical activation in organic synthesis under solvent-free reaction conditions. *Molecules (Basel, Switzerland)*, *25*(16), 3579. doi:10.3390/molecules25163579 PMID:32781678

Bates, E. D., Mayton, R. D., Ntai, I., & Davis, J. H. (2002). CO2 capture by a task-specific ionic liquid. *Journal of the American Chemical Society*, *124*(6), 926–927. doi:10.1021/ja017593d PMID:11829599

Bényei, A., & Joó, F. (1990). Organometallic catalysis in aqueous solutions: The biphasic transfer hydrogenation of aldehydes catalyzed by water-soluble phosphine complexes of ruthenium, rhodium and iridium. *Journal of Molecular Catalysis*, *58*(2), 151–163. doi:10.1016/0304-5102(90)85035-G

Breslow, R. (1991). Hydrophobic effects on simple organic reactions in water. *Accounts of Chemical Research*, *24*(6), 159–164. doi:10.1021/ar00006a001

Buhler, D. R., & Reed, D. J. (1990). *Ethel Browning's toxicity and metabolism of industrial solvents*. Elsevier.

Byrne, F. P., Jin, S., Paggiola, G., Petchey, T. H., Clark, J. H., Farmer, T. J., ... Sherwood, J. (2016). Tools and techniques for solvent selection: green solvent selection guides. *Sustainable Chemical Processes, 4*(1), 1-24.

Bystrzanowska, M., Pena-Pereira, F., Marcinkowski, Ł., & Tobiszewski, M. (2019). How green are ionic liquids?–A multicriteria decision analysis approach. *Ecotoxicology and Environmental Safety*, *174*, 455–458. doi:10.1016/j.ecoenv.2019.03.014 PMID:30852310

Camp, J. E. (2018). Bio-available Solvent Cyrene: Synthesis, Derivatization, and Applications. *ChemSusChem*, *11*(18), 3048–3055. doi:10.1002/cssc.201801420 PMID:30044553

Capello, C., Fischer, U., & Hungerbühler, K. (2007). What is a green solvent? A comprehensive framework for the environmental assessment of solvents. *Green Chemistry*, *9*(9), 927–934. doi:10.1039/b617536h

Chen, J., Spear, S. K., Huddleston, J. G., & Rogers, R. D. (2005). Polyethylene glycol and solutions of polyethylene glycol as green reaction media. *Green Chemistry*, *7*(2), 64–82. doi:10.1039/b413546f

Cintas, P., Tabasso, S., Veselov, V. V., & Cravotto, G. (2020). Alternative reaction conditions: Enabling technologies in solvent-free protocols. *Current Opinion in Green and Sustainable Chemistry*, *21*, 44–49. doi:10.1016/j.cogsc.2019.11.007

Clarke, C. J., Tu, W. C., Levers, O., Brohl, A., & Hallett, J. P. (2018). Green and sustainable solvents in chemical processes. *Chemical Reviews*, *118*(2), 747–800. doi:10.1021/acs.chemrev.7b00571 PMID:29300087

Cortes-Clerget, M., Yu, J., Kincaid, J. R., Walde, P., Gallou, F., & Lipshutz, B. H. (2021). Water as the reaction medium in organic chemistry: From our worst enemy to our best friend. *Chemical Science (Cambridge)*, *12*(12), 4237–4266. doi:10.1039/D0SC06000C PMID:34163692

Cvjetko Bubalo, M., Vidović, S., Radojčić Redovniković, I., & Jokić, S. (2015). Green solvents for green technologies. *Journal of Chemical Technology and Biotechnology (Oxford, Oxfordshire)*, *90*(9), 1631–1639. doi:10.1002/jctb.4668

Deepthi, A., & Sathi, V. (2021). Supercritical Fluids as Green Solvents. In *Green Organic Reactions* (pp. 323–338). Springer. doi:10.1007/978-981-33-6897-2_17

Denmark, S. E., & Wynn, T. (2001). Lewis base activation of lewis acids: Catalytic enantioselective allylation and propargylation of aldehydes. *Accounts of Chemical Research*, *2002*(35), 209. doi:10.1021/ar000145a

Dolzhenko, A. V. (2021). Green solvents for eco-friendly synthesis of bioactive heterocycles. In *Green Synthetic Approaches for Biologically Relevant Heterocycles* (pp. 393–470). Elsevier. doi:10.1016/B978-0-12-820792-5.00007-X

Feng, G. R., Peng, J. J., Qiu, H. Y., Jiang, J. X., Tao, L., & Lai, G. Q. (2007). Synthesis of novel greener functionalized ionic liquids containing appended hydroxyl. *Synthetic Communications*, *37*(16), 2671–2675. doi:10.1080/00397910701465230

Filly, A., Fabiano-Tixier, A. S., Louis, C., Fernandez, X., & Chemat, F. (2016). Water as a green solvent combined with different techniques for extraction of essential oil from lavender flowers. *Comptes Rendus. Chimie*, *19*(6), 707–717. doi:10.1016/j.crci.2016.01.018

Gómez-López, P., Puente-Santiago, A., Castro-Beltrán, A., do Nascimento, L. A. S., Balu, A. M., Luque, R., & Alvarado-Beltrán, C. G. (2020). Nanomaterials and catalysis for green chemistry. *Current Opinion in Green and Sustainable Chemistry*, *24*, 48–55. doi:10.1016/j.cogsc.2020.03.001

Gu, Y., & Jérôme, F. (2010). Glycerol as a sustainable solvent for green chemistry. *Green Chemistry*, *12*(7), 1127–1138. doi:10.1039/c001628d

Gupta, P. K. (2020). Solvents, Gasses, and Vapors. In *Problem Solving Questions in Toxicology* (pp. 181–193). Springer. doi:10.1007/978-3-030-50409-0_13

Häckl, K., & Kunz, W. (2018). Some aspects of green solvents. *Comptes Rendus. Chimie, 21*(6), 572–580. doi:10.1016/j.crci.2018.03.010

Hansen, B. B., Spittle, S., Chen, B., Poe, D., Zhang, Y., Klein, J. M., Horton, A., Adhikari, L., Zelovich, T., Doherty, B. W., Gurkan, B., Maginn, E. J., Ragauskas, A., Dadmun, M., Zawodzinski, T. A., Baker, G. A., Tuckerman, M. E., Savinell, R. F., & Sangoro, J. R. (2020). Deep eutectic solvents: A review of fundamentals and applications. *Chemical Reviews, 121*(3), 1232–1285. doi:10.1021/acs.chemrev.0c00385 PMID:33315380

Hartonen, K., & Riekkola, M. L. (2017). Water as the first choice green solvent. In *The Application of Green Solvents in Separation Processes* (pp. 19–55). Elsevier. doi:10.1016/B978-0-12-805297-6.00002-4

Hauk, P., Wencel-Delord, J., Ackermann, L., Walde, P., & Gallou, F. (2021). Organic Synthesis in Aqueous Multiphase Systems-Challenges and Opportunities ahead of Us. *Current Opinion in Colloid & Interface Science, 56*, 101506. doi:10.1016/j.cocis.2021.101506

Hume, A. S., & Ho, K. (2019). Toxicity of solvents. In *Basic Environmental Toxicology* (pp. 157–184). CRC Press. doi:10.1201/9781315138091-6

Jahangirian, H., Lemraski, E. G., Webster, T. J., Rafiee-Moghaddam, R., & Abdollahi, Y. (2017). A review of drug delivery systems based on nanotechnology and green chemistry: Green nanomedicine. *International Journal of Nanomedicine, 12*, 2957–2978. doi:10.2147/IJN.S127683 PMID:28442906

Jimenez-Gonzalez, C. (2019). Life cycle considerations of solvents. *Current Opinion in Green and Sustainable Chemistry, 18*, 66–71. doi:10.1016/j.cogsc.2019.02.004

Lomba, L., Giner, B., Bandrés, I., Lafuente, C., & Pino, M. R. (2011). Physicochemical properties of green solvents derived from biomass. *Green Chemistry, 13*(8), 2062–2070. doi:10.1039/c0gc00853b

Loupy, A. (2004). Solvent-free microwave organic synthesis as an efficient procedure for green chemistry. *Comptes Rendus. Chimie, 7*(2), 103–112. doi:10.1016/j.crci.2003.10.015

Malolan, R., Gopinath, K. P., Vo, D. V. N., Jayaraman, R. S., Adithya, S., Ajay, P. S., & Arun, J. (2021). Green ionic liquids and deep eutectic solvents for desulphurization, denitrification, biomass, biodiesel, bioethanol and hydrogen fuels: A review. *Environmental Chemistry Letters, 19*(2), 1001–1023. doi:10.100710311-020-01113-7

Oklu, N. K., Matsinha, L. C., & Makhubela, B. C. (2019). *Bio-solvents: synthesis, industrial production and applications*. Solvents, Ionic Liquids and Solvent Effects.

Okuhara, T. (2002). Water-tolerant solid acid catalysts. *Chemical reviews, 102*(10), 3641-3666.36. *Chemical Communications (Cambridge), 268*. Advance online publication. doi:10.1039/b108851n

Otto, S., Bertoncin, F., & Engberts, J. B. (1996). Lewis acid catalysis of a Diels– Alder reaction in water. *Journal of the American Chemical Society, 118*(33), 7702–7707. doi:10.1021/ja960318k

Pacheco-Fernández, I., & Pino, V. (2019). Green solvents in analytical chemistry. *Current Opinion in Green and Sustainable Chemistry, 18*, 42–50. doi:10.1016/j.cogsc.2018.12.010

Pena-Pereira, F., & Tobiszewski, M. (Eds.). (2017). *The application of green solvents in separation processes*. Elsevier.

Ratti, R. (2014). Ionic liquids: Synthesis and applications in catalysis. *Adv. Chem, 2014*(3), 1–16. doi:10.1155/2014/729842

Savitskaya, T., Kimlenka, I., Lu, Y., Hrynshpan, D., Sarkisov, V., Yu, J., & Wang, L. (2021). Principle of Green Chemistry. In *Green Chemistry* (pp. 1–14). Springer.

Sheldon, R. A. (2019). The greening of solvents: Towards sustainable organic synthesis. *Current Opinion in Green and Sustainable Chemistry, 18*, 13–19. doi:10.1016/j.cogsc.2018.11.006

Soh, S. H., & Lee, L. Y. (2019). Microencapsulation and nanoencapsulation using supercritical fluid (SCF) techniques. *Pharmaceutics, 11*(1), 21. doi:10.3390/pharmaceutics11010021 PMID:30621309

Talele, G. S., & Shahare, H. V. (2021). Green Solvents for Pharmaceuticals. In *Natural Pharmaceuticals and Green Microbial Technology* (pp. 1–9). Apple Academic Press. doi:10.1201/9781003003229-1

Tebby, C., Mombelli, E., Pandard, P., & Péry, A. R. (2011). Exploring an ecotoxicity database with the OECD (Q) SAR Toolbox and DRAGON descriptors in order to prioritise testing on algae, daphnids, and fish. *The Science of the Total Environment, 409*(18), 3334–3343. doi:10.1016/j.scitotenv.2011.05.029 PMID:21684579

Uzma, N., Khaja Mohinuddin Salar, B. M., Kumar, B. S., Aziz, N., David, M. A., & Reddy, V. D. (2008). Impact of organic solvents and environmental pollutants on the physiological function in petrol filling workers. *International Journal of Environmental Research and Public Health*, *5*(3), 139–146. doi:10.3390/ijerph5030139 PMID:19139531

Virot, M., Tomao, V., Ginies, C., & Chemat, F. (2008). Total lipid extraction of food using d-limonene as an alternative to n-hexane. *Chromatographia*, *68*(3), 311–313. doi:10.136510337-008-0696-1

Wagare, D. S., Shirsath, S. E., Shaikh, M., & Netankar, P. (2021). Sustainable solvents in chemical synthesis: A review. *Environmental Chemistry Letters*, *19*(4), 1–20. doi:10.100710311-020-01176-6

Wu, J., Liang, Q., Yu, X., Lü, Q. F., Ma, L., Qin, X., Chen, G., & Li, B. (2021). Deep Eutectic Solvents for Boosting Electrochemical Energy Storage and Conversion: A Review and Perspective. *Advanced Functional Materials*, *31*(22), 2011102. doi:10.1002/adfm.202011102

Xing, H., Wang, T., Zhou, Z., & Dai, Y. (2005). Novel Brønsted-acidic ionic liquids for esterifications. *Industrial & Engineering Chemistry Research*, *44*(11), 4147–4150. doi:10.1021/ie0488703

Yaseen, G., Ahmad, M., Zafar, M., Akram, A., Sultana, S., Kilic, O., & Sonmez, G. D. (2021). Current status of solvents used in the pharmaceutical industry. In *Green Sustainable Process for Chemical and Environmental Engineering and Science* (pp. 195–219). Elsevier. doi:10.1016/B978-0-12-821885-3.00004-9

Young, P. J., Harper, A. B., Huntingford, C., Paul, N. D., Morgenstern, O., Newman, P. A., Oman, L. D., Madronich, S., & Garcia, R. R. (2021). The Montreal Protocol protects the terrestrial carbon sink. *Nature*, *596*(7872), 384–388. doi:10.103841586-021-03737-3 PMID:34408332

Zimmerman, J. B., Anastas, P. T., Erythropel, H. C., & Leitner, W. (2020). Designing for a green chemistry future. *Science*, *367*(6476), 397–400. doi:10.1126cience.aay3060 PMID:31974246

Chapter 6
Recycled Raw Banana Peels for Bioplastics

Suchismita Satapathy
https://orcid.org/0000-0002-4805-1793
KIIT University, India

Divyalok Samal
KIIT University, India

Partha Pratim Sarmah
KIIT University, India

Tushar Mandal
KIIT University, India

Abhinav Sinha
KIIT University, India

Simranpreet Singh
KIIT University, India

ABSTRACT

Petroleum-based plastics may not be sustainable in the long run as crude oil will get scarce. Hence, it is imperative to find alternative sources of material from which plastics can be prepared. Starch, found abundantly in raw banana peel, which is a well-known kitchen waste, can be used to prepare biodegradable plastics. The preparation and some of its properties including its biodegradability have been studied here. A homemade DIY bioplastic sample has been prepared and its biodegradability tested. The world today is suffering an unprecedented menace of plastic pollution. The ubiquitous presence of plastic has sent alarm bells all across the globe. So bioplastics may be the solution.

DOI: 10.4018/978-1-7998-9851-1.ch006

INTRODUCTION

Petrochemical based plastics are used in many industries including the packaging industry because of their affordability, high-strength, lightweight, good barrier properties and easy preparation processes. The characteristics of high-strength combined with their hydro-phobic nature causes them to be non-biodegradable as natural microorganisms present in soil are unable to degrade them using their enzymes. The issue is aggravated by plastics having low surface area and high molecular weight. The yearly production of widely used petrochemical plastic can be more than 350 million tonnes. Their usage, especially of disposable plastic items is responsible for severe environmental damage. Greenhouse gases are emitted because of the production of petroleum-based plastics. Waste management options of plastic are not adequate. Increase in amounts of plastic have caused landfills to overflow and this problem of shrinking landfill space poses a threat to environment. The problem of disposal of plastics is faced by nations round the world and is a concern of global proportions. It affects nations across economies and sizes with enormous environmental ramifications. Their combustion leads to release of several pulmonary irritants. Plastic pollution can be carcinogenic to humans and it is known to cause disruption of thyroid hormone levels. Plastic packaging of food can contaminate foodstuff which will lead to adverse health impacts. Recycling rates of plastic is very low.[9] Plastic fragments reaching the oceans are harmful to the marine ecosystem. If current trends on plastic pollution continues, there could be more plastic than fishes in oceans by 2050. Marine organisms can think of plastic waste as food and ingest them which can lead to injuries and suffocation. This can be harmful for humans too when they consume sea-food. Plastics which have a diameter of less than 5 mm are called microplastics and they are formed by macroplastics. They have been found in food, water, air, and even in Antarctica and in the Arctic. Also, with the depletion of fossil fuel reserves, it will be necessary to find alternatives of petroleum-based plastics. Sufficient amount of cheap crude oil which serves as a raw material for plastics will not be available on the future. With oil reserves depleting, it is becoming increasingly difficult and expensive to extract oil for plastic manufacturing. Substitutes of petroleum-based plastics must be looked at. Various biodegradable, eco-friendly and economically viable options that can replace plastic has come up in the last few years. Renewable biomass can be used to prepare biodegradable plastic. Bio-plastic can be starch-based, cellulose-based or protein-based. Bioplastic is a biodegradable material that come from renewable sources and can be used to reduce the problem of plastic waste that is suffocating the planet and polluting the environment. Bioplastics are used for disposable items, such as packaging, crockery, cutlery, pots, bowls, and straws. Few commercial applications exist for bioplastics. Starch is commonly used for preparation of bioplastic. Starch is richly found in banana peels, a common household waste. Raw banana has more starch than ripe banana.[8]

Although a lot of study has been done on bio-plastics produced from ripe banana peels, not much has been done on those prepared from peels of raw banana. Raw bananas are consumed widely and their peels are discarded. This study aims to study bio-plastics prepared from raw banana peels.

The raw bananas used in this experiment were sources from a local market in Guwahati, Assam.

LITERATURE REVIEW

Kanoujiya and Khanna (2019), have explored the manufacturing of a biodegradable plastic sample and conducted its biodegradability testing. The bio-plastic was made using banana peels sourced from local markets. The sample was buried in soil and reduction in its weight was noted every seven days by exhuming it. The bio-degradation of the specimen was calculated by the following formula:

Degradation percent = [(Initial Dry Weight – Final Dry Weight)/Final Dry Weight] x 100 (1)

The protein and carbohydrate content of the prepared bio-plastic was also estimated.

The paper mentions that the bio-plastic is biodegradable with a weight reduction of about 80% in 70 days. It was found that the bio-plastic was non-toxic. Bioplastics called drop-in bioplastics are chemically identical to their fossil-fuel counterpart but made from renewable resources. Thermoplastic starch represents the most widely used bioplastic, constituting about 50 percent of the bioplastics market. Cellulose bioplastics are mainly the cellulose esters, (including cellulose acetate and nitrocellulose) and their derivatives, including celluloid. Cellulose can become thermoplastic when extensively modified. An example of this is cellulose acetate, which is expensive and therefore rarely used for packaging. However, cellulosic fibers added to starches can improve mechanical properties, permeability to gas, and water resistance due to being less hydrophilic than starch

Abdullah et al (2019), explored the morphological characteristics, mechanical and physical characteristics and biodegradability of plastics prepared from starch of sweet potato. Starch and glycerol were mixed in different ratios and bio-plastics prepared from them were studied. It was seen that irrespective of starch to glycerol ratio, bio-plastics showed a characteristic yellow colour. It was observed that biodegradation was faster in specimens with higher concentration of starch.

Jayachandra Yaradoddi et al (2016), has investigated biodegradability of samples of bio-plastic prepared from banana peels both untreated and treated with Sodium metabisulphite. Solubility of both kinds of samples in various solvents namely Ammonia, Acetic acid, Chloroform, Acetone, Methanol, Sulfuric acid, Orth phosphoric acid, Ethhy alcohol and water has been tested. It was observed that 2 grams of bio-plastic which was treated with Sodium metabisulfite degraded in soil and 0.46 grams remained after a period of 13 days. Whereas only 0.04 grams remained out of the original quantity of 2 grams remained in case of bio-plastic not treated with Sodium metabisulfite.

Manimaran et al (2016), have produced a bio-plastic from banana peel starch and have performed biodegradability and elongation tests on it. The plastic sample did not produce characteristics of plastics but biodegraded rapidly in 6 days and completely biodegraded in 90 days. It was observed that the amount of glycerine used is an important factor in terms of biodegradability as increased amount of glycerine leads to slower biodegradability. The bio-plastic was stretched and it was found that it is stretchable. The bio-plastic was moulded into various shapes and it was concluded that it is mouldable.

N A Azieyanti et al (2020), compares the characteristics of banana starch based bio-plastics using two kinds of techniques. One technique encompasses usage of chemicals like Sodium Hydroxide and Hydrochloric acid; and glycerol as plasticizer. The other technique involves usage of natural-based materials like potato-starch and corn-starch; and glycerol as plasticizer.

Folino et al (2020), concludes that the appropriate environment for bio-plastics to degrade are aerobic and not anaerobic. This is because rapid degradation happens in aerobic conditions.

Rizwana Beevi et al (2020), have synthesized bio-plastic using banana peels and potato starch. Their solubility test and swelling tests have been carried out.

Sultan and Johari (2017), have studied films made of banana peels with varying concentrations of corn-starch added. Arikan and Bilgen (2019), compares the biodegradability and water absorption capacity of Potato peel bio-plastic(PPB) and commercial bio-plastic(CB) which was procured from an online retailer. Chandrana and Chandra (2021), have prepared a bio-plastic from banana peels and mention that it can sustain a weight of 2 kilograms Huzaisham and Marsi (2020), studies properties of banana peel and tapioca starch composite. A total of eight samples with weight of banana peel extract in them ranging from 5 to 40% of weight were studied. Arasaretnam S (2020), mentions possible usage of banana peel bio-plastic to purify wastewater since banana peel bio-plastic absorbs considerable amount of cationic heavy metal pollutants and cationic dye pollutants. The Royal Society of Chemistry portrays the nonexclusive cycle for the assembling of starch based bioplastics. This includes hydrolysis of the starch by utilizing a corrosive. Abdorreza et al (2011) have depicted in their paper the physiological, warm and rheological properties of corrosive hydrolyzed starch. This paper shows that the amylose content increments at first yet consistent hydrolysis causes a reduction in the amylose content. This reality is additionally validated in the paper by Karntarat Wuttisela et al (2008). The amylose content is answerable for the plastic development in starch. Plasticizers are utilized to give adaptability and mouldability to the bioplastic tests. Thawien Bourtoom, of the Prince of Songkla University, Thailand, in his paper (2007) examines the impacts of the normal kinds of plasticizers utilized and their consequences for different properties like elasticity, stretching at break and water fume porousness of the bioplastic film. Utilizations of bioplastics, particularly in the bundling business have been examined in the paper by Nanou Peelman et al (2013) where biobased polymers utilized as a segment in (food) bundling materials is thought of, various methodologies for further developing obstruction properties of biobased bundling and porousness esteems and mechanical properties of multifaceted biobased plastics is additionally examined. Recent trends indicate the biocompatible and biodegradable polyhydroxyalkanoates (PHAs) as alternatives to conventional plastics which has wide variety of thermal and mechanical characteristics (Khatami et al. 2021). PHAs are linear polyesters, produced by microbiological, enzymatic, or chemical processes, but their industrial production is still not cost-competitive (Medeiros Garcia Alcântara et al. 2020). Renewable and inexpensive carbon sources—such as macroalgae, peanut oil, crude glycerol, and whey—have been studied to reduce production costs (El-malek et al. 2020). Innovative research proposed the production of PHAs by a three-step process consisting of CO2 reduction to acetate and butyrate by microbial electrosynthesis, extraction/concentration of acetate and butyrate, and PHAs production from volatile fatty acids. This process meets the demand to decrease CO2 emissions and convert a greenhouse gas to bioplastics (Pepè Sciarria et al. 2018).

A residual product of crude oil palm production is an empty fruit bunch, composed of cellulose, hemicellulose, and lignin. Having high cellulose content (36.67%), this abundant waste could be used to produce bioplastics (Isroi and Panji 2016; Isroi et al. 2017). Microcrystalline cellulose and glycerol were added to keratin from waste chicken feathers to produce biopolymeric films (Ramakrishnan et al. 2018; Sharma et al. 2018). Microcrystalline cellulose was a reinforcing additive in bioplastic production also from avocado seeds (Sartika et al. 2018), jackfruit seeds (Lubis et al. 2018), and cassava peels (Maulida and Tarigan 2016). Waste cassava peels were investigated in combination with kaffir lime essential oil for future applications in industry and medicine (Masruri et al. 2019).

METHODOLOGY

The methodology consists of extraction of starch from raw banana peel, preparing bio-plastic from it, conducting its biodegradability test by soil burial method, studying its behaviour under water and preparing useful articles from it.

The peels of banana were removed using a stainless steel knife and cut into small sizes. The peels were them kept immersed in a solution of Sodium metabisulfite. The peels were then removed and boiled for 30 minutes in 800 mL of distilled water. After the boiling process, the water was decanted off and the peels were dried for 30 minutes. The peels were then ground to a paste which was filtered to obtain banana peel starch.25 mL of banana peel starch was measured and 3 mL of acetic acid was added to it and stirred. After this, 3 mL of glycerol was added to it and stirred. The mixture was then spread on a dish and was sun-dried.

Figure 1. Photograph showing Bio-plastic prepared from raw banana peel

The sample was weighed and its weight noted. The sample was then buried in soil and removed every two days, gently cleaned and weighed again. The reduction in weight was noted.

The rate of biodegradation was measured using the formula mentioned in equation (1).

The bio-plastic was immersed in regular water, hot water, cold water and saltwater to study their impact on it. The bio-plastic can be moulded to prepare articles like cups and plates. The paste can be put in a mould and dried to make it usable.

Figure 2. Bioplastics degrade in soil

Figure 3. Sequence of experiments done

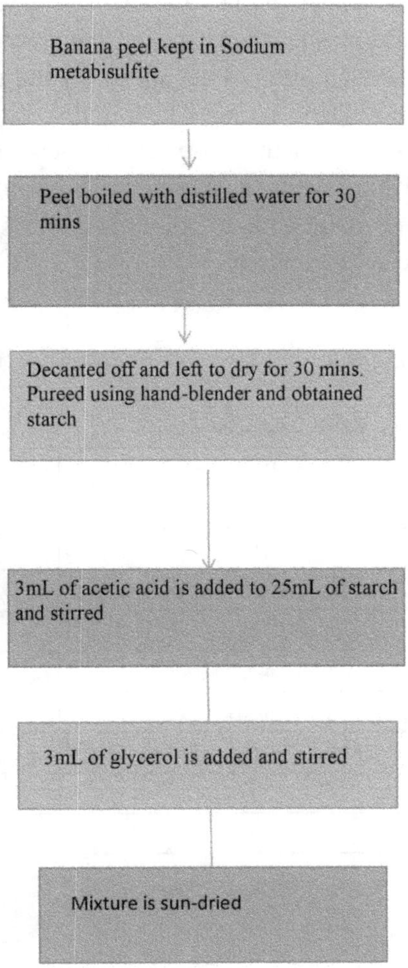

RESULTS AND DISCUSSION

Materials such as starch, cellulose, wood, sugar and biomass are used as a substitute for fossil fuel resources to produce bioplastics; this makes the production of bioplastics a more sustainable activity compared to conventional plastic production. Although bioplastics save more nonrenewable energy than conventional plastics and emit less GHG compared to conventional plastics, bioplastics also have negative environmental impacts such as eutrophication and acidification. So bioplastics are preferred .It was observed that the weight of the bio-plastic sample reduced from 15.57 grams to 1.54 grams in a period of 19 days, by virtue of its biodegradability in soil.

Biodegradability is dependent on a lot of uncontrollable factors like soil composition, microorganisms present, temperature etc. From the data it can be seen that biodegradation rate remained high for the first five days and reduced thereafter.

It was observed that bio-plastic when exposed to hot water fragmented and small pieces were formed. Other samples did not show any immediate fragmentation. It was observed that the bio-plastic samples kept under water developed fungi in a matter of few days. This shows that the bio-plastic is consumable by microorganisms thriving under moisture and asserts the biodegradable nature of the bio-plastic.

The bio-plastic can me moulded into articles like bowls and tumblers which are biodegradable and disposable in nature.

Table 1. Reduction in weight and percentage degradation with time

Day	Weight of sample(grams)	% Degradation
1st	15.57	-
3rd	8.94	74.16
5th	5.96	50.00
7th	5.57	7.00
9th	5.19	7.32
11th	4.08	27.21
13th	3.30	23.64
15th	2.57	28.40
17th	2.17	18.43
19th	1.54	40.91

Figure 4. Graph showing percentage of degradation of sample with the number of days

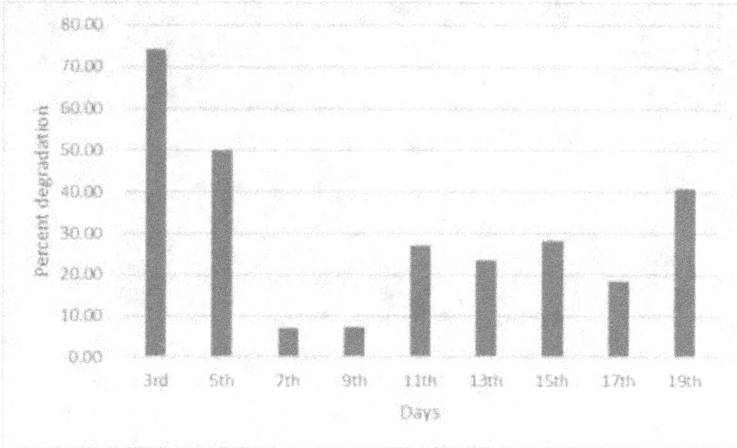

Figure 5. Graph showing reduction in weight of sample with number of days

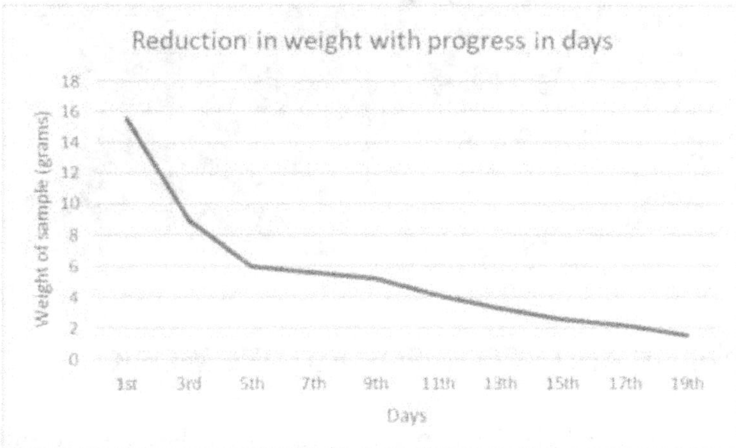

Figure 6. Bio-plastic moulded to form disposable utensils

CONCLUSION AND SCOPE FOR FUTURE WORK

Use of Petroleum-based plastics is extremely taxing for the environment and is not sustainable. Hence, there is a need for biodegradable plastic today. Raw bananas are consumed at many places and their peel which is a waste product is rich in starch. Biodegradable plastic can be made from this banana peel. The bio-plastic produced here rapidly degrades in soil in the first five days, which is consistent with (Manimaran et al, 2016), where rapid degradation was observed for first six days. This indicates that bioplastic prepared from raw banana peel might be less taxing on our landfills. The bio-degradation in this case differs from the one observed in (Supriya Nandlal Kanoujiya and Shivani Kakkar Khanna, 2019). Here, percent bio-degradation on the sample is not uniform whereas there the percent bio-degradation was seen to increase consistently. To further advance the appliance of bioplastic, it's very necessary to manage carefully the waste disposal. Recycling appears the simplest solution from that time, for disposal of the bio-based product to maximise the environmental footprint also as reduce the renewable resources consumption. Recycling of a bioplastic results in an overall decrease of environmental impact which can related to the assembly and disposal of the bioplastic itself. it's worth noting that thanks to the improper management and applications of bioplastics, the knowledge reported during this paper are often useful for the environmental reliability. PHA materials are the most resource to substitute conventional plastic use in most of the engineering applications fields. Nowadays, the PHA costs of production are too high, but further research on technology and sourcing can reduce manufacturing costs for a versatility and heterogeneity and strengthen the applications of bioplastic. Further the bio-plastic acts as a food for fungi under fresh and saltwater conditions which asserts its biodegradability. Since this bio-plastic is made with natural and non-toxic ingredients, it can be used as Do-it-yourself (DIY) disposable utensils. This is demonstrated by the preparation of bowls. Because of its tendency to decompose, degrade and fragment; the disposable utensils should not be re-used. They should be used within the shelf life of 3 days. It should also not be exposed to hot fluids.

Further studies can be done to find out tensile strength of the bio-plastic. Other mechanical properties can also be explored and compared with bio-plastics prepared from ripe banana peel.

REFERENCES

Kanoujiya & Khanna. (2019). *Production of Bioplastic from banana peels*. Academic Press.

Akbar, H. D. A., Pudjiraharti, S., Karina, M., Putri, O. D., & Fauziyyah, R. H. (2019). Fabrication and characterization of sweet potato starch-based bioplastics plasticized with glycerol. *The Journal of Biological Sciences*, *19*, 57–64.

Yaradoddi. (2016). *Biodegradable plastic production from fruit waste material and its sustainable usage for green applications*. Academic Press.

Manimaran. (2016). *Production of Biodegradable plastic from banana peel.* Academic Press.

Azieyanti. (2020). *Mechanical and Morphology Studies of Bioplastic-Based Banana Peels*. Academic Press.

Folino. (2020). *Biodegradation of Wasted Bioplastics in Natural and Industrial Environments: A Review.* Academic Press.

Beevi. (2020). *Bioplastic Synthesis Using Banana Peels And Potato Starch And Characterization*. Academic Press.

Sultan & Johari. (2017). *The Development of Banana Peel/Corn Starch Bioplastic Film: A Preliminary Study*. Academic Press.

Arikan & Bilgen. (2019). *Production of bioplastic from potato peel waste and investigation of its biodegradability*. Academic Press.

Chandarana & Chandra. (2021). *Production of Bioplastics from Banana Peels*. Academic Press.

Huzaisham & Marsi. (2020). *Utilization of banana (Musa paradisiaca) peel as bioplastic for planting bag application*. Academic Press.

Arasaretnam, S. (2020). *Preparation of biobased plastic from banana peel and application in industrial waste water purification*. Academic Press.

Khatami, K., Perez-Zabaleta, M., Owusu-Agyeman, I., & Cetecioglu, Z. (2021). Waste to bioplastics: How close are we to sustainable polyhydroxyalkanoates production? *Waste Management (New York, N.Y.)*, *119*, 374–388. doi:10.1016/j.wasman.2020.10.008 PMID:33139190

Medeirosgarciaalcântara, J., Distante, F., Storti, G., Moscatelli, D., Morbidelli, M., & Sponchioni, M. (2020). Current trends in the production of biodegradable bioplastics: The case of polyhydroxyalkanoates. *Biotechnology Advances*, *42*, 107582. doi:10.1016/j.biotechadv.2020.107582 PMID:32621947

El-malek, F. A., Khairy, H., Farag, A., & Omar, S. (2020). The sustainability of microbial bioplastics, production and applications. *International Journal of Biological Macromolecules*, *157*, 319–328. doi:10.1016/j.ijbiomac.2020.04.076 PMID:32315677

Pepè Sciarria, T., Batlle-Vilanova, P., Colombo, B., Scaglia, B., Balaguer, M. D., Colprim, J., Puig, S., & Adani, F. (2018). Bio-electrorecycling of carbon dioxide into bioplastics. *Green Chemistry*, *20*(17), 4058–4066. doi:10.1039/C8GC01771A

Isroi, A. C., & Panji, T. (2016). Bioplastic production from oil palm empty fruit bunch. *International Conference on Biomass*.

Isroi, A. C., Panji, T., Wibowo, N. A., & Syamsu, K. (2017). Bioplastic production from cellulose of oil palm empty fruit bunch. *IOP Conference Series. Earth and Environmental Science*, *65*, 11. doi:10.1088/1755-1315/65/1/012011

Maulida, M. S., & Tarigan, P. (2016). Production of starch based bioplastic from cassava peel reinforced with microcrystalline cellulose avicel PH101 using sorbitol as plasticizer. *Journal of Physics: Conference Series*, *710*, 012012. doi:10.1088/1742-6596/710/1/012012

Masruri, M., Azhar, A. Z., Rosyada, I., & Febrianto, A. (2019). The effect of kaffir lime (*Citrus hystrix* DC) essential oil on bioplastic derived from cassava peel waste. *Journal of Physics: Conference Series*, *1374*(1), 012015. doi:10.1088/1742-6596/1374/1/012015

Ramakrishnan, N., Sharma, S., Gupta, A., & Alashwal, B. Y. (2018). Keratin based bioplastic film from chicken feathers and its characterization. *International Journal of Biological Macromolecules*, *111*, 352–358. doi:10.1016/j.ijbiomac.2018.01.037 PMID:29320725

Sartika, M., Lubis, M., Harahap, M. B., Afrida, E., & Ginting, M. H. S. (2018). Production of bioplastic from avocado seed starch as matrix and microcrystalline cellulose from sugar palm fibers with Schweizer's reagent as solvent. *Asian Journal of Chemistry*, *30*(5), 1051–1056. doi:10.14233/ajchem.2018.21155

Chapter 7
Scientomeric Evaluation and Visual Aanalytics of the Scientific Literature Production on Bioplastics

Anchal Rana
DAV College, Chandigarh, India

Anjali Sharma
DAV College, Chandigarh, India

Nidhi Gupta
https://orcid.org/0000-0003-4244-4470
Panjab University, Chandigarh, India

Rupak Chakravarty
https://orcid.org/0000-0001-5046-1663
Panjab University, Chandigarh, India

ABSTRACT

Incessant use of plastics in the past decades has been a serious issue of environmental concern. The aim of this study is to examine the current status of the bioplastic research field by identifying the key publications or scientific productions, the most productive researchers in terms of publications and citations, latest hotspots (keywords), the countries with the most active research, collaborations between the countries, highly cited journals, highly contributing organizations, and co-cited references with their cluster analysis. This chapter considers whole literature study of bioplastics (i.e., documents published from 1894 to 2021). One thousand eight hundred sixty-three documents were identified using the Scopus database. The largest number of papers published were during the year 2020 with 308 records. Bibliometric analysis shows author "Guerrero, A." was the highly productive author and USA was the most productive region of the world. The Journal of Polymers and the Environment was the most common outlet for publication.

DOI: 10.4018/978-1-7998-9851-1.ch007

INTRODUCTION

Petrochemical-based plastics have become increasingly popular in last few decades due to easy manufacturing, low production cost, better strength to weight ratio, durability and diverse applicability (Narancic *et al.,* 2020). However, the accumulation of conventional plastics is a cause of major environmental concern (Saharan *et al.,* 2012). These have an extremely slow biodegradability, may produce toxins during degradation, increase carbon dioxide emission and over utilization of fossil fuels (Narancic *et al.,* 2020).

In the present scenario of reckless environmental deterioration, there is an enhanced need and demand for alternatives to petrochemically derived plastics. Bioplastics have gathered spotlight in recent years owing to their faster degradability and low toxicity. Bioplastics can be categorized as bio based biodegradable plastics and fossil based biodegradable plastics. Petroleum based biodegradable plastics include polycaprolactone, polybutylene succinate etc. However, their use still faces biodegradability issues. Bio based plastics are synthesized from natural resources including cellulose, starch, Poly -Beta- Hydroxyalkanoates, Polyhydroxybutyrate, chitin etc. (Shah *et al.,* 2021). These employ plasticizers along with the aforementioned polymers. Plasticizers including glycerin and sorbitol are organic molecules that reduce brittleness and crystallinity of polymers thus enhancing their durability and toughness upon lowering melting points (Groote *et al.,* 2002). Bioplastics synthesized by fermentation and action of microorganisms on carbohydrates for the production of polyhydroxyalkanoates and polyhydroxy fatty acids has generated immense research interest at present. Polyhydroxybutyric acid (PHB) is one of the major types of polyhydroxyalkanoate. Polylactic acid (PLA) is produced from lactic acid, a natural acid has excellent biocompatibility and processing ability.

Packaging holds the largest segment in application of bioplastic in 2019. There is escalating demand for packaging made from bioplastic to be used in organic food wrapping as well as premium and branded products with specific requirement. Bioplastics can be used in rigid packing included Bio-Pet bottles, flexible packing for perishable products and service packing such as carrier bags (Coppola *et al.,* 2021). The present article is an endeavor to employ bibliometric methods for reviewing bioplastic research since its inception in 1894 till early 2021. A bibliometric analysis of bioplastics research can substantiate the knowledge in this research field, as it allows a reader to obtain a quick overview of the types of publications over time and information regarding the bioplastics development.

Bibliometrics or scientometrics represents a tool to evaluate scientific output of an individual, institution or country employing specific parameters such as quantity, impact factor and citation of published articles during a course of specific time span (Wang *et al.,* 2016). Bibliometrics has witnessed significant developments recently owing to advancements in computers enhanced internet connectivity, availability of databases, algorithms and programs required for bibliometric studies (Zhang and Yuan 2019). In this article, different bibliometric tools have been employed to review the global growth in research interest towards bioplastics. Since, earliest studies available to the latest publications in this area.

LITERATURE REVIEW

Abejon (2018) used Scopus database for bibliometric analysis to identify the research trends related to hemicellulose valorization from 2000 to 2016. The analysis was done on 109 articles. The bibliometric analysis was based only on the Scopus database and only two keywords i.e., "hemicellulose" and "valorization". European countries with Spain, France and Portugal as leaders, must be considered reference points for the most innovative research in hemicellulose valorization. USA in this research occupied only the fourth position and other countries that were great contributors to specific research (China, Japan, Canada or South Korea) did not appear in ranking for research regarding hemicellulose valorization. This paper also proposed the simultaneous valorization of three main components of lignocellulosic biomass i.e., hemicellulose, cellulose & lignin. So, researchers could view data for these compounds also.

Wang *et al* (2019) used Web of Science database to conduct a bibliometric analysis on Resource Recycling Industry (RRI) from 1996 to 2018. Study was conducted on 7041 articles. CiteSpace, a visualization tool was used to classify and analyze the research status of RRI field. The number of documents increased from 94 in 1996 to a peak of 963 in 2018. The number of papers published by "Chinese Academy of Science" ranks first among all researcher institutions. A publication by M.C. Monte on waste management is the most cited paper. Within this area, latest emerging subject categories include "Green and Sustainable technology" and "Science and Technology". Furthermore, "e-waste", "reverse logistics" and "lean manufacturing" were emerging research trends for RRI and "carbon emissions", "policy", "demolition waste", "supply chain management" and "compressive strength" had become hot topics. The study concluded the findings which provide inspiration for scholars to search for new research directions and ideas.

Perea *et al* (2020) used the Web of Science database to conduct a bibliometric analysis on Bioeconomy from 2006 to 2019. The aim of the study was to understand bioeconomy through the exploration of scientific studies related to the concept in academic literature. Bioeconomy is a model of production in full evolution. It addresses global challenges such as scarcity of natural resources and its conservation in products with added value (food, bio-plastics, bio-fertilizers) and the management of available resources for sustainable economic development. The study was conducted on 2,321 publications. The analysis used Bibliometrix R Package (Biblioshiny) as a bibliometric analysis tool for quantitative research on bioeconomy. VOS Viewer was used as visualization program to show the scientific landscape. The first report on the use of word bioeconomy in publication was from 1979. Bioeconomy was promoted worldwide in 2018. The structural analysis of the work in bioeconomy showed that there is a high rate of co-operation with a rich research network between institutions of European countries and comparatively a low level of collaboration with other nation.

Obileke *et al.* (2020) retrieved data from Web of Science database and perform bibliometric analysis on 1153 publication (from 1998 to 2018) regarding bioenergy from biowaste. The essence of the study was to analyze the state of research evolution in bioenergy from bio-waste and its trends. VOS Viewer was used as visualization tool. The study depicted 84 countries participating in this area. China proved to be most productive country based on number of publications and citations. This is the first bibliometric analysis study on bioenergy from bio-waste in the literature. The study concluded the need for further investigation in the research field and also emphasizes the need for more collaboration and more countries and more institutional participation.

Qin *et al.* (2020) performed bibliometric analysis on Microplastic Research from 2004 to 2019 at global level. Web of Science database was employed in the study and data was analyzed by VOS Viewer visualization software. Developed countries were important contributors to Microplastic research, as England, the USA, and Germany occupied the top three positions, based on the criteria of NSC. China was the only developing country in top 10 contributors but more exchange and co-operation between these countries and others are needed. Thompson RC, who defined the term 'Microplastics' was the most productive author, as well as the most influential one. All the top 10 authors were ESI highly cited researchers in the Ecology/ Environment field.

Ndlovu *et al.* (2020) used the Scopus database to conduct a bibliometric analysis on Hemp Breeding. The aim of the study was to perform a bibliometric and scientometric analysis of Scopus – indexed papers covering the field of 'Hemp Breeding' between 1908 to 2020. Hemp is an herbaceous plant that is currently receiving undivided attention from researchers and policymakers owing to its industrial use (in production of textiles, biofuel and bioplastics). For visualization purpose, VOS Viewer software and for quantitative data Microsoft Excel is used. Bibliometric study was performed on 152 papers. Most research on this topic was conducted or reported in the USA, Italy & Netherlands. The highest number of papers over the studied period and topic were published by authors affiliated to Wageningen University & Research. Results indicated that there is need to scale up research initiatives targeting hemp trait improvements to cater for the projected high demand and climate change which can be achieved through synergistic partnerships & knowledge exchanges across the hemp value chain.

To the best of our knowledge, bibliometric analysis and mapping of the Bioplastic research have not been performed and published to date.

Research Problem: Evaluation of literature on bioplastics via bibliometric or scientometric studies.

Study Objectives

The main objective of this study is analyse and understand the results of Scientometric evaluation and visual analytics of the scientific literature production on Bioplastics. To accomplish this major objective, several objectives have been framed as mentioned below:

1. To undertake the Descriptive bibliometric analysis of the research output on Bioplastic.
2. To measure the Annual Scientific Production of Bioplastic research.
3. To calculate Average article citations per year.
4. To find out Most prolific authors.
5. To evaluate Author co-authorship network
6. To depict Author's Keyword Co-occurrence Analysis
7. To find analysis of Countries (Highly Productive Countries, Country Collaboration Map on Bioplastic Research, International Citation Analysis)
8. To evaluate Journal ranking.
9. To depict organizations contributions.
10. To relate Three Factor Analyses (Keyword, Organization & Country)
11. To evaluate Top Ten Highly Cited Articles
12. To depict Co-Cited References of Papers on Bioplastics

METHODOLOGY

In the present study, data pertaining to bioplastics was retrieved from Scopus online database (the largest database of abstracts and bibliographic references of peer-reviewed scientific literature). The Search string used for this study was: TITLE-ABS-KEY (bioplastic*). This study covers the time span of 128 years from 1894 to 2021. In total, 1863 documents were retrieved. All types of publications (articles, book reviews, conference papers etc.) and languages (English, Spanish,

German etc.) were considered for evaluation. Scopus database facilitates data in different data formats. For this analysis, the data was extracted in CSV file format. The basic data processing work was carried out using the CSV file formats and tables. Graphs were generated out of the processed data using Microsoft Excel. For developing the network visualization maps from the data, three freely available softwares, VOSviewer (VV), Biblioshiny - a web-interface for bibliometrix), and ScientoPy were used. RStudio is an integrated development environment (IDE) for R. For data processing and analysis, the exported CSV file was imported in Biblioshiny and VV respectively. For ScientoPy, the records were exported from WoS by selecting the option "Full Record and Cited References" in Tab-delimited (Win, UTF-8) file format select.

THE ANALYSIS TOOLS

VOSviewer (VV)

VV software (freely available at www.vosviewer.com) was used to map the bibliographic material in this study. The tool specifically enables the creation, visualization and exploration of bibliometric maps in sciences (Mas-Tur *et al.,*2020, Selvavinayagam *et al.,* 2018) and provides visualizations of the results through a wide range of bibliometric indicators, such as co- authorship analysis of authors (Ke L. *et al*, 2020), co–occurrence analysis of keywords (Yan *et al*., 2021, Rahaman et al., 2020), citation analysis of sources and organizations (Zhao *et al.,* 2019), International Citation Analysis (Hassan *et al.,* 2020), and co–citation of cited references (Li *et al*., 2021).

RStudio (Biblioshiny)

Biblioshiny is the app used for bibliometrix study of scientific literature (Darvish 2019). It is based on the R language and run under the R Studio software (Version 1.3.1093). The data retrieved from Biblioshiny includes the Annual Scientific Production, Average article citation per year, three-factor analysis, highly productive countries, analysis of author's keywords, country collaboration map (Secinaro et al., 2020) and information about the highly cited articles.

The Impact Factor (IF) of the Journals was visualized from the Journal Citation Report (JCR) 2020. IF is a quantitative indicator, used to assess, compare and rank the scientific publication in different scientific areas (Malazy *et al.,* 2016).

ScientoPy is an Open-Source Python Based Scientometric Analysis Tool. It can process data set imported from Clarivate Web of Science (WoS) and Scopus data.

It has features including h-index extraction for the analyzed topics, country and institution extraction from author affiliations, top authors, countries, or institutions based on first document's authors or all document's authors, top/trending topics analysis. It provides five different visualization graphs: bar, bar trends, timeline, evolution, and word cloud through a graphical user interface (GUI). In this study some of the features of ScientoPy has been used (Ruiz-Rosero et al., 2019).

RESULTS AND DISCUSSION

The analysis of bibliometric results starts with a description of the main bibliometric statistics.

Descriptive Bibliometric Analysis

Table 1 shows the information on 1863 articles published between 1894 to early 2021. Total 6081 authors used 4006 keywords for literature work, out of which 138 authors did their work as single author and rest (5943) worked in collaboration. The collaboration index is 3.52.

Table 1. Main information

Main information	Explanation	Number
Documents	Total number of documents	1863
Timespan	Documents from that year span	1894-2021 (128 years)
Sources	The frequency distribution of sources as journals	688
Author's keywords	Total no. of keywords	4006
Authors	Total no. of authors	6081
Author's appearances	The author's frequency distribution	8467
Authors of single-authored documents	The number of single authors per articles	138
Authors of multi-authored documents	The number of authors of multi-authored articles	5943
Authors per document	Average number of authors in each document	3.26
Co-authors per document	Average number of co-authors in each document	4.54
Average citations per article	Average number of citations in Each article	18.81
Collaboration index		3.52

Annual Scientific Production

Analysis of overall growth trend showed an annual growth rate is 7.05%. The first article related to bioplastics was published in 1894 with no publication in this field for next 57 years. An article was published in 1952 but still there was no increase in production. From 1952 to the start of 21st century there was only 55 papers on bioplastic. Increase in production started from last 2 decades. The more production was in the last decade. The highest producing year was 2020 as shown in Figure.1 depicting annual scientific production. No peak is shown in starting years and a highest rise is in the year 2020 with 308 articles published on bioplastics.

Figure 1. Annual scientific production graph showing peak in 2020

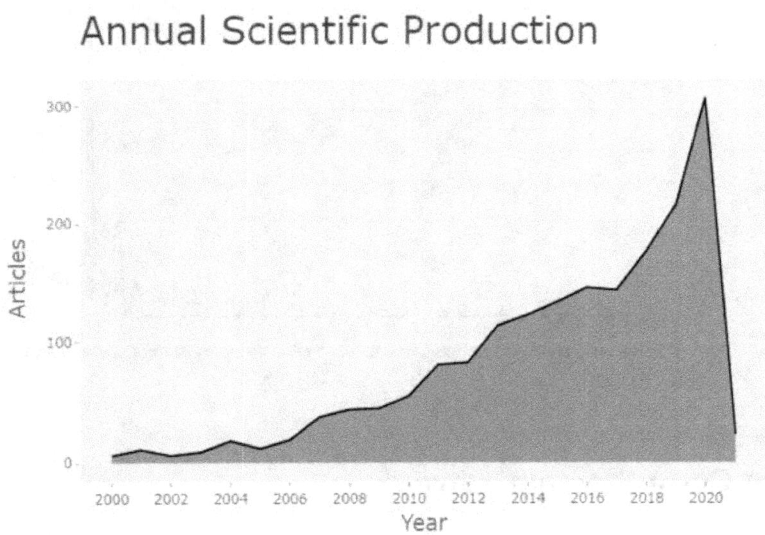

Average Article Citations Per Year

Figure.2 highlights that the highest cited articles were published in 2002 having 22.6 average citations per year. There are 1511 total citations of the data.

Most Prolific Authors

Table 2 and Figure 3 represent the top 10 most prolific authors researching on bioplastic, their total publications, total citations, h-index and first year of publication. The publications for these authors range from a maximum 41 to a minimum of 15 articles. 3 authors have more than 37 publications. The list shows that Guerrero, A. is the most productive author with 41 publications, 637 citations and 15 h-index followed by Misra, M. with 39 publications,3152 citations and 19 h-index. Mohanty, A.K. ranked 3rd according to the number of publications (37) but have highest citations (n=3467) and highest h-index (n=20). Partal P is at the bottom of the top 10 list with 15 publications.

Figure 2. Plot of average article citation per year. (From 1894-2021)

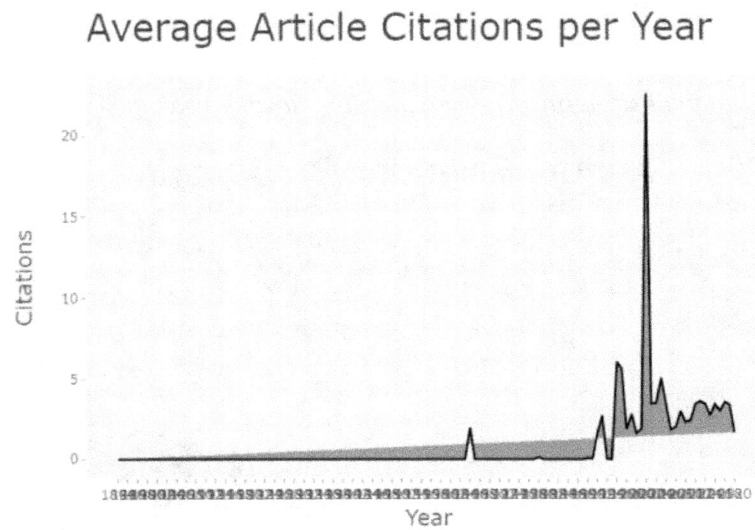

Author Co-Authorship Network

The link, collaboration or network between the authors is shown with the help of visualization map from VOSviewer. From a total of 5955 authors, 48 met the threshold by considering the author having atleast 8 numbers of documents. Out of 48, only 10 authors showed connections to each other. As highlighted in figure.4, the network contains 10 nodes, 30 co-authorship links and 3 clusters. Each node in the figure represents an author's productivity and the links between the authors denote the collaboration established through the co-authorship in the articles (Sudarsana and Baba, 2019). The total link strength is 198. The number of publications two researchers have co-authored is known as the link strength. All 10 authors are divided into 3 clusters. The color of the circles remains the same for the authors in the same cluster. Cluster 1 (Red) includes 5 authors Felix M., Guerrero A., Jimenez-Rosado M., Perez-Puyana V., Romero A. Cluster 2 (Green) has 3 members; Gallegos C.,

Martinez I., Partal P. Cluster 3 (Blue) includes 2 authors; Bengoechea C., Alvarez Castillo E.

Table 2. Top 10 most prolific authors

SN	Author	Total production	Total citations	h-index	First year of publication
1	Guerrero, A	41	637	15	2005
2	Misra, M	39	3152	19	2002
3	Mohanty, A.K	37	3467	20	2002
4	Romero, A	25	358	11	2011
5	Bengoechea, C	23	245	10	2013
6	Accinellic	17	264	10	2009
7	Verbeek, C.J.R	17	155	7	2009
8	Baimark, Y	16	47	4	2015
9	Martanez, I	16	357	10	2005
10	Partal, P	15	487	12	2005

Figure 3. Top 10 most prolific authors

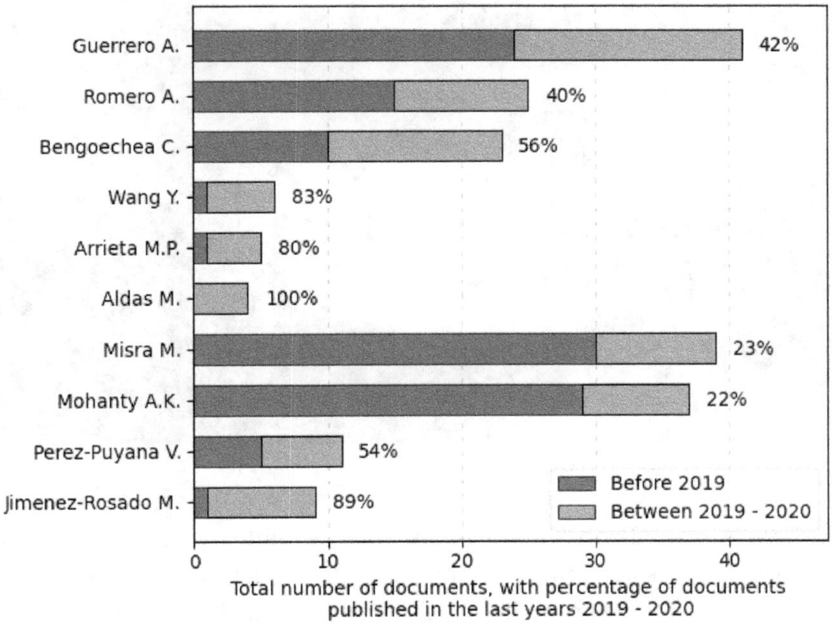

Figure 4. Network visualization for authors co-authorship network

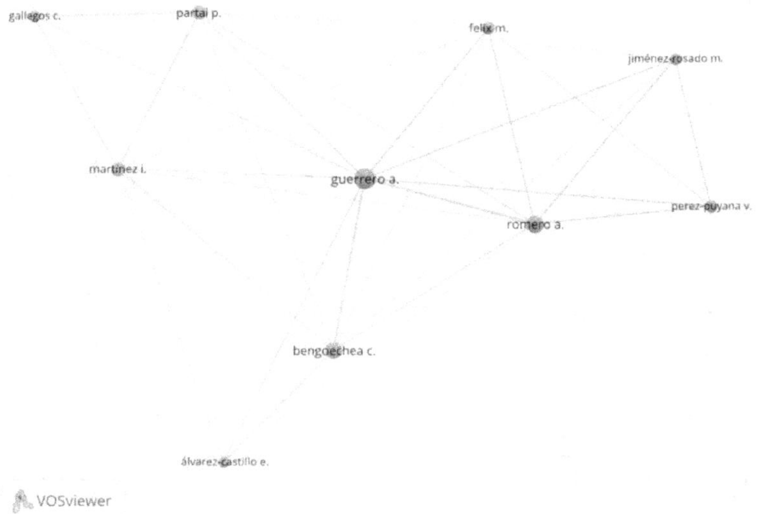

Figure 5. Density visualization for author co-authorship network

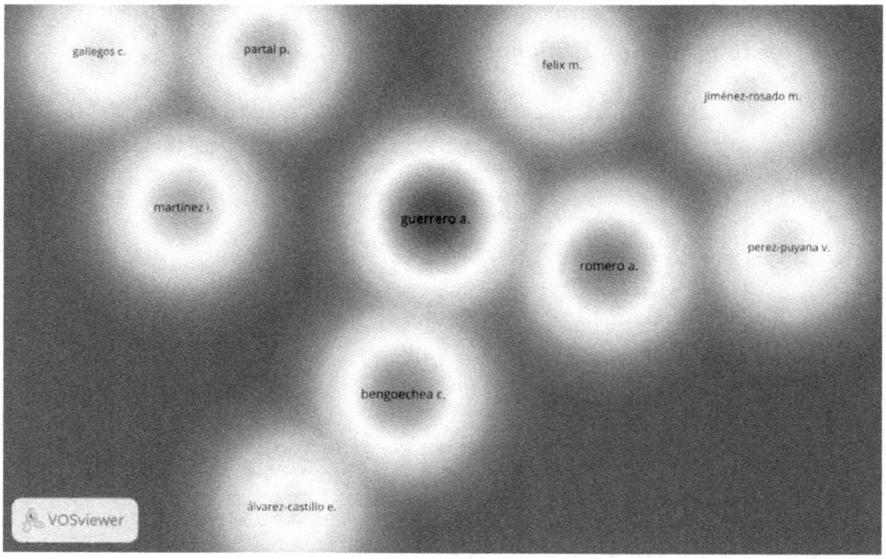

Guerrero A. in collaboration network was the highly productive author with 41 documents, 9 links and 96 total link strength present in Cluster 1 as shown in table 3. followed by Romero, A. having 25 documents with 8 links and 66 total link strength and also present in Cluster 1.

Table 3. List of most collaborative authors according to their total link strength

	Author	Documents	Citations	Total link Strength
1	Guerrero, A.	41	637	96
2	Romero, A.	25	358	66
3	Misra, M.	39	3152	47
4	Mohanty, A.K.	37	3467	47
5	Bengoechea, C.	23	245	45
6	Partal, P.	15	487	38
7	Accinelli, C.	17	264	37
8	Martinez, I.	16	357	33
9	Abbas, H.K.	12	132	29
10	Perez-Puyana, V.	10	68	29

In Figure5, Density Visualization of authors is shown. Denser yellow color shows higher collaborations of author depicting that the weight of the neighboring items is higher. Guerrero A. shows the dense yellow color due to more collaborations.

Author's Keyword Co-Occurrence Analysis

Figure.6 shows the word cloud of author's keywords. Word cloud is taken from the Biblioshiny with setting parameter at 50 words and 'bioplastics' word due to plurality is not included. According to figure, bioplastic is the most used keyword in documents.

Co-occurrence Analysis of author's keyword is shown by using VOSviewer. There are total 4007 keywords, but to keep the display interpretable, the resulting map is limited to keywords that occurred at 10 as minimum frequency. 71 keywords met the threshold. Figure.7. displays the Visualization of keyword co-occurrence network showing 71 nodes, 594 co-occurrence links and 8 clusters having total link strength of 1399. As can be seen in the map, nodes such as Bioplastics, Bioplastic, mechanical properties, polyhydroxyalkanoates, biodegradable have the highest frequency of occurrence and represent the research hotspots in the bioplastic field i.e., we can concentrate on biodegradability of plastics and synthesis of bioplastic by polyhydroxyalkanoate. Bioplastics is the most frequently used author's keyword with 297 occurrences. It is present in cluster 5. It has total 63 co-occurrences links. It is followed by its singular word which is also highly used i.e., 280 occurrences of keyword 'bioplastic'. Thick line exists in network between keyword, bioplastics and mechanical properties depicting most frequent keywords that occurred together. This line has 17 link strength.

Figure 6. Word cloud of author's keywords

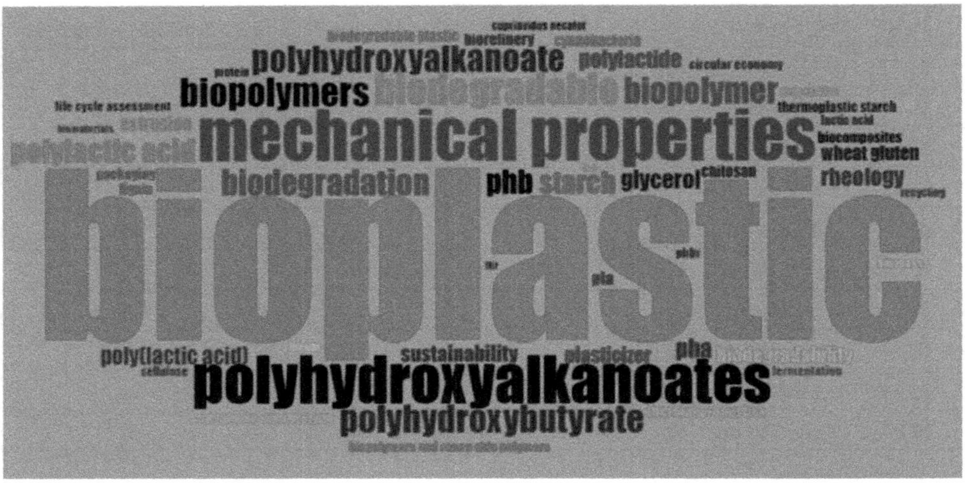

Figure 7. Author's keyword network analysis

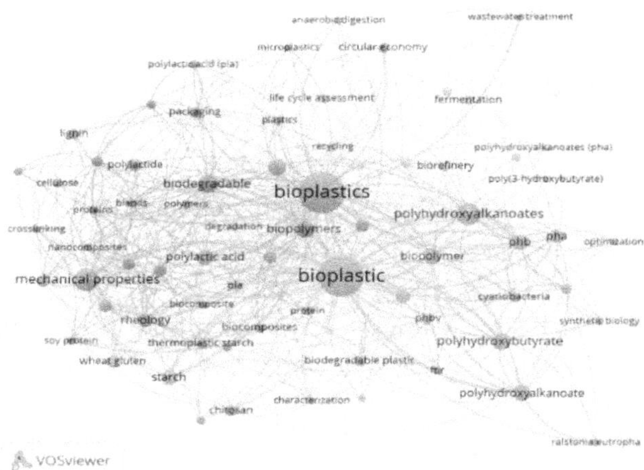

Overlay visualization is shown in Figure.8. which shows the new and trending research hotspots in this field. The colour gradient (blue to yellow) represents trends in keyword co-occurrence from 2014 to 2018. The colour is related to the time when the word appeared. The keywords with blue colour appeared early and the one with yellow later (Wang *et al.,* 2016). In early stage of bioplastic concept, use of wheat gluten for synthesis of bioplastic for better biodegradation was the main hotspot. After sometime synthesis of bioplastic from biopolymer Chitosan for better biodegradability was the research trend. The occurrence of chitosan is highest (20 times) and average publication in the year is 2017. The latest research hotspot in bioplastic study is the Microplastics (n=10) from last two years having average publication in the year 2019. However, in the last 2 years an increasing number of researchers have given attention to the concept of Microplastics (n=10), anaerobic digestion (n=10), circular economy (n=17) & food packaging (n=12) in bioplastic field.

Figure 8. Overlay visualization of the keywords showing trending hotspots for research in bioplastic. The average publication for most frequent used word bioplastics is in the year 2015

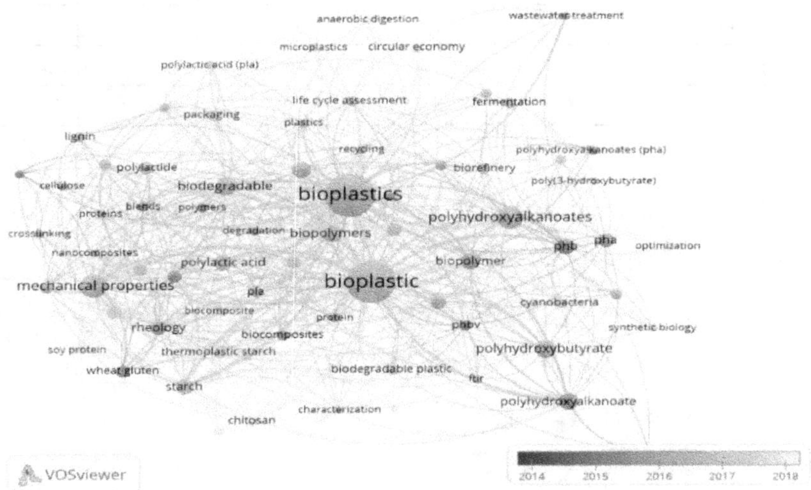

Top 10 keywords in table 4 represent total occurrences of keywords in the dataset, their links, total link strength and cluster in VOSviewer. In this list, singulars of keyword ‘biopolymers’ and ‘polyhydroxyalkanoates’ is not considered having 43 and 45 occurrences respectively. Abbreviation of Polyhydroxybutyrate was also excluded having 39 occurrences.

Figure.9 shows the horizontal bar graph of frequently used keywords according to their total number of documents published and percentage of documents published between 2019-2020. The mostly used keyword ‘Bioplastics’ contributes 33% of total (297) documents published in last year. The top 2nd keyword ‘bioplastic’ and the top 3rd keyword ‘mechanical properties’ contributes 31% and 28% documents published in between 2019-2020. It shows the mostly studied period of this research field i.e., 2019-2020.

Table 4. List of top 10 keywords their occurrences, links, total link strength and cluster in VOSviewer

	Keywords	Occurrences	cluster	Links	Total link strength
1	Bioplastics	297	5	63	313
2	Bioplastic	280	2	64	315
3	Mechanical properties	86	3	36	129
4	Polyhydroxyalkanoates	79	2	36	98
5	Biodegradable	53	1	36	93
6	Biopolymers	48	1	30	64
7	Polyhydroxybutyrate	47	2	21	50
8	Biodegradation	43	5	23	55
9	Polylactic acid	40	3	25	51
10	Starch	38	7	21	55

Figure 9. Mostly used keywords with their percentage of documents published

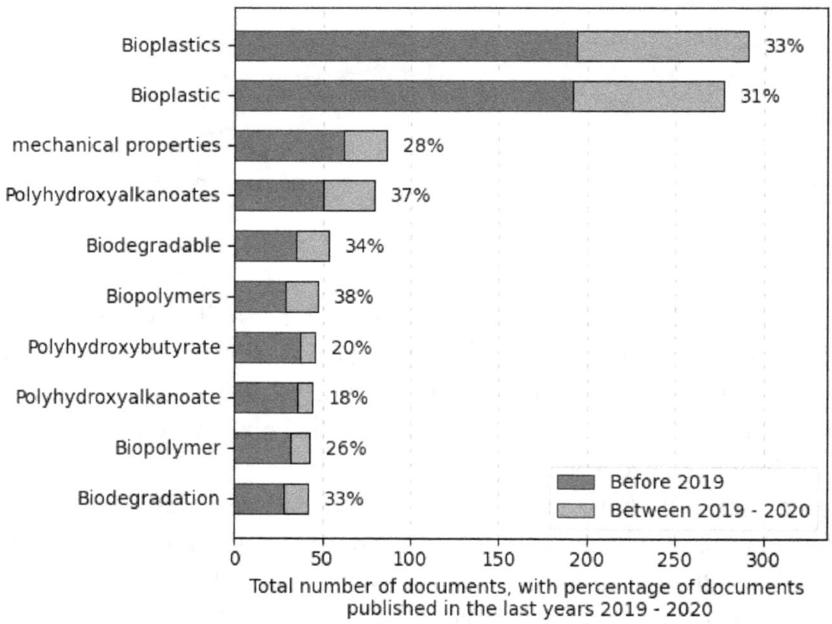

COUNTRIES

Highly Productive Countries

Three countries produced over 91 publications on 'bioplastic'. Figure10 and Table 5. depicts USA is on top of the list with 186 publications and 8913 citations. Italy holds 2nd rank with 111 publications and 1550 citations. Spain has 3rd rank with 91 publications and 2327 citations. Japan has produced 76 publications and received 1087 citations followed by the China with 66 publications and 2167 citations. India is on the 6th rank with 55 publications and 875 citations. Netherlands is at the bottom of the top 10 with 39 publications and 2340 citations. USA contributes more percent documents from whole literature on bioplastic. The percent is calculated by the formula;

Total Percent = production of a country / Total number of documents × 100
Production of a country = number of documents contributed by a country.
Total documents = number of total documents of the taken dataset (1863).

Table 5. Ranking of top 10 highly productive countries

Rank	Country	Number of documents	Percent	Total citations
1	USA	186	9.98	8913
2	Italy	111	5.96	1550
3	Spain	91	4.89	2327
4	Japan	76	4.08	1087
5	China	66	3.55	2167
6	India	55	2.96	875
7	Germany	54	2.89	1302
8	Canada	53	2.85	1465
9	Korea	48	2.57	551
10	Netherlands	39	2.09	2340

Country Collaboration Map on Bioplastic Research

Figure.11 highlights the top 10 country collaboration map on bioplastic study. USA has more collaborations with countries i.e., with Italy (17 publications), China (n=14), Germany (n=9), Korea (n=9), Australia (n=8) followed by Spain with Italy

(13 publications) and with Germany (n=9). The least collaborator countries among 10 collaborators are Saudi Arabia and Egypt with seven publications (Jabali et al., 2020). Frequency of collaborations is shown in table 6.

Figure 10. Ranking of top 10 highly productive countries

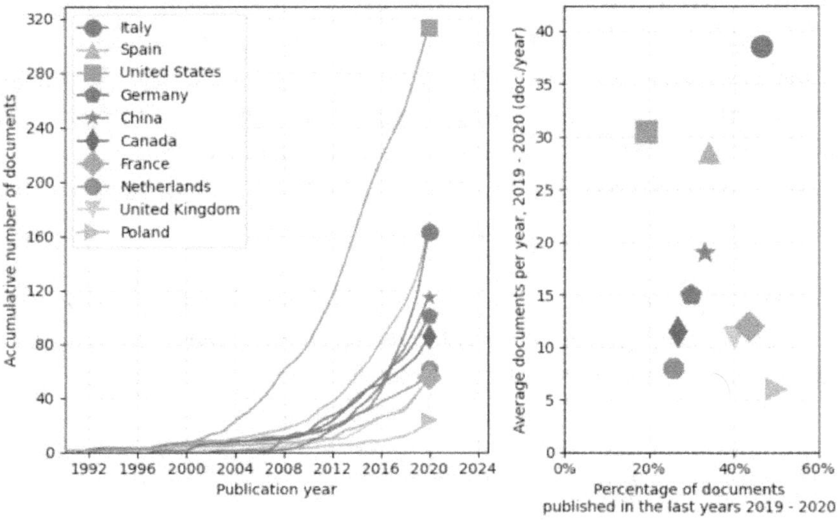

International Citation Analysis

Each country appearing in Figure12, met a minimum of productivity threshold of five publications. Total 36 countries are selected out of 89 each having minimum 100 citations. There are 5 clusters, 357 links between countries with 2130 total link strength (TLS). Clusters were limited to a minimum of five members. Each node represents the country's productivity and the links between the countries denote the collaborations established through the co-citations in the articles (Sudarsana and Baba 2019). As a representative of developed countries, 'United States' is the most active in producing 'bioplastic' studies (Figure9). According to VOSviewer data researchers from US contributed 319 documents accounting for 17.13% with 11360 citations.US is present in cluster 1 (red colour). Researchers from the Italy contributed 169 papers (9.08%) having 2133 citations and 359 total link strength. Spain contributes 168 documents with 3388 citations and 431 total link strength. Philippines has lowest number of documents published(n=5) with 134 citations. The lowest cited documents are from Czech Republic having 17 documents published with 119 citations.

Figure 11. Map showing connections or collaboration between countries

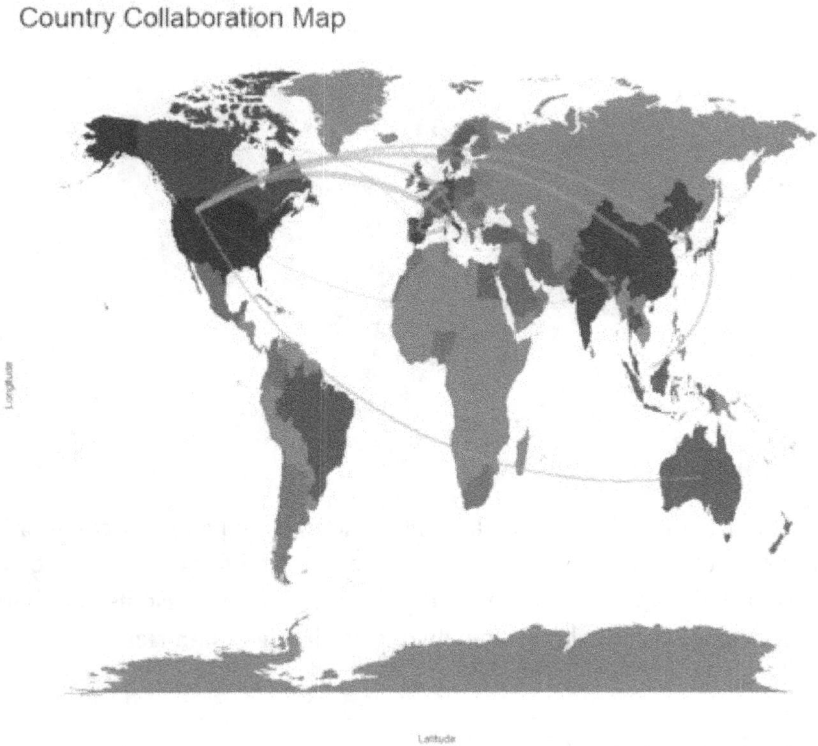

Table 6. Collaborations between countries with frequency and total link strength

Rank	From	To	Frequency	Total link strength
1	Italy	USA	17	17
2	China	USA	14	13
3	Spain	Italy	13	12
4	Germany	USA	9	10
5	Malaysia	Japan	9	10
6	Korea	USA	9	9
7	Germany	Spain	9	8
8	Italy	United Kingdom	9	8
9	USA	Australia	8	8
10	Saudi Arabia	Egypt	7	8

Figure 12. Visualization of network of countries

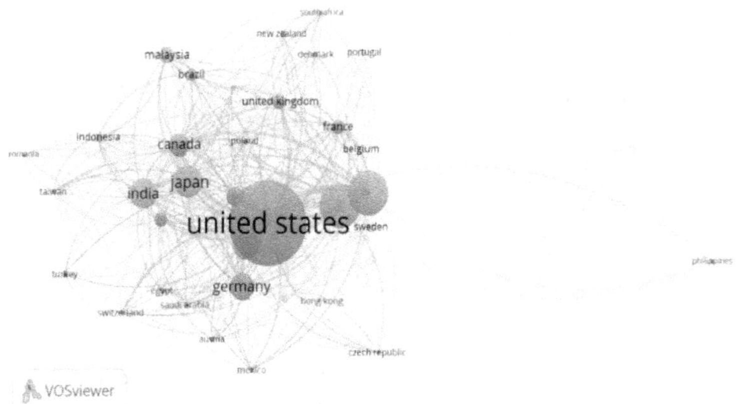

USA played a core role in the collaboration network & had good collaborations with other countries having 675 total link strength. Italy and Spain have strong co-operative relations with the USA. Philippines need to increase its collaborative strength or enhance its collaboration networks with other countries.

JOURNALS (SOURCES)

There are total 688 journals that published research documents on Bioplastics. For clarity visualization, a frequency threshold was setup at more than 5 documents and citations of a source. 83 journals met the threshold. Out of these only 81 journals are connected to each other. Figure.13 shows the network visualization of journals. It depicts the network of 81 nodes and 383 collaboration links. There are 10 clusters on the map with '2' minimum cluster size. Journals centring on the same investigative field tend to be within the same cluster (Fang et al., 2019). Different colours show diverse clusters and the line between the journal shows the co-citation of journals with each other. Journal of Polymer and the Environment, Journal of Applied polymer Science, Bioresource Technology are the top three journals with the most elevated number of distributed papers.

Density Visualization: Journal of Polymers and the Environment have highest article density showing intense yellow colour (Figure.14). It is present in the second cluster having 39 links with other journals and 152 is its total link strength. Most papers are published in this journal between the year 2014-2015. (Average Publication in Year 2014.84).

Figure 13. Citation network and cluster of highly cited journals based on VOS viewer

Figure 14. Density visualization of journals

The top 10 journals are shown in table 7. The top 10 most productive journals distributed 338 articles accounting for 18.15% of total publications.

Journal of Polymers and the Environment (1st rank) is highly influential journal that produced the largest number of articles related to Bioplastic. It produced maximum 67 published documents and also have highest citations i.e., 2432. The

Journal of Applied Polymer Science has 2nd rank with 55 publications and 1025 citations. But it has lowest impact factor among top 10 journals. i.e., 2.52, followed by "Bioresource Technology" on 3rd rank having 38 publications and 1265 citations. The "International Journal of Biological Macromolecules" is at the bottom of the list and has produced 20 publications and received 222 citations. Four journals in top 10 are from the Netherlands, three from United States, two from United Kingdom and one from Switzerland. H-index of journals is taken from Biblioshiny (Source Impact). The journal with the highest h-index is of the journal of "Bioresource technology" (n=22).

Journal Impact Factor is taken from the JCR 2020 Report. According to Journal Citation Report 2020, the range of impact factor given in table 7.is 2.52 to 7.632. The 2020 impact factor of "ACS Sustainable Chemistry & Engineering" has the highest and that of Journal of Applied polymer Science has the lowest.

Among the top 10 sources, included in the Journal Citation Report JCR database, 8 journals ranked first in quartile and 2 ranked second in quartile.

Organizations

According to VOSviewer analysis, 1863 documents have been published by 3718 organizations. The visualization of organization is performed by selecting those organizations that have at least 2 documents and 2 citations each. 247 organizations met the threshold. After excluding unconnected organizations, the remaining 120 organizations are used for the visualization map. Figure.15 shows a network of organizations of 120 nodes and 421 collaboration links. The size of the node indicates the publication frequency of the organization. The analysis is divided into 10 clusters.

The most prolific organization is "School of Engineering, Thornbrough building, University of Guelph" that contributed 10 documents. This was followed by "Usda-ars, Biological Control of Pests Research Unit, Stoneville, US" having 7 documents with 82 citations published related to Bioplastic. The organization "Composite Materials and Structures Centre, Michigan State University" have highest citations (1887) with only 4 documents published and have higher number of links (32) in the network with good number of collaborations with many of the organizations with 46 total link strengths. The most collaborative organization i.e., 'University of Guelph' is from Canada. It can be seen that the organizations on the left are close together, indicating strong connection, while those on the right are farther apart have a weaker co-operation with respect to bioplastic research.

Table 7. Top 10 journals based on total productions

SN	Journal	Country	Documents	h-Index	Total citations	Avg. Citation per page	Impact factor	Quartile of JCR	Publisher
1.	Journal of Polymers and the Environment.	United States	67	17	2432	36.29	2.572	Q2	Springer New York
2.	Journal of Applied polymer Science.	United States	55	18	1025	18.63	2.52	Q2	John Wiley and Sons Inc.
3.	Bioresource Technology	Netherlands	38	22	1265	33.28	7.539	Q1	Elsevier ltd.
4.	Journal of Cleaner Production	Netherlands	31	16	994	32.06	7.246	Q1	Elsevier ltd.
5.	ACS Sustainable Chemistry & Engineering	United States	29	16	793	27.34	7.632	Q1	American chemical society
6.	Carbohydrate Polymers	United Kingdom	28	17	1006	35.92	7.182	Q1	Elsevier ltd.
7.	Polymers	Switzerland	25	6	115	4.6	3.426	Q1	MDPI AG
8.	Industrial Crops and Products	Netherlands	23	13	790	34.34	4.244	Q1	Elsevier
9.	Polymer degradation and stability	United Kingdom	22	14	786	35.72	4.032	Q1	Elsevier ltd.
10.	International journal of biological macromolecules.	Netherlands	20	9	222	11.1	5.162	Q1	Elsevier

Figure 15. Organizations network visualization on the basis of citations

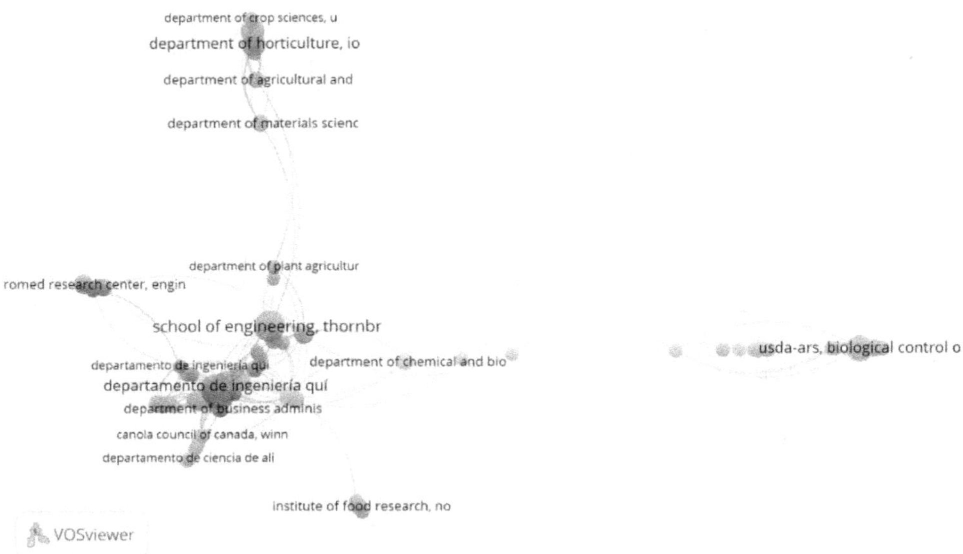

Three Factor Analyses (Keyword, Organization & Country)

A three-field plot (Figure.16) has been generated for top 10 keywords, countries and organizations based on literature of bioplastic. The size of block shows the associational relationship amongst each factor (Jabali *et al.*, 2020). The main keywords i.e., bioplastics, bioplastic and polyhydroxyalkanoate have strong relation with top 10 countries (Spain, Italy, USA, Korea, Canada, India, Japan, China, Thailand & Germany). Keywords also show associational links with organizations of different countries. The highly associated country Spain has associational links with the Universities: Universidad de Sevilla, University of Naples Federico ii, Delft University of Technology and Michigan State University. The second most associated country Italy is connected with University of Bologna, University of Naples Federico ii.

Not reported (organizations not reported by the software) organizations have more associational links with seven countries i.e., Italy, Germany, Japan, Canada, USA, China & Thailand.

Top Ten Highly Cited Articles

Table 8. shows the bibliographic information of top ten most cited articles. The publication years of top 10 highly cited articles range between 1996 to 2016. The article entitled "Sustainable bio-composites from Renewable Resources: opportunities and challenges in the Green Materials World" by A.K. Mohanty published in 2002 in "Journal of Polymers and the Environment" is on the top of the list with 1511 citations followed by article entitled "Complete genome sequence and comparative analysis of the metabolically versatile *Pseudomonas putida* KT2440" by K.E. Nelson in 2002 with 923 publications. The article "Replacing fossil-based PET with biobased PEF; process analysis, energy and GHG balance written by Eerhart A.J.J.E. *et al.* is at the bottom of the list in top ten highly cited articles with 270 citations.

Figure 16. Three Factor Analysis. (Keyword, Country, Organizations)

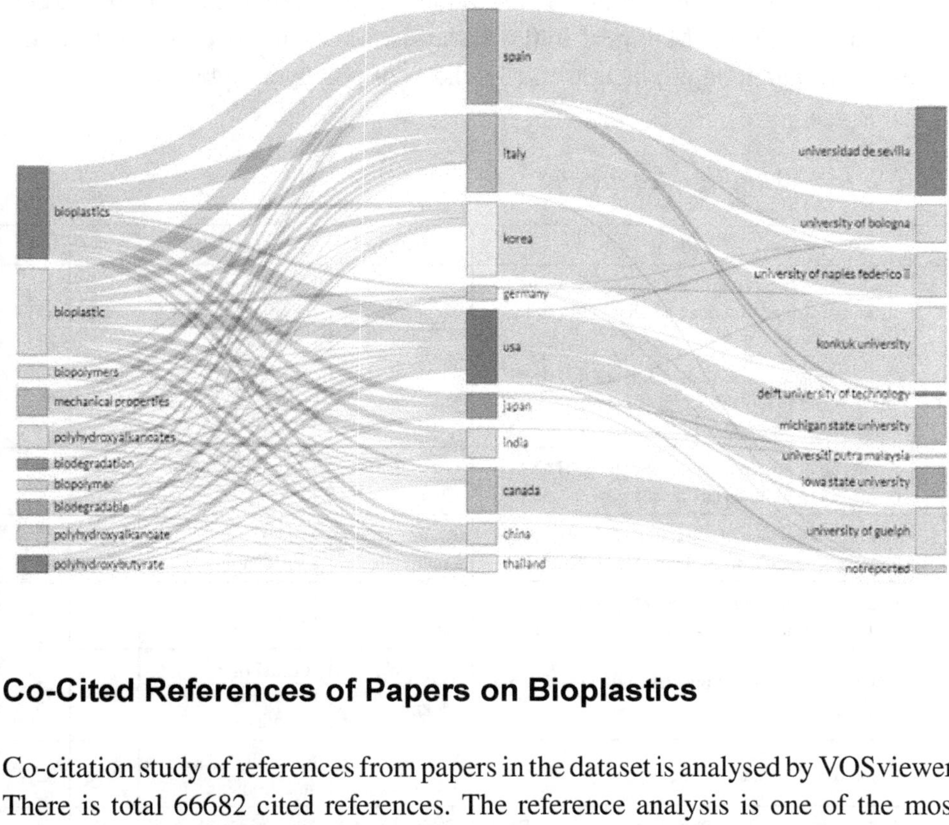

Co-Cited References of Papers on Bioplastics

Co-citation study of references from papers in the dataset is analysed by VOSviewer. There is total 66682 cited references. The reference analysis is one of the most important indicators of the bibliometrics (Wang et al., 2016). Considering the large number of cited references, those papers are selected which were cited more than 4 times for analysis. 246 met the threshold, out of which 210 showed connections to each other. There were a total 9 clusters visible upon setting the minimum cluster size at 10 i.e., a cluster must contain more than 10 items. Figure.17 shows 210 points with different colours represent 210 cited papers and the line between every 2 points depicts that both were cited in one paper. Cluster 3 mainly consisting of papers related to polyhydroxyalkanoate and polyhydroxybutyrate. There are 4 references in top 10 from 3rd cluster.

Table 9 shows top 10 references by co-citation. Sudesh *et al.* (2000) is the only reference co-cited more than 14 times. Two references are cited more than 13 times, including Anderson *et al.*, (1990) and Lagrain *et al.,* (2010). Among the top 10 references by co-citation, 'Biomacromolecules' Journal contributes 2 references. There are 2 articles by author 'Yves Poirier' on the study of 'Polyhydroxybutyrate'. The most commonly cited reference is "Synthesis, structure and properties of polyhydroxyalkanoates: biological polyesters; Sudesh *et al.,* 2000. This paper reported the production of Polyhydroxyalkanoate with various physical properties by genetically engineered micro-organisms. Sudesh K is the first author of this article from 'Universiti sains Malaysia' in the Malaysia. Maximum papers of this author are on the research on production and synthesis of polyhydroxybutyrate.

Table 8. Top 10 highly cited articles

	Title	Author	Source Title	Year	TC
1	Sustainable Bio-composites from Renewable Resources: opportunities and challenges in the Green Materials World.	A.K. Mohanty	Journal of Polymers and the Environment	2002	1511
2	Complete genome sequence and comparative analysis of the metabolically versatile Pseudomonas putida KT2440.	K.E. Nelson	Environmental Microbiology	2002	923
3	A microbial polyhydroxyalkanoates (PHA) based bio- and materials industry.	Guo-Qiang Chen	Chemical Society Reviews	2009	777
4	Genome Sequence of the bioplastic producing "knallgas" bacterium Ralstonia eutropha H16.	Anne Pohlmann	Nature Biotechnology	2006	370
5	Lignin Valorization through integrated biological funnelling and chemical catalysis.	Jeffrey G. Linger	PNAS	2014	355
6	Progress in bio-based plastics and plasticizing modifications.	Tizazu Mekonnen	Journal of Materials Chemistry A	2013	329
7	Crystallinity in starch bioplastics.	Van Soest JJG	Industrial Crops and Products	1996	327
8	Cannabis Sativa: The Plant of the Thousand and one molecules.	Christelle M. Andre	Frontiers in plant Science.	2016	305
9	Production of polyhydroxyalkanoates by mixed culture: recent trends and biotechnological importance.	H. Salehizadeh	Biotechnology Advances	2004	280
10	Replacing fossil-based PET with biobased PEF; process analysis, energy and GHG balance.	A.J.J.E. Eerhart	Energy and Environmental Science.	2012	270

Figure 17. Visualization of Reference's network by co citation on Bioplastics

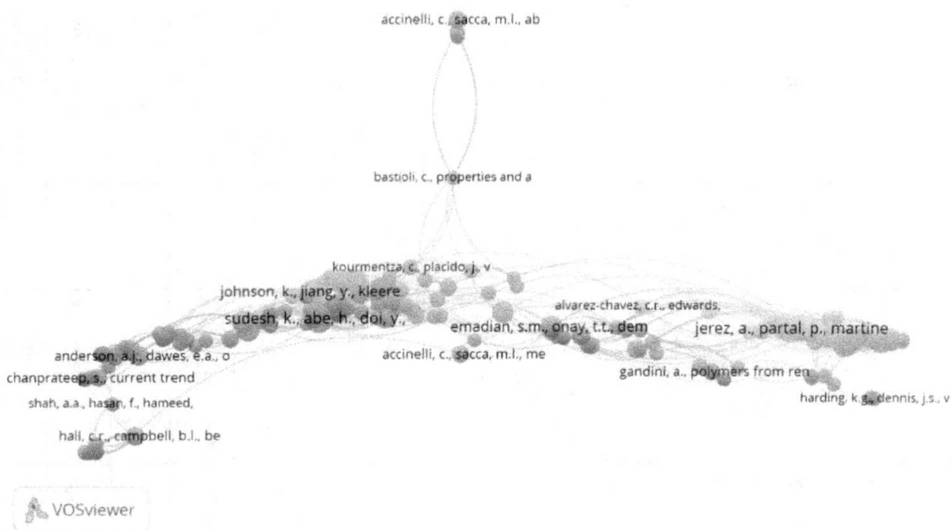

Table 9. Top ten References by Co-Citation

	Reference	Author	Source	Local Citations	Links	Total link strength	Cluster number
1	Synthesis, structure and properties of polyhydroxyalkanoates: biological polyesters (Sudesh k, *et al.*,2000)	Sudesh K, Abe H, Doi Y	Progress in Polymer Science	14	26	40	3
2	Occurrence, metabolism, metabolic role, and industrial uses of bacterial polyhydroxyalkanoates (Anderson *et al.*,1990)	Anderson A J, Dawes E A	Microbiology and Molecular biology Reviews	13	27	41	3
3	Molecular basis of Processing Wheat Gluten toward biobased Materials (Lagrain *et al.*, 2010)	Lagrain B, Goderis B, Brijs K and Delcour J A	Biomacromolecules	13	24	51	7
4	Rheology and processing of gluten-based bioplastics (Jerez, A. *et al.*, 2005)	Jerez A, Partal P, Martinez I, Gallegos C, Guerrero A.	Biochemical Engineering Journal	11	27	41	4
5	Enrichment of a mixed bacterial culture with a high polyhydroxyalkanoate storage capacity (Johnson *et al.*, 2009)	Johnson K, Jiang Y, Kleerebezem R, Muyzer G and Loosdrecht M V.	Biomacromolecules	11	25	47	6

continues on following page

Table 9. Continued

	Reference	Author	Source	Local Citations	Links	Total link strength	Cluster number
6	Polyhydroxybutyrate, a biodegradable Thermoplastic, produced in Transgenic plants (Poirier *et al.*, 1992)	Poirier Y, Dennis D E, Klomparens K, Somerville C	Science	11	18	29	3
7	Transgenic Arabidopsis plants can accumulate polyhydroxybutyrate to up to 4% of their fresh weight (Bohmert *et al.*, 2000)	Bohmert K, Balbo I, Kopka J, Mittendorf V, Nawrath C, Poirier Y, Tischendorf G, Trethewey R N & Willmitzer L.	Planta	10	17	31	3
8	Morphologies and properties of thermomolded biodegradable plastics based on glycerol – plasticized wheat gluten (Sun *et al.*, 2007)	Sun S, Song Y, Zheng Q.	Food Hydrocolloids	10	27	51	7
9	Production of polyhydroxyalkanoates by activated sludge treating a paper mill wastewater (Bengtsson *et al.*, 2008)	Bengtsson S, Werker A, Christensson M, Welander T.	Bioresource Technology	9	22	49	6
10	Biodegradation of bioplastics in natural environments (Emadian *et al.*, 2017)	Emadian S M, Onay T T, Demirel B	Waste Management	9	17	17	8

SUMMARY OF FINDINGS

1. A bibliometric study was conducted to evaluate the literature on Bioplastic, publication in the field, author working in this research area and research impact.
2. Various softwares including VOS Viewer, R Studio, ScientoPy and Databases (Scopus, Scimago) were employed for the study.
3. Journal of Polymer and Environment showed maximum number of publications related to bioplastics.
4. Most productive year was 2020.
5. USA has maximum number of publications.
6. USA played a core role in the collaboration network & had good collaborations with other countries.
7. The author with maximum research impact in this field was Guerrero, A.
8. Bioplastics is the most used keyword in documents.

9. The most prolific organization is "School of Engineering, Thornbrough building, University of Guelph" that contributed 10 documents related to Bioplastics.
10. In Bioplastics literature Sudesh *et al.* (2000) is the reference co-cited more than 14 times.

CONCLUSION

Plastics, pose a great threat to the environment owing to massive production in order to meet demands of ever-increasing population. Plastics have a slow degradation, increase carbon footprint and escalate burden on landfill sites and oceans. Hence bioplastics as an alternative to conventional plastics are inviting the research endeavours globally at present. This paper shows a bibliometric or scientometric study on whole literature on Bioplastic. It encompasses data of 128 years from 1894 to 2021. The study was conducted on total 1863 documents. Most productive and network analysis of authors, countries, keywords, journals, organizations, articles and references were identified in the said period. The collaboration between countries and associational relationship between keywords, countries and organizations was analysed. The most productive author is Guerrero A. The most productive country is USA which also played a core role in the collaboration network & had good collaborations with other countries. Journal of Polymers and the Environment is the most productive source, providing a platform for high quality publications in this area. This study shows that the Wheat Gluten for synthesis of bioplastic for better biodegradation was the main hotspot. Synthesis of bioplastic from Chitosan was the research trend and latest research hotspot in bioplastic is the Microplastics.

REFERENCES

Abejon, R. (2018). A Bibliometric Study of Scientific Publications regarding Hemicellulose Valorization during the 2000–2016 Period: Identification of Alternatives and Hot Topics. *ChemEngineering*, *2*(1), 7. doi:10.3390/chemengineering2010007

Coppola, G., Gaudio, M. T., Lopresto, C. G., Calabro, V., Curcio, S., & Chakraborty, S. (2021). *Bioplastic from Renewable Biomass: A Facile Solution for a Greener Environment.* Earth Systems and Environment. doi:10.100741748-021-00208-7

Darvish, H. (2019). Bibliometric Analysis using Bibliometrix an R Package. *Journal of Scientometric Res.*, *8*(3), 156–160. doi:10.5530/jscires.8.3.32

Fang, J., Pan, L., Gu, Q. X., Juengpanich, S., Zheng, J. H., Tong, C. H., Wang, Z. Y., Nan, J. J., & Wang, Y. F. (2020). Scientometric analysis of mTOR signaling pathway in liver disease. *Annals of Translational Medicine*, *8*(4), 93. doi:10.21037/atm.2019.12.110 PMID:32175386

Groote, P. D., Devaux, J., & Godard, P. (2002). Effect of Benzenesulfonamide Plasticizers on the glass- transition temperature of semicrystalline polydodecamide. *Journal of Polymer Science. Part B, Polymer Physics*, *40*(19), 2208–2218. doi:10.1002/polb.10281

Hassan, M. D., Castanha, R. C. G., & Wolfram, D. (2020). Scientometric analysis of global trypanosomiasis research: 1988–2017. *Journal of Infection and Public Health*, *13*(4), 514–520. doi:10.1016/j.jiph.2019.10.006 PMID:31831393

Jabali, K. A., Ashiq, M., Ahmad, S., & Rehman, S. U. (2020). A Bibliometric Analysis of Research Productivity on Diabetes Modeling and Artificial Pancreas 2001 to 2020. *Library Philosophy and Practice (e-journal),* 4305. https://digitalcommons.unl.edu/libphilprac/4305

Ke, L., Lu, C., Shen, R., Lu, T., Ma, B., & Hua, Y. (2020). Knowledge Mapping of Drug-Induced Liver Injury: A Scientometric Investigation (2010-2019). *Frontiers in Pharmacology*, *11*, 842. doi:10.3389/fphar.2020.00842 PMID:32581801

Li, Z., Chen, Z., Yang, N., Wei, K., Ling, Z., Liu, Q., Chen, G., & Ye, B. H. (2021). Trends in research on the carbon footprint of higher education: A bibliometric analysis (2010-2019). *Journal of Cleaner Production*, *289*, 125642. doi:10.1016/j.jclepro.2020.125642

Malazy, O. T., Ramezani, A., Atlasi, R., Larijani, B., & Abdollahi, M. (2016). Scientometric study of academic publications on antioxidative herbal medicines in type 2 diabetes mellitus. *Journal of Diabetes and Metabolic Disorders*, *15*(1), 48. doi:10.118640200-016-0273-3 PMID:27785446

Mas-Tur, A., Roig-Tierno, N., Sarin, S., Haon, C., Sego, T., Belkhouja, M., Porter, A., & Merigo, J. M. (2021). Co-citation, bibliographic coupling and leading authors, institutions and countries in the 50 years of Technological Forecasting and Social Change. *Technological Forecasting and Social Change*, *165*, 120487. doi:10.1016/j.techfore.2020.120487

Narancic, T., Cerrone, F., Beagan, N., & O'Connor, K. E. (2020). Recent Advances in Bioplastics: Application and Biodegradation. *Polymers*, *12*(4), 920. doi:10.3390/polym12040920 PMID:32326661

Ndlovu, N., Banda, L., Muitire, C., Munyengwa, N., & Mayaya, T. (2020). Bibliometric and Scientometric Analysis of Scopus-Indexed Literature on Hemp (Cannabis Sativa) Breeding. International Journal of Plant & Soil Science, 32(17), 29-43.

Obileke, K., Onyeaka, H., Omoregbe, O., Makaka, G., Nwokolo, N., & Mukumba, P. (2020). Bioenergy from Bio-waste: A bibliometric analysis of the trend in scientific research from 1998-2018. *Biomass Conversion and Biorefinery*. Advance online publication. doi:10.100713399-020-00832-9

Perea, L.N., Gaviria, D., & Redondo, M.I. (2020). Bioeconomy: Bibliometric analysis from 2006 to 2019. *Revista ESPACIOS*, *41*(43).

Qin, F., Du, J., Gao, J., Liu, G., Song, Y., Yang, A., Wang, H., Ding, Y., & Wang, Q. (2020). Bibliometric Profile of Global Microplastics Research from 2004 to 2019. *International Journal of Environmental Research and Public Health*, *17*(16), 5639. doi:10.3390/ijerph17165639 PMID:32764298

Rahaman, M. S., Ahmad, S., Ansari, K. M. N., & Abdel-Magid, I. M. (2020). Scientometrics Profile of the Banasthali Vidyapith: A Deemed University of Rajasthan, India. *Library Philosophy and Practice (e-journal)*. 4329. https://digitalcommons.unl.edu/libphilprac/4329

Ruiz-Rosero, J., Ramirez-Gonzalez, G., & Viveros-Delgado, J. (n.d.). *Software survey: ScientoPy, a scientometric tool for topics trend analysis in scientific publications*. Academic Press.

Saharan, B.S., & Ankita, S. D. (2012). Bioplastics- for Sustainable Development: A Review. *International Journal of Microbial Resource Technology*, *1*(1), 11–23.

Secinaro, S., Brescia, V., Calandra, D., & Biancone, P. (2020). Employing bibliometric analysis to identify suitable business models for electric cars. *Journal of Cleaner Production*, *264*, 121503. Advance online publication. doi:10.1016/j.jclepro.2020.121503

Selvavinayagam, A. V., Muthu, M., & Boopalan, E. (2018). Scientometrics, Techniques, Sources And Their Key Points To Analysis Of Lis Research: An Overview. *International Journal of Science & Technology, 8*(1). www.ijst.co.in

Shah, M., Rajhans, S., Pandya, H.A., & Mankand, A.U. (2021). Bioplastic for future: A review then and now. *World Journal of Advanced Research and Reviews, 9*(2), 56-67. doi:10.30574/wjarr.2021.9.2.0054

Sudarsana, D., & Baba, M. S. (2019). Global nuclear fuel research during 2000 to 2017: A scientometric analysis. *Annals of Library and Information Studies*, *66*, 85–93.

Wang, M., Liu, P., Gu, Z., Cheng, H., & Li, X. (2019). A Scientometric Review of Resource Recycling Industry. *International Journal of Environmental Research and Public Health*, *16*(23), 4654. doi:10.3390/ijerph16234654 PMID:31766728

Wang, Y., Wang, Q., Zhu, R., Yang, C., Chen, Z., Bai, Y., Li, M., & Zhai, X. (2016). Trends of spinal tuberculosis research (1994-2015) A bibliometric study. *Medicine*, *95*(38), e4923. doi:10.1097/MD.0000000000004923 PMID:27661044

Yan, W. T., Lu, S., Yang, Y. D., Ning, W. Y., Cai, Y., Hu, X. M., Zhang, Q., & Xiong, K. (2021). Research trends, hot spots and prospects for necroptosis in the field of neuroscience. *Neural Regeneration Research*, *16*(8), 1628–1637. doi:10.4103/1673-5374.303032 PMID:33433494

Zhang, W., & Yuan, H. (2019). A Bibliometric Analysis of Energy Performance Contracting Research from 2008 to 2018. *Sustainability*, *11*(13), 3548. doi:10.3390u11133548

Zhao, L., Tang, Z. Y., & Zou, X. (2019). Mapping the Knowledge Domain of Smart-City Research: A Bibliometric and Scientometric Analysis. *Sustainability*, *11*(23), 6648. doi:10.3390u11236648

KEY TERMS AND DEFINITIONS

Bibliometrix: bibliometrix package provides a set of tools for quantitative research in bibliometrics and scientometrics.

Biodegradation: Is the degradation of the materials into environmentally acceptable products such as water, carbon dioxide, and biomass by the action of naturally available microorganisms under normal environmental conditions.

Biomass: Is organic, meaning it is made of material that comes from living organisms, such as plants and animals.

Bioplastic: A type of biodegradable plastic derived from biological substances rather than petroleum. A bioplastic is a substance made from organic biomass sources, unlike conventional plastics which are made from petroleum.

R: R is a free, open source software program for statistical analysis, based on the S language.

RStudio: RStudio is a free, open source IDE (integrated development environment) for R. R is a programming language used for statistical computing while RStudio uses the R language to develop statistical programs.

Scientometrics: Scientometrics is concerned with the quantitative features and characteristics of science and scientific research.

VOSviewer: VOSviewer is a software tool for constructing and visualizing bibliometric networks.

Chapter 8
Magnetic Nanoparticles for Environmental Management

Mustafa Akgün
Ardahan University, Turkey

ABSTRACT

Magnetic nanoparticles are an emerging technique that has attracted attention in recent years in nanotechnology, biomedical, electronics, environmental science, and engineering applications. Nanoparticles have optical, electrical, catalytic, and thermal properties with their supermagnetic properties, large surface area, and biocompatibility. The major benefit of using nanoparticles is that due to their size, they can be accurately oriented and can be targeted and interacted with a specific biological entity or marker. In addition, it is easy to separate the magnetic property from the aqueous solution with the application of an external magnetic field. From an environmental perspective, MNPs have been used as catalysts in the purification of whey; removal of heavy toxic metals such as Arsenic (As), Lead (Pb), and toxic pollutants such as Fluoride (F) from contaminated water; and photocatalytic degradation of dyes and pollutants in water. In this study, types of magnetic nanoparticles, synthesis methods, properties, and environmental science and engineering applications are included.

INTRODUCTION

Many of the most pressing issues facing the world in the coming years will be related to environmental quality. Global warming, energy, material and food security, and clean water, soil and air are the main environmental issues that need to be addressed. The direct or indirect use of magnetic nanoparticles is among the options offered as

DOI: 10.4018/978-1-7998-9851-1.ch008

a new technology. Magnetic nanoparticles are used in a wide range of fields, such as the simultaneous monitoring and treatment of various diseases in the field of Health, the removal of industrial pollutants from underground and marine environments or the purification of aqueous environments by improving the quality of drinking water resources. Over two billion people live in countries experiencing water stress (UNESCO, 2021). In 2016, it is estimated that approximately 829,000 people die from diarrhoea as a result of unsafe drinking water, sanitation and hand hygiene. These causes represent 60% of all deaths due to diarrhoea globally, including nearly 300,000 children under the age of five, 5.3% of all deaths in this age group (Prüss-Ustün et al., 2019). Furthermore an estimated 2.86 million cholera cases occur annually in endemic countries. Among these cases, there are an estimated 95,000 deaths (M. Ali, Nelson, Lopez, & Sack, 2015). Organic pollutants and heavy metals are common pollutants that enter water systems as a result of anthropogenic activities and can cause various cancers, respiratory and skin diseases even at low concentrations. Successful removal of chemical and biological contaminants (bacteria, fungi, algae, viruses) allows soil regeneration, cleaning of marine and freshwater environments, and hygienic treatment of wastewater for reuse, with great benefits to human health and aquatic biota.

Magnetic nanoparticles combine with natural magnetic phenomena, with their enhanced activity and superparamagnetism resulting from their enormous surface-to-volume ratio. This powerful combination enables magnetic nanoparticles not only to act as efficient absorbers for contaminants, but also to magnetically separate from the surrounding aqueous medium for further processing and removal. The coating and functionalization of a magnetic nanoparticle creates a core-shell structure that enables specific chemical and metallic contaminants to be targeted with impressive selectivity and ultra-high sensitivity. Recently, increasing interest is environmental chemosensors using functionalized magnetic nanoparticles. These molecules form a complex that provides rapid detection and in situ detection of a target pollutant, and as a result of this complex, they also undergo some physicochemical changes. The natural reactivity of the nanoscale zero-valent (nZVI) form of iron can also be used to treat soil and groundwater pollution by reducing pollutants to less harmful compounds. Therefore, the environmental applications of magnetic nanoparticles for pollutants removal can be divided into three main categories: removal by chemical reaction, removal by physical adsorption, and detection by chemical sensing. (Pratt, 2014).

TYPES OF MAGNETIC NANOPARTICLES

In general, we can divide magnetic nanoparticles into 2 classes: Oxides: ferrites and Metallic with a Shell (Sreenivasulu et al., 2013).

Oxides: Ferrit

Ferrite nanoparticles are the most researched magnetic nanoparticles to date. Once ferrite nanoparticles become smaller than 128 nm (A. H. Lu, Salabas, & Schüth, 2007), they become superparamagnetic, preventing self-agglomeration as they exhibit their magnetic behavior only when an external magnetic field is applied. When the external magnetic field is turned off, the redundancy drops to zero. Just like non-magnetic oxide nanoparticles, the surface of ferrite nanoparticles is often modified by surfactants, silicones or phosphoric acid derivatives to increase their stability in solution (Sreenivasulu et al., 2013).

Metallic with a Shell

The metallic core of magnetic nanoparticles can be passive through mild oxidation, surfactants, polymers and precious metals. In an oxygen environment, Co nanoparticles form an anti-ferromagnetic Co O layer on the surface of the Co nanoparticle. Recently, the study investigated the effect of synthesis and exchange bias in these Co-core CoO shell nanoparticles with a golden outer shell. Nanoparticles with a magnetic core made of elemental Iron or Cobalt with a non-reactive shell made of graphene have been recently synthesized. Its advantages over ferrite or basic nanoparticles are:

- Higher magnetization
- Higher stability in acidic and basic solutions as well as organic solvents
- Chemistry on grapheme surface using methods already known for carbon nanotubes

(Sreenivasulu et al., 2013).

SYNTHESIZE METHODS FOR MAGNETIC NANOPARTICLES

Magnetic nanoparticles are produced by three methods: physical, chemical and biological. These production methods are briefly summarized in Figure 1.

Figure 1. Synthesize methods for magnetic nanoparticles

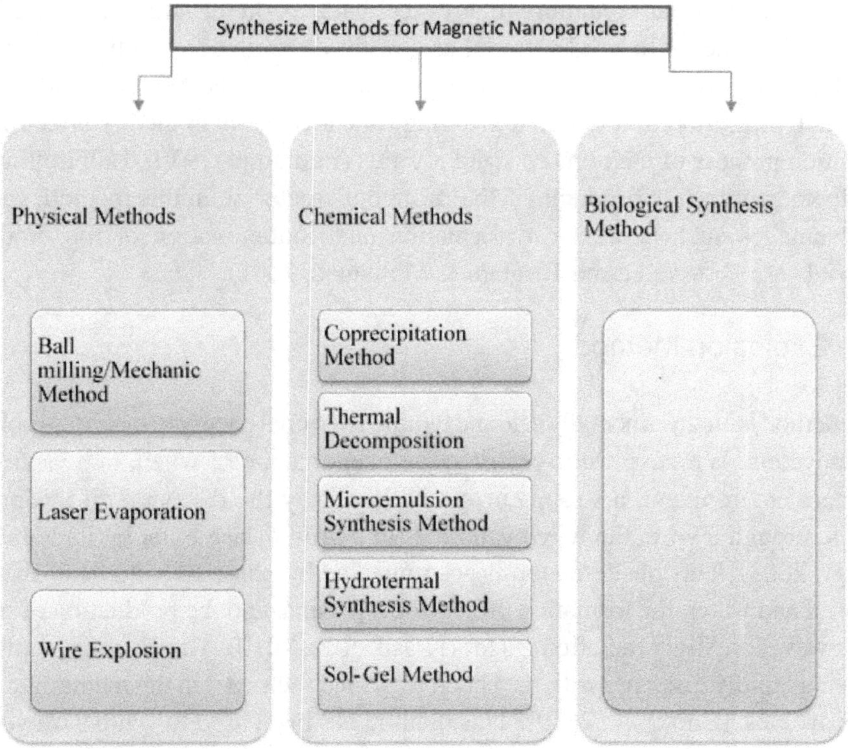

Physical Methods

Physical methods are divided into "top-down" and "bottom-up" two procedures. Top-down methods are based on the reduction of macroscopic magnetic materials to the nanometer range by grinding. Bottom-up methods, on the other hand, use condensation of nanoparticles from the liquid or gas phase. (Biehl, von der Lühe, Dutz, & Schacher, 2018). In MNP production by physical method; Three physical methods, namely ball milling, laser evaporation and wire blasting, will be briefly explained.

Ball Milling Method/Mechanical Method

Ball milling is a simple, fast, cost-effective green technology with enormous potential. Ball milling is a mechanical process that enables physical and chemical transformations to be carried out in powder form in accordance with the purpose (Delogu, Gorrasi, & Sorrentino, 2017). The working principle of this method, which was developed in 1970, is quite simple. The raw materials are enclosed in a small, hollow cylindrical jar containing many steel balls as the grinding medium. Balls apply kinetic energy to solid material as a result of continuous collisions between steel balls and solid materials resulting in nano/micro sized powder. Ball/powder ratio, ball size, vibration rate and grinding time are the main factors affecting the formation process of nano/micro sized crystals (Benjamin, 1970). Ball milling is a top-down approach to producing MNP from bulk material. In this method, various disadvantages such as plastic deformation and contamination of the produced materials are observed (Abd-Elsalam & Mohamed, 2001).

Wire Explosion Method

This method is environmentally safe and requires minimal energy to make less polluted nanopowders. Wire explosion refers to the phenomenon of when, in a vacuum or a certain environment, a strong current produced by the discharge of a capacitor passes through a wire, the wire evaporates in a bright flash burst by joule heating (Kotov, 2003). This phenomenon is accompanied by phase transitions of the wire material and water, the formation of non-ideal plasma, and the production of strong shock waves and light radiation fluxes (Krasik et al., 2010). The ultra-hot explosion products rapidly dissipate with the shock wave and then cool in the medium to form nanoparticles. (Gao et al., 2019). This method has been used to synthesize various nanoparticles, including metal nanoparticles, metal compound nanoparticles, and multiple carbon nanomaterials. (Yilmaz et al., 2013).

Laser Evaporation

Laser evaporation, which we can also call laser ablation, is a bottom-up approach in which nanoparticles are formed from the liquid or gas phase by condensation. Compared to traditional methods, laser ablation in liquids is the most effective physical method for nano-production (Mostafa, Mwafy, & Toghan, 2021). According to (Awwad, El-Kader, Ibrahium, Asnag, & Morsi, 2021), laser ablation (LA) of metallic samples is the process of removing material from a solid surface by irradiating it with a laser beam to form micro/nano structures. Laser ablation of solids in liquids and laser deposition of solid metal in air is a new and rapid way to produce nanoparticles with distinct advantages over other means.

Chemical Methods

There are many ways to produce MNP by chemical method. In this section, some of the common ones such as Sol-Gel method, coprecipitation method, hydrothermal synthesis method, thermal decomposition method and microemulsion synthesis method will be mentioned.

Coprecipitation Method

The co-precipitation technique is probably the simplest and most effective chemical way to obtain magnetic particles. Iron oxides are usually prepared by stoichiometric mixture of iron and iron salts in aqueous medium (Laurent et al., 2008). Among the various techniques for magnetite synthesis, the co-precipitation method is a convenient way to synthesize magnetite nanoparticles from an aqueous solution of iron salt (Fe^{2+}, Fe^{3+}). In this process, a base is simply added at room temperature under an inert atmosphere. Co-precipitation does not produce or use any toxic intermediates or solvents, does not require precursor complexes, and takes place at temperatures below 100 °C. This process is recognized for its industrial importance due to its scalability, reproducibility and environmentally friendly reaction conditions. (A. H. Lu et al., 2007). The main advantage of the co-precipitation process is that a large amount of nanoparticles can be synthesized. However, control of the particle size distribution is limited, as only kinetic factors control the growth of the crystal (Laurent et al., 2008).

Thermal Decomposition Method

Thermal separation methods are among the most successful methods for obtaining magnetic nanoparticles with a high degree of size control and narrow particle size distribution (Effenberger et al., 2017). The thermal decomposition method has advantages such as simple operation, low cost, and ease of obtaining high purity products. It is therefore quite promising and easy to apply to industry (Tang, Claveau, Corcuff, Belkacemi, & Arul, 2008).

Microemulsion Synthesis Method

Microemulsions are defined as clear thermodynamically stable dispersions of two immiscible liquids containing an appropriate amount of surfactant or surfactant and co-surfactant. The dispersed phase consists of small droplets with a diameter in the range of 100-1000 A°. Because of these properties, such systems have several advantages over macroemulsions for industrial applications. The small droplet size in microemulsions also results in a large surface-to-volume ratio in an oil-water system. This is important for chemical reactions where the reaction rate is dependent on the interfacial area. Microemulsion can also be classified as water/oil (W/0) or oil/water (0/W), similar to macroemulsion systems (Sharma & Shah, 1985). Synthesis of nanoparticles by microemulsion method is an area of great interest today. Microemulsion technique promises to be one of the versatile preparation methods that allows controlling particle properties such as particle size control mechanisms, geometry, morphology, homogeneity and surface area (Malik, Wani, & Hashim, 2012).

Hydrothermal Synthesis Method

Hydrothermal synthesis is one of the most widely used methods for the preparation of nanomaterials. It is basically a solution reaction based approach. The formation of nanomaterials in hydrothermal synthesis can occur over a wide temperature december, from room temperature to very high temperatures. Many types of nanomaterials have been successfully synthesized using this approach. The hydrothermal synthesis method has significant advantages over others. Hydrothermal synthesis can produce unstable nanomaterials at high temperatures. Nanomaterials with high vapor pressure can be produced by the hydrothermal method with minimal material loss. The compositions of the nanomaterials to be synthesized can be well controlled in hydrothermal synthesis through liquid phase or multiphase chemical reactions. (Gan, Jayatissa, Yu, Chen, & Li, 2020).

Sol-Gel Method

Polycondense in liquids is the formation of inorganic compounds through interactions. Sol is a colloidal distribution of particles in a solution and concentrated in the form of a gel. The starting raw materials are usually metal salts, dissolved together in distilled water or other solvents, and mixed to homogenize the dispersion. The left particles interact through van der Waals forces, sometimes it is necessary to add an organic complex agent as ethylene glycol and increase the interaction temperature by heating it on a hot plate. Then it is necessary to dry and heat the gel at a minimum temperature of 400 ° C for several hours to make sure that all organic compounds have completely evaporated. Then the resulting result is ready for use or it can be subjected to another processing according to the prepared material. A schematic summary for the sol-gel method is presented in Figure 2 (Abd-Elsalam & Mohamed, 2001).

Figure 2. Schematic diagram for the sol-gel method (Abd-Elsalam & Mohamed, 2001)

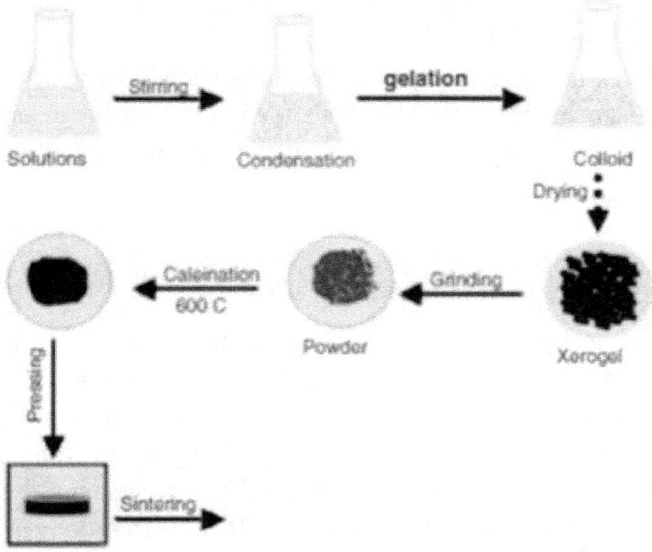

Biological Synthesis Method

We can also call the biological synthesis method as green synthesis. It has been reported that MNPs can be synthesized in various shapes and sizes using different living organisms including plants, microorganisms, fungi, bacteria, actinomycetes and even algae (Barabadi et al., 2019; Nath & Banerjee, 2013). When nanoparticles are produced using green synthesis methods, they are especially preferred in production because they do not contain harsh working conditions (high temperature and pressure) and the addition of dangerous chemicals. A large number of plants and microorganisms are tested to ensure that this method is environmentally friendly, cost-effective and biologically safe (Kumari, Dhand, & Padma, 2021).

ENVİRONMENTAL APPLICATION OF MAGNETIC NANOPARTICLES

As a result of excessive use of natural resources, population growth and industrialization, environmental pollution has increased and has become an important problem today.

Nanomaterials used for the treatment of organic pollutants can be generaly classified as inorganic nanomaterials (for example, transition metal/metal oxide/

metal sulfide nanoparticles, carbon-based nanomaterials) and organic molecule-based nanomaterials (for example, metal-organics, nanomembranes, and organic polymer-based nanomaterials). (F. Lu & Astruc, 2020).

Nanoparticles; Since they are environmentally friendly, reusable and comply with the logic of green chemistry, they have found use in many disciplines today. In this section, the main applications of magnetic nanoparticles in environmental engineering will be discussed.

Water Remediation

Unlike inorganic heavy metals, which are pollutants of a limited type, organic molecules are in the millions with a variety of functional groups, properties, and applications. Organic pollutants often include pharmaceuticals, personal care products, endocrine disruptors, pesticides, detergents, organic dyes and common industrial organic waste (F. Lu & Astruc, 2020). Li et al. (2018), in their study, established a model wastewater system using methylene blue and tried to treat this wastewater with nanoparticles. In this purification process using nanoparticles, after a waiting period of 4 hours, they managed to reduce the methylene blue concentration from 20.82 mg/lt to 1.92 mg/lt with a purification efficiency of 90.79%. In a study conducted for the treatment of groundwater contaminated with heavy metals, Cr(VI) was removed from groundwater with a yield of 94.7% at pH 5, 303 K and 0.4 g/L dosage (Zhu, Ma, Liu, & Deng, 2018). Again, a study on the removal of Cr(VI) from water was carried out at pH 2 and it was reported that the yield increased as the contact time and dose amount increased. In the same study, it was reported that by applying a dose of 0.2 g/L and a contact time of 10 minutes, a purification of over 90% was obtained with a contact time of 30 minutes (Fazlzadeh et al., 2017). Sadak et al. (2020), In their study, they used nanoparticles coated with azo dye for lead removal from water. reported that the removal efficiency of lead was highest at pH 6.5 and reaction time of 45 minutes, and they predicted a maximum absorption of 195.3 mg/g according to the Langmuir isotherm.

Soil Remediation

Engineered nanomaterials have greater potential for environmental pollution control as they are more cost-effective and more reactive than standard methods. These materials also enhance the option of in-situ treatment. Some examples of such engineered nanomaterials used in soil remediation are:

- Nanoscale calcium peroxide – used to break down organic compounds such as gasoline

- Nanoscale zerovalent iron – used to destroy halogenated organic compounds
- Nanoscale metal oxides – used for metal adsorption
- Carbon nanotubes, bionanoparticles, polymeric nanoparticles etc. other nanoparticles such as – used for the removal of aromatic and heavy metal contaminants (Sarkar, Sengupta, & Sen, 2019).

Although remediation with nanomaterials offers a more effective and less expensive approach than traditional methods due to the increased reactivity of nanoparticles and the possibility of in-situ treatment, more research is needed on the mobility of these particles in soil and their effects on the environment, as it is a new technology (Mueller & Nowack, 2010).

Waste Management

Anand et al. (2017), in their study; prepared Ag nanoparticles simply using the aqueous leaf extract of Ekebergia capensis and used these nanoparticles to catalyze the degradation of Allura and Congo red through an electron relay mechanism. This result; showed that these nanoparticles can be used in the degradation of industrial azo-type paints in wastewater. In another study on waste management; The effects of iron oxide nanoparticles on composting, organic matter degradation, dehydrogenase and urease activities and the quality of the final compost product were investigated. The results showed that composting enriched with Fe2O3 nanoparticles was more effective in facilitating organic matter degradation. (Zhang et al., 2019). Ali et al. (2017), in their study to increase the efficiency of energy obtained from urban solid waste; They added the nanoparticles they produced to the anaerobic reactor, which took place under mesophilic temperature ($37 \pm 0.5°C$), at concentrations of 50 mg/L, 75 mg/L, 100 mg/L and 125 mg/L. As a result, they observed that the nanoparticle added reactor produced maximum CH_4, followed by the 50 mg/L added reactor, and concluded that the addition of 75 mg/L nanoparticles biologically stimulated anaerobic treatment, which resulted in increased methane production.

Environmental Sensing

In order to obtain measurements of contamination levels at the site of environmental events such as oil spills, natural disasters, nuclear, industrial and chemical accidents; There is an idea to develop sensors that are fast, accurate and field usable. As outlined, magnetic nanoparticles are extremely effective at removing a wide variety of contaminants through chemical reaction or adsorption. The separation of the nanoparticles from the contaminants and their removal from the solution occurs subsequently, thanks to the natural magnetism of the nanoparticles (Jung, Lee, &

Shinkai, 2011). There is also a study in the literature emphasizing that both the detection tendency and the pollutant concentration contribute to the signal strength in the detection of pollutants. This result; The use of nanoparticles in studies for the detection of heavy metals such as Cd, Pb, Cu shows that detection can be made easier due to the increase in signal strength (Pratt, 2014). There is also a study in the literature on municipal solid waste to determine the effect of TiO_2 and Ag nanoparticles on aerobic storage processes. In this study, TiO_2 and Ag nanoparticles at a concentration of 100 mg/kg were added to the wastes and the bioreactor was operated with an aeration of approximately 0.03 L/min kg for 250 days. The results showed that leachate properties, gas components, solid quality parameters and temperature changes, which are the most important indicators of landfill operations, did not differ significantly in overall aerobic degradation performance between the control reactor and the reactors containing TiO2 and Ag nanoparticles. (Yazici Guvenc, Alan, Adar, & Bilgili, 2017).

FUTURE RESEARCH DIRECTIONS

In future studies, with the developing technology, magnetic nanoparticles can be coated with different types of surface coatings and after this process, the application results of these nanoparticles in different areas of use such as water treatment, soil and water remediation, the effect of biodegradation of solid wastes and perhaps the purification of air pollutants from the air can be examined. In addition, different production techniques can be studied to ensure that nanoparticles are produced in requested sizes and in a more economical way.

CONCLUSION

Magnetic nanoparticles can be broadly classified as iron salts and surface coated metals. These nanoparticles can generally be produced by physical, chemical and biological means. Produced nanoparticles are used in environmental engineering applications, water and soil remediation, and in the diagnosis and treatment of diseases in the field of health.

Nowadays, as a result of increasing environmental pollution, there is an increasing demand for reliable, non-toxic, highly efficient and eco-sustainable techniques that can replace traditional methods. Magnetic nanoparticles; due to they are non-toxic in green chemistry applications, they have been popularly used in environmental engineering in areas such as waste management, water treatment, soil and water remediation. Magnetic nanoparticles give acceptable and reliable results when compared with conventional purification methods. Thus, the negative effects of environmental pollution on humans and the ecosystem are reduced. In addition, as a result of using these nanoparticles many times, less chemicals will be used in treatment processes, so less waste will be generated and more economic gain will be achieved.

REFERENCES

Abd-Elsalam, K. A., & Mohamed, M. A. (2001). Magnetic nanostructures. *Anais da Academia Brasileira de Ciências*. Advance online publication. doi:10.15900001-37652001000300026

Ali, A., Mahar, R. B., Soomro, R. A., & Sherazi, S. T. H. (2017). Fe3O4 nanoparticles facilitated anaerobic digestion of organic fraction of municipal solid waste for enhancement of methane production. *Energy Sources. Part A, Recovery, Utilization, and Environmental Effects*, *39*(16), 1815–1822. doi:10.1080/15567036.2017.1384866

Ali, M., Nelson, A. R., Lopez, A. L., & Sack, D. A. (2015). Updated global burden of cholera in endemic countries. *PLoS Neglected Tropical Diseases*, *9*(6), 1–13. doi:10.1371/journal.pntd.0003832 PMID:26043000

Anand, K., Kaviyarasu, K., Muniyasamy, S., Roopan, S. M., Gengan, R. M., & Chuturgoon, A. A. (2017). Bio-Synthesis of Silver Nanoparticles Using Agroforestry Residue and Their Catalytic Degradation for Sustainable Waste Management. *Journal of Cluster Science*, *28*(4), 2279–2291. doi:10.100710876-017-1212-2

Awwad, N. S., El-Kader, M. F. H. A., Ibrahium, H. A., Asnag, G. M., & Morsi, M. A. (2021). Green synthesis of different ratios from bimetallic gold: Silver nanoparticles core@shell via laser ablation scattered in Chitosan-PVA matrix and its electrical conductivity behavior. *Composites Communications*, *24*(February), 100678. doi:10.1016/j.coco.2021.100678

Barabadi, H., Najafi, M., Samadian, H., Azarnezhad, A., Vahidi, H., Mahjoub, M. A., … Ahmadi, A. (2019). A systematic review of the genotoxicity and antigenotoxicity of biologically synthesized metallic nanomaterials: Are green nanoparticles safe enough for clinical marketing? *Medicina (Lithuania)*, *55*(8). doi:10.3390/medicina55080439

Benjamin, J. S. (1970). Dispersion strengthened superalloys by mechanical alloying. *Metallurgical Transactions*, *1*(10), 2943–2951. doi:10.1007/BF03037835

Biehl, P., von der Lühe, M., Dutz, S., & Schacher, F. H. (2018). Synthesis, characterization, and applications of magnetic nanoparticles featuring polyzwitterionic coatings. *Polymers, 10*(1), 91. Advance online publication. doi:10.3390/polym10010091 PMID:30966126

Delogu, F., Gorrasi, G., & Sorrentino, A. (2017). Fabrication of polymer nanocomposites via ball milling: Present status and future perspectives. *Progress in Materials Science, 86*, 75–126. doi:10.1016/j.pmatsci.2017.01.003

Effenberger, F. B., Couto, R. A., Kiyohara, P. K., Machado, G., Masunaga, S. H., Jardim, R. F., & Rossi, L. M. (2017). Economically attractive route for the preparation of high quality magnetic nanoparticles by the thermal decomposition of iron(III) acetylacetonate. *Nanotechnology, 28*(11), 115603. Advance online publication. doi:10.1088/1361-6528/aa5ab0 PMID:28192283

Fazlzadeh, M., Rahmani, K., Zarei, A., Abdoallahzadeh, H., Nasiri, F., & Khosravi, R. (2017). A novel green synthesis of zero valent iron nanoparticles (NZVI) using three plant extracts and their efficient application for removal of Cr(VI) from aqueous solutions. *Advanced Powder Technology, 28*(1), 122–130. doi:10.1016/j.apt.2016.09.003

Gan, Y. X., Jayatissa, A. H., Yu, Z., Chen, X., & Li, M. (2020). Hydrothermal Synthesis of Nanomaterials. *Journal of Nanomaterials, 2020*, 1–3. Advance online publication. doi:10.1155/2020/8917013

Gao, X., Hiraoka, T., Ohmagari, S., Tanaka, S., Sheng, Z., Liu, K., Xu, M., Chen, P., & Hokamoto, K. (2019). High-efficiency production of large-size few-layer graphene platelets via pulsed discharge of graphite strips. *Nanomaterials (Basel, Switzerland), 9*(12), 1785. Advance online publication. doi:10.3390/nano9121785 PMID:31888135

Jung, J. H., Lee, J. H., & Shinkai, S. (2011). Functionalized magnetic nanoparticles as chemosensors and adsorbents for toxic metal ions in environmental and biological fields. *Chemical Society Reviews, 40*(9), 4464–4474. doi:10.1039/c1cs15051k PMID:21607241

Kotov, Y. A. (2003). Electric explosion of wires as a method for preparation of nanopowders. *Journal of Nanoparticle Research, 5*(5–6), 539–550. doi:10.1023/B:NANO.0000006069.45073.0b

Krasik, Y. E., Fedotov, A., Sheftman, D., Efimov, S., Sayapin, A., Gurovich, V. T., Veksler, D., Bazalitski, G., Gleizer, S., Grinenko, A., & Oreshkin, V. I. (2010). Underwater electrical wire explosion. *Plasma Sources Science & Technology, 19*(3), 034020. Advance online publication. doi:10.1088/0963-0252/19/3/034020

Kumari, S. C., Dhand, V., & Padma, P. N. (2021). Green synthesis of metallic nanoparticles: A review. *Nanomaterials (Basel, Switzerland)*, 259–281. doi:10.1016/B978-0-12-822401-4.00022-2

Laurent, S., Forge, D., Port, M., Roch, A., Robic, C., Vander Elst, L., & Muller, R. N. (2008). Magnetic iron oxide nanoparticles: Synthesis, stabilization, vectorization, physicochemical characterizations and biological applications. *Chemical Reviews*, *108*(6), 2064–2110. doi:10.1021/cr068445e PMID:18543879

Li, Z. L., Wang, W., Li, M., Zhang, M. J., Tang, M. J., Su, Y. Y., Liu, Z., Ju, X.-J., Xie, R., & Chu, L. Y. (2018). Facile Fabrication of Bubble-Propelled Micromotors Carrying Nanocatalysts for Water Remediation. *Industrial & Engineering Chemistry Research*, *57*(13), 4562–4570. doi:10.1021/acs.iecr.7b04941

Lu, A. H., Salabas, E. L., & Schüth, F. (2007). Magnetic nanoparticles: Synthesis, protection, functionalization, and application. *Angewandte Chemie International Edition*, *46*(8), 1222–1244. doi:10.1002/anie.200602866 PMID:17278160

Lu, F., & Astruc, D. (2020). Nanocatalysts and other nanomaterials for water remediation from organic pollutants. *Coordination Chemistry Reviews*, *408*, 213180. doi:10.1016/j.ccr.2020.213180

Malik, M. A., Wani, M. Y., & Hashim, M. A. (2012). Microemulsion method: A novel route to synthesize organic and inorganic nanomaterials. 1st Nano Update. *Arabian Journal of Chemistry*, *5*(4), 397–417. doi:10.1016/j.arabjc.2010.09.027

Mostafa, A. M., Mwafy, E. A., & Toghan, A. (2021). ZnO nanoparticles decorated carbon nanotubes via pulsed laser ablation method for degradation of methylene blue dyes. *Colloids and Surfaces. A, Physicochemical and Engineering Aspects*, *627*(July), 127204. doi:10.1016/j.colsurfa.2021.127204

Mueller, N. C., & Nowack, B. (2010). Nanoparticles for Remediation: Solving Big Problems with Little Particles. *Elements*, *6*(6), 395–400. doi:10.2113/gselements.6.6.395

Nath, D., & Banerjee, P. (2013). Green nanotechnology - A new hope for medical biology. *Environmental Toxicology and Pharmacology*, *36*(3), 997–1014. doi:10.1016/j.etap.2013.09.002 PMID:24095717

Pratt, A. (2014). Environmental applications of magnetic nanoparticles. In *Frontiers of Nanoscience* (pp. 259–307). Elsevier Ltd. doi:10.1016/B978-0-08-098353-0.00007-5

Prüss-Ustün, A., Wolf, J., Bartram, J., Clasen, T., Cumming, O., Freeman, M. C., Gordon, B., Hunter, P. R., Medlicott, K., & Johnston, R. (2019). Burden of disease from inadequate water, sanitation and hygiene for selected adverse health outcomes: An updated analysis with a focus on low- and middle-income countries. *International Journal of Hygiene and Environmental Health*, *222*(5), 765–777. doi:10.1016/j.ijheh.2019.05.004 PMID:31088724

Sadak, O., Hackney, R., Sundramoorthy, A. K., Yilmaz, G., & Gunasekaran, S. (2020). Azo dye-functionalized magnetic Fe3O4/polyacrylic acid nanoadsorbent for removal of lead (II) ions. *Environmental Nanotechnology, Monitoring & Management*, *14*(October), 100380. doi:10.1016/j.enmm.2020.100380

Sarkar, A., Sengupta, S., & Sen, S. (2019). Nanoparticles for Soil Remediation. In *Nanoscience and Biotechnology for Environmental Applications* (pp. 249–262). doi:10.1007/978-3-319-97922-9_9

Sharma, M. K., & Shah, D. O. (1985). *Introduction to Macro- and Microemulsions*. doi:10.1021/bk-1985-0272.ch001

Sreenivasulu, K., Satyaveni, M., Rajwardanreddy, M., Sowjanya, M. S., Nama, S., & Baburao, C. (2013). A Revıew On Magnetıc Nano Partıcles. *International Journal of Current Pharmaceutical & Clinical Research*, *3*(2), 45–49.

Tang, Z. X., Claveau, D., Corcuff, R., Belkacemi, K., & Arul, J. (2008). Preparation of nano-CaO using thermal-decomposition method. *Materials Letters*, *62*(14), 2096–2098. doi:10.1016/j.matlet.2007.11.053

United Nations. (n.d.). The United Nations World Water Development Report 2021. *Valuing Water. Author.*

UNESCO. (2021). *Valuing water*. United Nations Educational, Scientific and Cultural Organization. doi:10.4324/9781315627250-3

Yazici Guvenc, S., Alan, B., Adar, E., & Bilgili, M. S. (2017). The impact of nanoparticles on aerobic degradation of municipal solid waste. *Waste Management & Research*, *35*(4), 426–436. doi:10.1177/0734242X17695884 PMID:28367755

Yilmaz, F., Lee, D. J., Song, J. W., Hong, H. S., Son, H. T., Yoon, J. S., & Hong, S. J. (2013). Fabrication of cobalt nano-particles by pulsed wire evaporation method in nitrogen atmosphere. *Powder Technology*, *235*, 1047–1052. doi:10.1016/j.powtec.2012.10.024

Zhang, L., Zhu, Y., Zhang, J., Zeng, G., Dong, H., Cao, W., Fang, W., Cheng, Y., Wang, Y., & Ning, Q. (2019). Impacts of iron oxide nanoparticles on organic matter degradation and microbial enzyme activities during agricultural waste composting. *Waste Management (New York, N.Y.)*, *95*, 289–297. doi:10.1016/j.wasman.2019.06.025 PMID:31351614

Zhu, F., Ma, S., Liu, T., & Deng, X. (2018). Green synthesis of nano zero-valent iron/Cu by green tea to remove hexavalent chromium from groundwater. *Journal of Cleaner Production*, *174*, 184–190. doi:10.1016/j.jclepro.2017.10.302

Chapter 9
An Overview to Green Nanotechnology for Environmental Issues

Naveen Kumar Chourasia
Government College, Bichhua, India

ABSTRACT

Environmental pollution is a serious issue that has a negative impact on human health. The situation is exacerbated by the gradual development of industrialization and urbanization. The removal of toxins from the environment is characterized as environmental remediation. Environmental remediation refers to all of the approaches used to lessen the risks of pollution in the environment. Nanotechnology has the potential to significantly reduce existing environmental issues. Green nanotechnology is the science and technology for developing nanoparticles using environmentally benign approaches. Green nanotechnology makes use of nature's biological qualities (through a variety of activities). Another area where nanotechnology can play a critical role is in environmental remediation because the properties of nanoparticles can be designed by designing their size and shape. Due to their huge surface area and, as a result, strong reactivity, nanoparticles perform significantly better than bulk materials for environmental remediation.

DOI: 10.4018/978-1-7998-9851-1.ch009

INTRODUCTION

Environmental contamination is one of the world's significant challenges in this period of industrialization. For the removal of contaminants from both terrestrial and aquatic ecosystems, many technologies are used. Filtration, adsorption, chemical reactions, photocatalysis, and other methods are extensively employed for environmental clean-up. Because of their unique mechanical qualities, increased surface area, chemical reactivity, regeneration efficiency, and cost-effectiveness, nanotechnology-based materials are becoming increasingly desirable for environmental remediation. Different environmental (organic or inorganic) contaminants, such as dyes (anionic and cationic) from industrial sources, heavy metals in various forms, gaseous hazardous chemicals, and so on, are removed using nanomaterials (Singh, 2022).

New innovative nanomaterials created using a variety of processes have raised the hazard to the environment and, as a result, human health. As a result, the development of nanomaterials-based environmentally friendly approaches has gained importance, attracting the attention of the scientific community. Researchers began producing nanoparticles using various biomaterials and claimed that their method was environmentally friendly. These nanoparticles can be classified according to how they were made (inorganic, organic, carbon-based, etc.) (Kumar et al., 2020). Producing eco-friendly, nontoxic nanomaterials that can be easily disposed of following treatment of air, water, or soil is currently a difficulty.

PHYTOGENIC-MEDIATED NANOPARTICLES

Water contamination is a global problem that is worsening by the day. Direct discharge of wastewater comprising organic and inorganic, dissolved and suspended pollutants could endanger terrestrial and aquatic life, disrupting the natural ecosystem. As a result, the treatment of such toxic wastewater is critical. Several physicochemical and biological wastewater treatment technologies are being developed in this field. However, several of these traditional wastewater treatment technologies have proven to be costly, time-consuming, and need a large amount of labor and infrastructure with low treatment efficiency (Chahal & Mishra, 2021).

For the effective removal of toxins from wastewater, nanotechnology appears to be an emerging and diversified field. Because of their enormous specific surface areas and strong reactivity, nanomaterials are excellent adsorbents and catalysts. Furthermore, nanomaterials in solution have a great degree of flexibility, and even a small quantity of nanomaterials is sufficient to break down contaminants in wastewater. Chemical and biological methods can be used to make nanoparticles.

Figure 1. Methods for remediation treatment

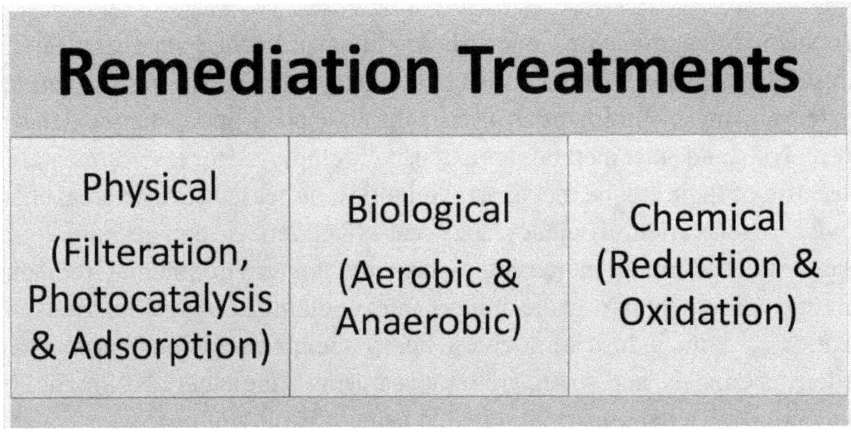

However, "green synthesis/phytogenic-mediated synthesis of nanoparticles" utilizing plants and plant products is an efficient, cost-effective, safe, and environmentally friendly method that does not require high pressure, temperature, or energy and does not produce harmful compounds. Furthermore, a wide range of plants is readily available for large-scale nanoparticle production.

GREEN NANOPARTICLES

Nanoparticles (NPs) are particles having a size of fewer than 100 nanometers in at least one dimension. Nanoparticles have different shapes and properties than bulk materials with the same chemical makeup and can be manufactured and tailored for specific uses and applications by selecting the appropriate synthesis, production technique, or technology. Many NPs can be manufactured via biological and/or biomimetic synthesis, which is regarded greener than chemical or physical synthesis. Metallic NPs are among the most studied NPs. These come in a variety of morphologies, including nanocubes, nanoflowers, nanorods, nanostars, nanoreefs, nanoboxes, nanowhiskers, and nanofibers, and can be spherical or nonspherical. Nonspherical gold, platinum, and silver NPs have a variety of intriguing features that are being investigated for medicinal, nanotechnical, and environmental applications (trace metal detection). Some insights into green NP biosynthesis are discussed here, with a focus on intracellular, extracellular, and bio-based techniques (Singh & Dhiman, 2021). The majority of biosynthetic approaches use reduction agents to promote metallic ions and tune MNP forms, while stabilising agents ensure that MNPs do not combine and that their sizes are controlled. Biological synthesis frequently relies on mild circumstances and all-in-one reduction/oxidation and stabilisation; however, some researchers look at MNP production in two or more phases.

Intracellular syntheses use fungus or bacteria to reduce MNPs, whereas extracellular syntheses use reduction, biosorption, enzyme- or latex-mediated processes and are less time-consuming since MNPs do not need to be recovered or purified. Bio-based approaches use isolated chemicals from plants, bacteria, and fungi to simulate biological synthesis. Other NPs, such as metal-oxide particles, can be made using biosynthetic techniques and have attracted attention due to potential uses, such as magnetic NPs. All of the methods mentioned use non-hazardous reagents, and the vast majority are safe and clean (Verma, 2017).

NANOMATERIALS FOR SUSTAINABLE CEMENTITIOUS COMPOSITES

Nanomaterials have been popular in recent decades due to their ability to solve a variety of engineering difficulties. Algal cells, for example, have been utilised to generate biomaterials (optoelectrical materials, tissue-engineering scaffolds, and supercapacitors), which are then used in medicine, heavy metal absorbents, and biosensors. The numerous designed uses of nanoparticles for sustainable cementitious composites are discussed in this chapter. The use of nanoparticles in building was highlighted, with a comprehensive look at the many forms and applications (Bharmoria & Ventura, 2021; Nyika, 2022).

The building industry has benefited greatly from nanotechnology. The interfacial transition zone of recycled aggregate and recycled aggregate concrete is modified using nanomaterials such as ureolytic or nonureolytic bacteria and nano-silica. Such changes reduce the composite matrix's water permeability while also increasing the specific gravity of aggregates. Microscale characteristics of cementitious composites are also altered by this technology. Biogenic calcite improves the hydration mechanism. This method has been used to create nanoengineered cementitious composites with good mechanical and durability properties (Li, 2019a; Li, 2019b).

NANOPARTICLES IN THE AGRICULTURE INDUSTRY

The successful application of nanostructures in biomedicine and agriculture has emerged as a significant and emerging field of study. Crop production and improvement are critical for maintaining a self-sustaining world and feeding the world's ever-growing population. This can be accomplished in a large scale using "agri-nanotechnology," which promises natural resource management through novel tools and technology platforms within limited land and water resources. Nanotechnology has the potential to improve agricultural output by inventing and creating nanoscale materials, devices, and systems that can exploit the desired qualities (physical, chemical, and biological) (Kushwah & Verma, 2022).

Nanoparticles (NPs) can be used to manage plant growth promotion in a variety of ways, depending on their occurrence, kind, and composition. These recent developments have opened up new possibilities for the advancement of genetically modified crops, such as site-directed delivery of various macromolecules such as genes and drugs, smart agrochemical delivery systems, early disease and pathogen detection, plant-protecting chemicals, and precision farming techniques. However, this can only be done successfully in plants if we improve our understanding of NP absorption, transit, and accumulation in plants. Because data on the effects of NPs on plants are conflicting, it is necessary to optimise experimental settings such as plant growth medium, NP size and shape, and so on. This would boost the plant management capability of nanoformulations, nanogenetic manipulations, and nanoarray-based technologies. Other notable achievements of nanotechnology include the control of plant infections, their prevention, and disease treatment utilising various nanocides (Carissimi et al., 2022; Sapna, 2018).

Understanding the interactions between plants and their surroundings (soil, water, and atmosphere) makes it necessary to investigate the influence of nanotechnology on agriculture, particularly with regard to toxicity concerns. Furthermore, plant growth and metabolic processes fluctuate in response to a variety of nanoparticles. They provide NPs with avenues to bioaccumulation in food chains, allowing them to reach higher species in ecosystems. The Indian economy is heavily reliant on agriculture, hence agricultural nanotechnology research and development is critical.

Agriculture can benefit from the development of more efficient and low-contaminant agrochemicals (nanoformulations), devices that help detect biotic or abiotic stresses before they affect production (nanosensors), or new genetic manipulation techniques that allow for greater efficiency during plant breeding programmes. Nanodevices can help deliver agrochemicals to the correct place at the right time while also reducing the impact of external agents that cause losses due to degradation, leaching, run-off, volatilization, and other factors. The predicted result should be a decrease in the amount of active chemicals integrated into plants and soils, resulting in less detrimental environmental impact (Saran et al., 2017).

PLANT ABSORPTION AND UPTAKE OF NANOPARTICLES

Plant uptake of nanoparticles is influenced by a number of aspects, including the nanoparticle's nature, plant physiology, and the interaction of nanomaterials with the environment. It is apparent that nanoparticle characteristics will have a significant impact on its activity and, as a result, on the plant's ability to absorb it. Size appears to be one of the key barriers to nanoparticle penetration into plant tissues, and there have been some reports regarding the maximum dimensions that plants allow nanoparticles to migrate and accumulate inside cells, with a size exclusion of 40–50 nm limit. Another aspect that influences uptake is the type of nanoparticle and its chemical composition. The features of nanomaterial absorption and accumulation by plants can be considerably changed and altered by functionalizing and coating the nanomaterial surface (Venzhik et al., 2022).

Plant species differ in their physiology, which leads to disparities in nanoparticle uptake. These studies demonstrated that crops from various botanical families, when exposed to magnetic carbon-coated, titanium dioxide, or gold nanoparticles, displayed varied absorption and accumulation patterns inside the plants. However, the methods of administration are also important in determining how well a plant would internalise nanomaterials: roots are designed for nutritional and water absorption, whereas leaves are designed for gas exchange and have a cuticle that prevents things from penetrating (Rahmat, 2022).

However, nanoparticles interact with other elements in the environment, which can alter their properties and features for plant absorption. Humic acids and other organic matter in the soil, for example, can improve the stability of nanomaterials and hence their bioavailability, but salt ions can cause precipitation and have the opposite effect (Xiong & Li, 2022).

MOVEMENT OF NANOPARTICLES INSIDE PLANTS

There are two methods for nanoparticles to migrate through tissues once they have penetrated the plant: the apoplast and the symplast. Apoplastic transport occurs outside the plasma membrane, in extracellular gaps, neighbouring cell walls, and xylem arteries. Water and chemicals are transported between the cytoplasms of neighbouring cells by plasmodesmata, and sieve plates. The apoplastic channel is vital for radial movement within plant tissues, since it allows nanoparticles to reach the central cylinder of the root and vascular tissues, allowing them to travel upwards to the aerial region. Nanoparticles can migrate toward the aerial section of the plant through the xylem, following the transpiration stream. To reach the xylem through the root, however, a barrier to the apoplastic pathway, the Casparian strip, must be crossed by endodermal cells in a symplastic approach (Salatin et al., 2015).

At the Casparian strip, some nanomaterials can be halted and accumulated. Another essential symplastic movement is feasible, thanks to the phloem's sieve tube elements, which allow distribution to non-photosynthetic tissues and organs.

Nanomaterials must pass through the cuticle barrier in foliar treatments, either via the lipophilic or hydrophilic pathway. The lipophilic system involves diffusion through cuticular waxes, while the hydrophilic pathway uses polar aqueous pores in the cuticle and/or stomata. Because cuticular pores are predicted to be roughly 2 nm in diameter, the stomatal pathway appears to be the most plausible route for nanoparticle penetration, with a size exclusion limit of greater than 10 nm.

The way nanomaterials migrate within plants is crucial because it can reveal which areas of the plant they can reach and where they can end up and aggregate. If a type of nanoparticle is transported mostly through the xylem rather than the phloem, it will likely go primarily from root to shoot and leaves, rather than downwards, hence it should be given to the roots to ensure a good distribution in the plant (Dachineni, 2016).

FUTURE RESEARCH

Nanotechnology's potential uses in various fields must not be underestimated. However, there is a need to take seriously the uncertainties and negative perceptions of nanotechnological interventions in agriculture. As a result, significant efforts must be made to advance and improve futuristic research based on identified knowledge gaps. Future research should focus on finding solutions to avoid the risks associated with nanoparticle use. The study of nanoparticle synthesis and the provision of a few applications limited to laboratory conditions could not contribute to nanotechnology's full acceptance in the agriculture sector. As a result, the scientific community must collaborate to better future research using a more realistic approach (Maruyama, n.d.).

It is necessary to investigate and clarify the allowed level of nanoparticle dose within safety limits. This might be accomplished by conducting a concentration-dependent investigation in a natural soil environment to determine the precise active and non-toxic nanoparticle dose.

To get a thorough understanding of nanotoxicity, an understanding of the transgenerational and trophic chain transfer effects of nanoparticles applications on plants is required. Surprisingly, choosing a permitted level and researching transgenerational and trophic chain transfer effects together could yield appropriate safety assessments. A comprehensive understanding of the physico-chemical features of the soil in agricultural fields where nanoparticles will be used may aid in lowering the risk of nanoparticles harming plant and soil biota (Palit, 2019a).

A comprehensive understanding of the soil physico-chemical features of agricultural fields where nanoparticles will be used may aid in lowering the danger of nanoparticles harming plant and soil biota. Changing the soil environment to change the fate, transport, and bioavailability of nanoparticles in order to lessen their toxicity could considerably improve their safe and beneficial agricultural applications. Advanced soil management approaches, for example, might help reduce nanoparticle transport, bioavailability, and toxicity, all of which would have a large positive influence on the agroecosystem. Finally, and most critically, we highly suggest that biosynthesized nanoparticles be included as precondition for significant and in-depth research. Redeeming the environmentally beneficial approach of green nanoparticle synthesis, it is thought that biosynthesized nanoparticles may have lower or no toxicity, and that future study should concentrate on their practical utility. Furthermore, to accurately reflect the environmental impact of nanoparticles, the experimental design must be set in a natural context (growing the plants in soil) (Palit, 2019b).

CONCLUSION

Environmental remediation relies mainly on using various technologies (e.g., adsorption, absorption, chemical reactions, photocatalysis, and filtration) for the removal of contaminants from different environmental media (e.g., soil, water, and air). The enhanced properties and effectiveness of nanotechnology-based materials makes them particularly suitable for such processes given that they have a high surface area-to-volume ratio, which often results in higher reactivity.

REFERENCES

Bharmoria, P., & Ventura, S. P. M. (2021). *Bio-based aerogels for environmental remediation problems, Advances in Aerogel Composites for Environmental Remediation.* doi:10.1016/B978-0-12-820732-1.00017-5

Carissimi, G., Montalbán, M. G., Fuster, M. G., & Víllora, G. (2022). *Nanoparticles as Drug Delivery Systems, 21st Century Nanostructured Materials - Physics.* Chemistry, Classification, and Emerging Applications in Industry, Biomedicine, and Agriculture. doi:10.5772/intechopen.100253

Chahal, K., & Mishra, R. (2021). Challenges and Future Prospects of Nanotechnology for Antiviral Drugs. *International Journal of Applied Pharmaceutical Sciences and Research.* https://www.sierrajournals.com/index.php/IJAPSR/issue/view/91 doi:10.21477/ijapsr.6.2.2

Dachineni, R. (2016). *Cellular Uptake Mechanisms for Hydrophobically Modified Carbohydrate Polymer Based Nanoparticles*. doi:10.26226/morressier.57d6b2bfd462b8028d88ce64

Kumar, V., Kumar, P., & Singh, J. (2020). *Environmental Degradation: Causes and Remediation Strategies, Environmental Degradation: Causes and Remediation Strategies*. doi:10.26832/aesa-2020-edcrs

Kushwah, K. S., & Verma, D. K. (2022). *Biological Synthesis of Metallic Nanoparticles from Different Plant Species, 21st Century Nanostructured Materials - Physics*. Chemistry, Classification, and Emerging Applications in Industry, Biomedicine, and Agriculture. doi:10.5772/intechopen.101355

Li, V. C. (2019a). *Introduction to Engineered Cementitious Composites (ECC), Engineered Cementitious Composites (ECC)*. doi:10.1007/978-3-662-58438-5

Li, V. C. (2019b). *Sustainability of Engineered Cementitious Composites (ECC) Infrastructure, Engineered Cementitious Composites (ECC)*. doi:10.1007/978-3-662-58438-5

Maruyama, E. (n.d.). Nanotechnology research promotion at RIKEN - past, present and future. *Proceedings of the 2nd IEEE Conference on Nanotechnology*. 10.1109/NANO.2002.1032269

Nyika, J. M. (2022). *Nanotechnology and Its Applications in Environmental Remediation, Research Anthology on Emerging Techniques in Environmental Remediation*. doi:10.4018/978-1-6684-3714-8.ch004

Palit, S. (2019a). *Nanotechnology Research, Industrial Ecology, and the Visionary Future, Chemical Nanoscience and Nanotechnology*. doi:10.1201/9780429398254-7

Palit, S. (2019b). *Nanotechnology Research, Green Engineering, and Sustainability: A Vision for the Future, Chemical Nanoscience and Nanotechnology*. doi:10.1201/9780429398254-8

Rahmat, S. (2022). *Effects, uptake, and translocation of selenium-based nanoparticles in plants, Toxicity of Nanoparticles in Plants*. doi:10.1016/B978-0-323-90774-3.00006-4

Salatin, S., Dizaj, S. M., & Khosroushahi, A. Y. (2015). Effect of the surface modification, size, and shape on cellular uptake of nanoparticles. Cell Biology International, 39(8), 881-890. doi:10.1002/cbin.10459

Sapna, D. K. (2018). *Biodegradable polymer-based nanoadsorbents for environmental remediation, New Polymer Nanocomposites for Environmental Remediation*. doi:10.1016/B978-0-12-811033-1.00012-3

Saran, A., Boddula, R., & Sharan, S. R. (2017). *Biological Nanoparticles: Optical and Photothermal Properties, Fungal Biology, Fungal Nanotechnology*. doi:10.1007/978-3-319-68424-6_8

Singh, B., & Dhiman, M. (2021). *Carbon aerogels for environmental remediation, Advances in Aerogel Composites for Environmental Remediation*. doi:10.1016/B978-0-12-820732-1.00012-6

Singh, V. V. (2022). *Green nanotechnology for environmental remediation, Sustainable Nanotechnology for Environmental Remediation*. doi:10.1016/B978-0-12-824547-7.00017-5

Venzhik, Y., Sokolov, A., Sokolov, O., Moshkov, I., & Dykman, L. (2022). *Effects, uptake, and translocation of Au-based nanoparticles in plant, Toxicity of Nanoparticles in Plants*. doi:10.1016/B978-0-323-90774-3.00016-7

Verma, M. L. (2017). *Fungus-Mediated Bioleaching of Metallic Nanoparticles from Agro-industrial By-Products, Fungal Biology, Fungal Nanotechnology*. doi:10.1007/978-3-319-68424-6_5

Xiong, T., & Li, S. (2022). *Effects, uptake, and translocation of Cu-based nanoparticles in plants, Toxicity of Nanoparticles in Plants*. doi:10.1016/B978-0-323-90774-3.00012-X

Chapter 10
Crop Improvement Technology With *Lawsonia inermis*

Shweta Saxena
School of Studies in Botany, Jiwaji University, Gwalior, India

Sweta Prakash
Department of Botany, Rajmata Scindia Government Girl's P.G. College, Chhindwara, India

Sadhna Panday
Department of Botany, Government KRG P.G. College, Gwalior, India

ABSTRACT

Lawsonia inermis L. (family Lythraceae), often known as henna, is a dye-producing plant cultivated in various parts of the world for cosmetic purposes. Since time immemorial, leaf powder made from this little tree has been used to beautify skin, hair, fingernails, leather, silk, and wool. The plant's leaf, which contains an active dye (red orange pigment), lawsone, is widely employed in cosmetic and pharmaceutical industries (2-hydroxy-1,4 naphthoquinone). Lawsone, an essential secondary metabolite, accumulates in the plant's aerial parts, with the highest concentration of 1.0–1.4% in the young leaf petiole. Analgesic, hypoglycemic, hepatoprotective, immune stimulant, anti-inflammatory, antibacterial, anti-dermatophytic, protein glycation inhibition, anti-sickling, antioxidant, anti-fertility, tuberculostatic, wound healing, anticomplimentary, and anticancer properties have all been reported for the plant. Henna is currently recognised as a valuable source of unique natural ingredients for the creation of medications and commercial products for a variety of ailments.

DOI: 10.4018/978-1-7998-9851-1.ch010

NAME OF THE CROP WITH BOTANICAL NAME AND FAMILY

Lawsonia inermis L. – Lythraceae
Syn. *Lawsonia alba* Lamk.
Family: Lythraceae
Hindi: Mehndi, Henna
Sanskrit: Medhini, Madyantika
English: Egyptian proverb, Cypress shrub
Fruiting: Sep- Nov

ORIGIN

It is a biennial dicotyledonous herbaceous shrub. A native of North Africa and South-West Asia, the plant is now widely cultivated throughout the tropics as an ornamental and dye plant.

GEOGRAPHICAL DISTRIBUTION

Henna is an ancient dye, evidence being the Egyptian mummies found in the tombs that had their nails dyed with henna. Henna occurs wild from Iran to western India. From where it has been spread eastward to the rest of India and Indonesia and westward to the Middle east where it became one of the important plant of Islam. It later followed Islamic armies and traders from Arabia reaching as far as Spain, Madagascar, the Moluccas Indo-China and Japan. It is now distributed throughout the tropics and subtropics. *Lawsonia inermis* is commercially cultivated in Afghanistan, Bangladesh, Egypt, Iran, Libya, Morocco, Pakistan, Saudi Arabia, Somalia, Sudan, Tunisia, UAE, Western India and Yemen.

Kingdom: Plantae-Plants
Division: Angiospermae
Class: Dicotyledoneae
Order: Myrtales
Family: Lythraceae
Genus: *Lawsonia*
Species: *inermis*

GROWING AREAS IN THE COUNTRY

Lawsonia inermis is found in Asia, Australia, South-East Africa, the Middle East, and Arab and in many other parts of the world. In India, it is cultivated in South Indian states such as Tamil Nadu, Karnataka, Kerala, and Andhra Pradesh. It is traditionally used in countries like Algeria, India, Saudi- Arabia, Pakistan and Bangladesh, Israel, Somalia, Tunisia, and Turkey. It grows mostly in 35-45 degrees centigrade. Below 5-degree centigrade kills mehndi. A glabrous, much-branched, deciduous shrub with tetragonous branches often ending in spines; leaves simple, opposite, entire, lanceolate, petioles very short or absent; flowers white or rose colored, fragrant, in large terminal pyramidal panicled cymes, stamens 8, in 4 pairs inserted on the calyx tube; fruits globose capsules, tipped with the style and supported by the persistent calyx, seeds numerous, smooth, pyramidal.

CLIMATE AND SOIL

Distribution: Scarcely in dry deciduous forests, widely cultivated as a hedge plant. It can be grown on a wide variety of soil and climatic conditions. However, deep fine sandy is useful for henna cultivation. The plant thrives well under arid to tropical and warm temperate climatic conditions. It needs moderate rainfall of about 400 mm and a temperature of about 30-40°C during the active growth period during the rainy season and hot, dry, and open weather for a good harvest of quality leaves.

The soil required for henna is a bit different than other soil types for cultivation. The soil must not hold too much moisture. If the soil has more moisture, the leaves will also carry moisture. Hence, tannin production will be low. Soils that are semi-dry can be best used for the cultivation of henna. Too dry soils will not support the good growth of the plants. The ph of the soil must be balanced. Too acidic or alkaline soils are not good for the growth of henna plants. Organic matter must be present in sufficient quantities in the soil. The soil must be well aerated. Well-drained soils are preferred for the cultivation of henna.

MEDICINAL PROPERTIES AND USE OF DIFFERENT PARTS OF PLANTS FOR SPECIFIC PURPOSES

The roots are bitter, diuretic, and useful in burning sensation, leprosy, skin diseases, and premature graying of hair. The leaves are bitter, astringent, refrigerant, vulnerary, expectorant, anti-inflammatory, constipating, liver tonic, and febrifuge. They are useful in wounds, ulcers, cough, bronchitis vitiated conditions of *Kapha* and *pitta*, burning sensation, inflammations, diarrhea, dysentery, leprosy, leucoderma, scabies, boil, anemia, fever, ophthalmia, falling of hair, graying of hair and jaundice. The flowers are intellect promoting, cardiotonic, refrigerant, febrifuge, and tonic. They are useful in burning sensation, cardiopathy, insomnia, and fever. The seeds are antipyretic, constipating useful in intermittent fevers, insanity, diarrhea, dysentery, and gastropathy.

SKIN DISEASES

- Leaves are astringent and used as a prophylactic against skin diseases. They are applied locally to boils, burns, wounds, and skin diseases (Nayak *et al.*, 2007).
- In Asian folk medicine this plant was included in the treatment of dandruff, eczema, scabies, and ulcers (Jellin, 2002).
- *Bodo* tribals apply leaves on scabies.
- *Gond* tribals apply leaf paste externally to cure eczema.
- *Sahariya* tribals apply leaves on skin diseases.
- *Nagapattinam* tribals apply leaves is ground with *Cipadessa baccifera* root, leaf, and bark applied to leaves on psoriasis.
- *Sahariya* tribals apply leaves on skin diseases.
- *Tharu* tribals apply leaf pounded with mustard oil and applied on burns.
- *Valaiyans* tribals apply its leaves to reduce body heat, mixed with Onion bulb and made into a juice, taken internally.

Other uses:

- Among Muslims, henna perfume is applied to the dead to repel insects (Levng, 1980).
- Other products: The fragrant flowers are used as a perfume, while small twigs are used as toothbrushes in Indonesia.
- Henna is commercially used for fodder, timber, firewood, and dye for staining clothes, wood, leather, wool, and hair (Orwa *et al.* 2009).
- *Lawsonia inermis* tree has been extensively used in Ayurveda, Unani and Homeopathic medicine as a household remedy against various human ailments from antiquity making it a cynosure of modern medicine (Biswa *et al.* 2002).
- Unani medicinal uses - Over 27 uses are reported in the literature (Nadkarni, 1976).

- Use in Western herbal - *Lawsonia* is official in British Pharmacopoeia 1962. It is used in certain allergic, infective, and inflammatory conditions (Khare, 2004).
- Henna is an ingredient of Anna Pavala Sindhooram - a Siddha antiatherosclerotic formulation (Shanmugasundaram *et al*, 1987).
- Henna twigs are used in Kashmir by Gujjar Bakerwals for relieving toothache (Ganai and Nawchoo, 2003).

CROP IMPROVEMENT INCLUDING BIOTECHNOLOGY

Biotechnological approaches help in increasing the application of medicinal plants for secondary metabolite production. *In vitro* culture allows the production of secondary metabolites under controlled culture conditions. Two types of in vitro cultures are currently used for secondary metabolite production: dedifferentiated cultures (callus and suspension cultures) and differentiated cultures (transformed roots and shoots). Various approaches used for increasing the production of secondary metabolites include optimization of growth and production culture media, as well as elicitation and metabolic engineering (Giulietti and Ertola 1999).

PROPAGATION INCLUDING MICRO-PROPAGATION

Planting cuttings is a very common method for the propagation of henna. The stem must be cut above the nodes and planted in the field. A few leaves must be kept in the cuttings. Good cuttings from a good variety of plants must be used for the propagation of henna. Henna is also propagated through seeds. The selected seeds must be healthy and disease-free. The seeds must be soaked in water before sowing. Soaking them prior to sowing allows better germination.

CULTIVATION, LAND PREPARATION, TIME OF PLANTING, SPACING, MANURING AND FERTILIZATION, IRRIGATION, WEED CONTROL, SPECIAL CARE, IF ANY

The henna plant is native to northern Africa, western and southern Asia, and northern Australia in semi-arid zones and tropical areas. It produces the most dye when grown in temperatures between 35°C and 45°C (95 and 113°F. During the onset of precipitation intervals, the plant grows rapidly, putting out new shoots. Growth subsequently slows. The leaves gradually slow and fall during prolonged dry or cool intervals. It does not thrive where minimum temperatures are below 11°C (52°F). Temperatures below 5°C (41°F) will kill the henna plant. It grows best in water-retentive soils and areas. It is commercially cultivated in countries such as Western India, Pakistan, Morocco, Yemen, Afghanistan, Somalia, Sudan, and Libya. In India, Rajasthan is the most cultivated heavily region in Sonat city. In other states like Gujarat, Madhya Pradesh, and Punjab, henna is used as a major commercial crop where the climate is relatively hot and dry and the rains are scanty. Twice a year, henna is harvested, best after the monsoon when the crop is ready by the month of September and October while the February and March crop acts as a side crop. It is a perennial shrub, if planted holds a good promise if planted a short rotation, forestry plantation and will not only maintain continuous soil cover through foliage/ residue but also help in building a cleaner environment in air fringes. Henna based agroforestry system could be a good drought proofing strategy that may help farmers to buffer the effect of climate change.

LAND PREPARATION

For large-scale cultivation, a suitable area of land must be selected. The land must be clear of any weeds or matter from previous cultivations. After the land is clear, it must be ploughed. Tractors may be used for ploughing. Bullock carts may also be used for ploughing of the fields. Ploughing ensures that the soil has reached a fine tilth. This makes cultivation easier. Ploughing must be done till the soil reaches desired texture. Proper quantities of farmyard manure must be mixed with the soil. Proper drainage channels are also important to be constructed in the field.

TIME OF PLANTING

The cuttings can be raised in nurseries till they develop roots. After that, they can be transplanted back to the main field. In the case of seed planting, the seeds are allowed to germinate. After germination, they are grown in nurseries under controlled conditions. The germinated seeds are taken together with the fine sand and are planted in the nurseries. Proper watering and nutrition management is done so that healthy seedlings can be raised. After about 3-4 months, the seedlings are transplanted to the main field.

At the time of transplanting, the seedlings have gained a height of 30-40cms. For good foliage growth, proper quantities of nitrogen must be applied in the field. Light irrigation must be provided after transplanting. While transplanting the seedlings from the nursery to the main field, care must be taken so that the roots are not damaged.

While placing them in the main field, they must be put inside the already dug-out pits. The pits are filled with manure mixed with topsoil. After planting the seedlings or cuttings, they must be properly covered with soil and watered. The soil around the base of the plant must be pressed firmly to avoid entry into air.

SPACING

The transplanting is done in July-August. The germinated seedlings are raised in nurseries in the month of March. The seed rate for henna cultivation is around 7-10kg for one hectare of land. Normally, a spacing of 30cm*45cm is applied in the crop of henna. The spacing and seed rate may vary for different locations.

MANURING AND FERTILIZATION

Too much application of fertilizer is not recommended for the cultivation of henna. The soil test is essential for determining the nutrients need for the henna crop. After the soil test, proper quantities of nutrients must be provided to the crop. To avoid micronutrient deficiency, appropriate steps must be taken. The henna plants respond very well to nitrogenous fertilizers. It is recommended after sowing or planting. The fertilizers can also be applied in split doses. Farmyard manure must be applied at the time of field preparation.

IRRIGATION

Heavy irrigation is not required in the cultivation of henna. In fact, the henna plant requires dry and low irrigation. In dry conditions, the tannin is better produced. To get quality henna leaves, over-irrigation must be avoided at any cost. In case of excessive rainfall, the extra water must be drained out of the field. However, in too dry seasons, watering must be done properly to provide minimum moisture to the plants. Low moisture is recommended for henna plants. But, the plants must not be made to suffer from moisture stress. For saving water, drip or sprinkler irrigation can be used. Light irrigation can be provided after sowing of seeds and application of fertilizers.

INSECT, PESTS, DISEASES, AND THEIR MANAGEMENT

The henna plant is not quite attacked by a lot many diseases and pests. A check must be kept on the sap-sucking insects of henna. The common insects and diseases are black root rot and bacterial leaf spot. Proper management of crops can ensure that these diseases do not happen. In case of such an incidence, appropriate steps must be taken. Proper control measures can ensure the healthy growth of the plants. A black root rot caused by *Corticium koleroga* and a bacterial leaf spot caused by *Xanthomonas lawsoniae* has been reported from India.

HARVESTING, YIELD, POST-HARVEST MANAGEMENT, AND STORAGE

Plants are generally harvested twice a year from the second year onwards under intensive cultivation. Harvesting starts 1 or 2 years later under extensive management. During the first year plants are cut at about 5 cm above the ground, later they are cut at ground level. Harvesting is done when the flower buds start to form.

Yield

Few reliable statistics on yields are available. Under irrigation, henna may yield 2500-3000 kg/ h per year of leaves on a dry weight basis, reaching 4000 kg /ha under optimal conditions. Under rainfed conditions in northern India yields of 700- 1500 kg /ha are obtained. In India plantation of Lawsonia inermis is mainly confined to Gujarat, Haryana, Punjab, Madhya Pradesh, and Rajasthan (Duke *et al.* 1993). Rajasthan State has 41,450 ha (2010-2011) area under henna cultivation which Pali district of Rajasthan alone accounted for 39,400 ha (95%) largest cultivated area (Sukla *et al.* 2012). As per the FAO report, India exported 45, 00 tones to 76, 00 tonnes of henna per annum. The estimated export of henna, in all forms, was around 10,500 tonnes with a value of around Rs 91 crores.

POST-HARVEST MANAGEMENT

The branches after harvesting are dried in sun. After dried properly, they are beaten with sticks to separate the leaves from the branches. After the leaves have dried properly, they must be powdered. The fresh leaves are sold in the market sometimes. For making powder, the leaves also have to undergo some processing.

GERMPLASM MANAGEMENT

Seed storage behavior is orthodox; long-term storage is feasible. On average, there are about 700 000 seeds/kg

Utilization/ Processing (By-Products)

Marketed Unani Formulations

Arq Murakkab Musaffi Khoon
Benazir Hair Oil
Habbe Musaffi Khoon
Marham Henna
Roghan Amla Khas
Safoof Surkhbada
Sharbat Musaffi Khoon

Marketed Ayurvedic Formulations

B-Cuts
Keo Karpin
Livin
Medicon-V

Trade and Marketing (Name of the Company with Address who are Utilizing the Crop- if Known)

The market for henna is quite good. The fresh leaves are sold directly to the market. The powdered form must be packaged properly after packing them in proper packages. The powder henna can also be sold to different brands for marketing.

MARKETING

Marketed preparations - Over 20 Unani, 4 Ayurvedic, and several Cosmetic marketed preparations containing Lawsonia are available in the market (All, 1996).

FUTURE RESEARCH THRUST

The very low toxicity of henna and its strong roots in tradition make it one of the few natural dyes for which demand is still growing. The ongoing research for innocuous natural dye may add to its present uses. If more attention were given to the selection of cultivars with a high lawsone content and to the development of better drying and processing techniques, it should be possible to expand its commercial production to more humid areas.

Cultivation of henna is profitable under low rainfall conditions and is an important source of income for farmers in drought-prone arid and semi-arid regions. Other benefits associated with henna cultivation include drought proofing, improved land use through sustainable forestry (henna plantations have an average productive life of about 15 years and come under short rotation forestry practices), and economic uses of marginal and gravely lands or wastelands.

Figure 1. (xvii) Colored photograph of plant and plant parts which are used for medicinal purposes
Lawsonia inermis L.

REFERENCES

Biswas, K., Chattopadhyay, I., Banerjee, R. K., & Bandyopadhyay, U. (2002). Biological activities and medicinal properties of neem (*Azadirachta indica*). *Current Science*, *82*(11), 1336–1345.

Duke's Phytochemical and Ethnobotanical Databases. (n.d.). *National Germplasm Resources Laboratory*. Available from: http://wwwarsgringov/duke/ plantshtml

Ganaie, K. A., & Nawachoo, I. A. (2003). Traditional treatment of toothache by Gujjar and Bakerwal tribes of Kashmir in India. *Journal of Economic and Taxonomic Botany*, *27*, 105–107.

Khare, C. P. (2004). Indian Herbal remedies, Rational Western Therapy, Ayurvedic and other traditional Usage, Botany. *Springer–Verlag*, *5*, 6.

Leung, A. Y. (1980). *Encyclopeida on common natural ingredients used in food, drugs and cosmetics*. John Willey and Sons.

Nadkarni, K. M. (1976). *Indian Materia Medica* (3rd ed.). PopularPrakashan.

Shanmugasundaram, K. R., Seethapathy, P. G., & Shanmugasundaram, E. R. (1987). Anna Pavala Sindhooram—An antiatherosclerotic Indian drug. *Journal of Ethnopharmacology*, *7*(3), 247–265. doi:10.1016/0378-8741(83)90001-6 PMID:6876850

Sukla, M., Regar, R. L., & Jangid, B. L. (2012). Henna (*Lawsonia inermis* L.) Cultivation: Available agri-enterprise in arid fringes of western Rajasthan. *Den News, 14*(2).

Compilation of References

A European Strategy for Plastics in a Circular Economy. (2018). *European Union.*

Abánades, A., Rubbia, C., & Salmieri, D. (2013). Thermal cracking of methane into hydrogen for a CO2-free utilization of natural gas. *International Journal of Hydrogen Energy*, *38*(20), 8491–8496. doi:10.1016/j.ijhydene.2012.08.138

Abbasi, T., & Abbasi, S. A. (2011). 'Renewable' hydrogen: Prospects and challenges. *Renewable & Sustainable Energy Reviews*, *15*(6), 3034–3040. doi:10.1016/j.rser.2011.02.026

Abbas, N., Safdar, W., Ali, S., Choudhry, S., & Ilahi, S. (2016). Citric acid production from *Aspergillus niger* using banana peel. *International Journal of Scientific and Engineering Research*, *7*(1), 1580–1583.

Abd-Elsalam, K. A., & Mohamed, M. A. (2001). Magnetic nanostructures. *Anais da Academia Brasileira de Ciências*. Advance online publication. doi:10.15900001-37652001000300026

Abdullah, A., & Mat, H. (2008). Characterisation of solid and liquid pineapple waste. *Reaktor*, *12*(1), 48–52. doi:10.14710/reaktor.12.1.48-52

Abejon, R. (2018). A Bibliometric Study of Scientific Publications regarding Hemicellulose Valorization during the 2000–2016 Period: Identification of Alternatives and Hot Topics. *ChemEngineering*, *2*(1), 7. doi:10.3390/chemengineering2010007

Aboagye, E. A., Chea, J. D., & Yenkie, K. M. (2021). Systems Level Roadmap for Solvent Recovery and Reuse in Industries. *iScience*, *24*(10), 103114. doi:10.1016/j.isci.2021.103114 PMID:34622166

Abou-Zeid, A. Z. A., Baghlaf, A. O., Khan, J. A., & Makhashin, S. S. (1983). Utilization of date seeds and cheese whey in production of citric acid by *Candida lipolytica. Agricultural Wastes*, *8*(3), 131–142. doi:10.1016/0141-4607(83)90112-9

Agarwal, S. (2020). Biodegradable Polymers: Present Opportunities and Challenges in Providing a Microplastic-Free Environment. *Macromolecular Chemistry and Physics*, *221*(6), 2000017. doi:10.1002/macp.202000017

Ahmad, I., & Mahanwar, P. A. (2010). Mechanical properties of fly ash filled high density polyethylene. *Journal of Minerals & Materials Characterization & Engineering*, *9*(03), 183–198. doi:10.4236/jmmce.2010.93016

Ajanovic, A. (2008). On the economics of hydrogen from renewable energy sources as an alternative fuel in transport sector in Austria. *International Journal of Hydrogen Energy*, *33*(16), 4223–4234. doi:10.1016/j.ijhydene.2008.06.041

Ajorloo, M., Ghodrat, M., & Kang, W. H. (2021). Incorporation of recycled polypropylene and fly ash in polypropylene-based composites for automotive applications. *Journal of Polymers and the Environment*, *29*(4), 1298–1309. doi:10.100710924-020-01961-y

Akbar, H. D. A., Pudjiraharti, S., Karina, M., Putri, O. D., & Fauziyyah, R. H. (2019). Fabrication and characterization of sweet potato starch-based bioplastics plasticized with glycerol. *The Journal of Biological Sciences*, *19*, 57–64.

Akhtar, M. N., & Tarannum, N. (2018). Flyash as a resource material in construction industry: A clean approach to environment management. In *Sustainable construction and building materials*. IntechOpen.

Alazemi, J., & Andrews, J. (2015). Automotive hydrogen fuelling stations: An international review. *Renewable & Sustainable Energy Reviews*, *48*, 483–499. doi:10.1016/j.rser.2015.03.085

Alcantara, A. R., & de Maria, P. D. (2018). Recent advances on the use of 2-methyltetrahydrofuran (2-MeTHF) in biotransformations. *Current Green Chemistry*, *5*(2), 86–103. doi:10.2174/2213346105666180727100924

Ali, A., Mahar, R. B., Soomro, R. A., & Sherazi, S. T. H. (2017). Fe3O4 nanoparticles facilitated anaerobic digestion of organic fraction of municipal solid waste for enhancement of methane production. *Energy Sources. Part A, Recovery, Utilization, and Environmental Effects*, *39*(16), 1815–1822. doi:10.1080/15567036.2017.1384866

Ali, M., Nelson, A. R., Lopez, A. L., & Sack, D. A. (2015). Updated global burden of cholera in endemic countries. *PLoS Neglected Tropical Diseases*, *9*(6), 1–13. doi:10.1371/journal.pntd.0003832 PMID:26043000

Ali, S. R., Anwar, Z., Irshad, M., Mukhtar, S., & Warraich, N. T. (2016). Bio-synthesis of citric acid from single and co-culture-based fermentation technology using agro-wastes. *Journal of Radiation Research and Applied Sciences*, *9*(1), 57–62. doi:10.1016/j.jrras.2015.09.003

Almakki, A., Mirghani, M. E., & Kabbashi, N. A. (2019). Production of citric acid from sugarcane molasses by *Aspergillus niger* using submerged fermentation. *Biological and Natural Resources Engineering Journal*, *2*(1), 47–55.

Amato, A., Becci, A., & Beolchini, F. (2020). Citric acid bioproduction: The technological innovation change. *Critical Reviews in Biotechnology*, *40*(2), 199–212. doi:10.1080/07388551.2019.1709799 PMID:31903797

American Chemical Society Green Chemistry Institute. (n.d.). *Pharmaceutical Roundtable Solvent Selection Guide*. https://www.acs.org/content/acs/en/greenchemistry/industry-business/pharmaceutical.html

Anand, K., Kaviyarasu, K., Muniyasamy, S., Roopan, S. M., Gengan, R. M., & Chuturgoon, A. A. (2017). Bio-Synthesis of Silver Nanoparticles Using Agroforestry Residue and Their Catalytic Degradation for Sustainable Waste Management. *Journal of Cluster Science*, *28*(4), 2279–2291. doi:10.100710876-017-1212-2

Anastas, P. T., & Warner, J. C. (1998). Green chemistry. *Frontiers*, *640*, 1998.

Anastas, P., & Eghbali, N. (2010). Green chemistry: Principles and practice. *Chemical Society Reviews*, *39*(1), 301–312. doi:10.1039/B918763B PMID:20023854

Anderson, E. M., Larsson, K. M., & Kirk, O. (1998). One Biocatalyst–Many Applications: The Use of candida Antarctica B-Lipase in Organic Synthesis. *Biocatalysis and Biotransformation*, *16*(3), 181–204. doi:10.3109/10242429809003198

Andrady, A. L. (2003). *Plastics and the Environment* (1st ed.). Wiley-Interscience. doi:10.1002/0471721557

Andrady, A. L. (2011). Microplastics in the marine environment. *Marine Pollution Bulletin*, *62*(8), 1596–1605. doi:10.1016/j.marpolbul.2011.05.030 PMID:21742351

Andrady, A. L., & Neal, M. A. (2009). Applications and societal benefits of plastics. *Philosophical Transactions of the Royal Society of London. Series B, Biological Sciences*, *364*(1526), 1977–1984. doi:10.1098/rstb.2008.0304 PMID:19528050

Andress, R. J., Huang, X., Bequette, B. W., & Martin, L. L. (2009). A systematic methodology for the evaluation of alternative thermochemical cycles for hydrogen production. *International Journal of Hydrogen Energy*, *34*(9), 4146–4154. doi:10.1016/j.ijhydene.2008.11.118

Angumeenal, A. R., & Venkappayya, D. (2013). An overview of citric acid production. *Lebensmittel-Wissenschaft + Technologie*, *50*(2), 367–370. doi:10.1016/j.lwt.2012.05.016

Arasaretnam, S. (2020). *Preparation of biobased plastic from banana peel and application in industrial waste water purification*. Academic Press.

Aravantinos-Zafiris, G., Tzia, C., Oreopoulou, V., & Thomopoulos, C. D. (1994). Fermentation of orange processing wastes for citric acid production. *Journal of the Science of Food and Agriculture*, *65*(1), 117–120. doi:10.1002/jsfa.2740650117

Arikan & Bilgen. (2019). *Production of bioplastic from potato peel waste and investigation of its biodegradability*. Academic Press.

Arnason, B., & Sigfusson, T. I. (2000). Iceland—A future hydrogen economy. *International Journal of Hydrogen Energy*, *25*(5), 389–394. doi:10.1016/S0360-3199(99)00077-4

Arslan, N. P., Aydoğan, M. N., & Taşkın, M. (2016). Citric acid production from partly deproteinized whey under non-sterile culture conditions using immobilized cells of lactose-positive and cold-adapted *Yarrowia lipolytica* B9. *Journal of Biotechnology*, *231*, 32–39. doi:10.1016/j.jbiotec.2016.05.033 PMID:27234881

Atikler, U., Basalp, D., & Tihminlioğlu, F. (2006). Mechanical and morphological properties of recycled high-density polyethylene, filled with calcium carbonate and fly ash. *Journal of Applied Polymer Science*, *102*(5), 4460–4467. doi:10.1002/app.24772

Attarde, S., Marathe, S., & Sil, A. (2014). Utilization of fly ash in construction industries for environment management. *International Journal of Environmental*, *3*(2), 117–121.

Avila-Ortiz, C. G., & Juaristi, E. (2020). Novel methodologies for chemical activation in organic synthesis under solvent-free reaction conditions. *Molecules (Basel, Switzerland)*, *25*(16), 3579. doi:10.3390/molecules25163579 PMID:32781678

Awwad, N. S., El-Kader, M. F. H. A., Ibrahium, H. A., Asnag, G. M., & Morsi, M. A. (2021). Green synthesis of different ratios from bimetallic gold: Silver nanoparticles core@shell via laser ablation scattered in Chitosan-PVA matrix and its electrical conductivity behavior. *Composites Communications*, *24*(February), 100678. doi:10.1016/j.coco.2021.100678

Ayeni, A., Daramola, M. O., Taiwo, O., Olanrewaju, O. I., Oyekunle, D. T., Sekoai, P. T., & Elehinafe, F. B. (2019). Production of citric acid from the fermentation of pineapple waste by *Aspergillus niger. The Open Chemical Engineering Journal*, *13*(1), 88–96. doi:10.2174/1874123101913010088

Azieyanti. (2020). *Mechanical and Morphology Studies of Bioplastic-Based Banana Peels*. Academic Press.

Baglari, S., Kole, M., & Dey, T. K. (2011). Effective thermal conductivity and coefficient of linear thermal expansion of high-density polyethylene—Fly ash composites. *Indian Journal of Physics*, *85*(4), 559–573. doi:10.100712648-011-0059-x

Baker, C. R. (1980). *Production of hydrogen for the commercial market: current and future trends*. Academic Press.

Balat, M. (2008). Potential importance of hydrogen as a future solution to environmental and transportation problems. *International Journal of Hydrogen Energy*, *33*(15), 4013–4029. doi:10.1016/j.ijhydene.2008.05.047

Ball, M., & Weeda, M. (2015). The hydrogen economy–vision or reality? *International Journal of Hydrogen Energy*, *40*(25), 7903–7919. doi:10.1016/j.ijhydene.2015.04.032

Barabadi, H., Najafi, M., Samadian, H., Azarnezhad, A., Vahidi, H., Mahjoub, M. A., … Ahmadi, A. (2019). A systematic review of the genotoxicity and antigenotoxicity of biologically synthesized metallic nanomaterials: Are green nanoparticles safe enough for clinical marketing? *Medicina (Lithuania)*, *55*(8). doi:10.3390/medicina55080439

Barclay, F. J. (2006). *Fuel cells, engines and hydrogen: an exergy approach*. John Wiley & Sons. doi:10.1002/9780470030264

Bargigli, S., Raugei, M., & Ulgiati, S. (2004). Comparison of thermodynamic and environmental indexes of natural gas, syngas and hydrogen production processes. *Energy*, *29*(12-15), 2145–2159. doi:10.1016/j.energy.2004.03.018

Bates, E. D., Mayton, R. D., Ntai, I., & Davis, J. H. (2002). CO2 capture by a task-specific ionic liquid. *Journal of the American Chemical Society*, *124*(6), 926–927. doi:10.1021/ja017593d PMID:11829599

Beevi. (2020). *Bioplastic Synthesis Using Banana Peels And Potato Starch And Characterization*. Academic Press.

Beghi, G. (1983). An introduction to hydrogen technology. In *Hydrogen-energy vector of the future* (pp. 5–12). Conference.

Behera, B. C. (2020). Citric acid from *Aspergillus niger*: A comprehensive overview. *Critical Reviews in Microbiology*, *46*(6), 727–749. doi:10.1080/1040841X.2020.1828815 PMID:33044884

Behera, B. C., Mishra, R., & Mohapatra, S. (2021). Microbial citric acid: Production, properties, application, and future perspectives. *Food Frontiers*, *2*(1), 62–76. doi:10.1002/fft2.66

Benjamin, J. S. (1970). Dispersion strengthened superalloys by mechanical alloying. *Metallurgical Transactions*, *1*(10), 2943–2951. doi:10.1007/BF03037835

Bényei, A., & Joó, F. (1990). Organometallic catalysis in aqueous solutions: The biphasic transfer hydrogenation of aldehydes catalyzed by water-soluble phosphine complexes of ruthenium, rhodium and iridium. *Journal of Molecular Catalysis*, *58*(2), 151–163. doi:10.1016/0304-5102(90)85035-G

Bertel, E., Lee, K. S., & Nordborg, C. (2004). Nuclear energy in the hydrogen economy. *NEA News*, 10-13.

Bhandari, R., Trudewind, C. A., & Zapp, P. (2014). Life cycle assessment of hydrogen production via electrolysis–a review. *Journal of Cleaner Production*, *85*, 151–163. doi:10.1016/j.jclepro.2013.07.048

Bharmoria, P., & Ventura, S. P. M. (2021). *Bio-based aerogels for environmental remediation problems, Advances in Aerogel Composites for Environmental Remediation*. doi:10.1016/B978-0-12-820732-1.00017-5

Biehl, P., von der Lühe, M., Dutz, S., & Schacher, F. H. (2018). Synthesis, characterization, and applications of magnetic nanoparticles featuring polyzwitterionic coatings. *Polymers*, *10*(1), 91. Advance online publication. doi:10.3390/polym10010091 PMID:30966126

Biswas, K., Chattopadhyay, I., Banerjee, R. K., & Bandyopadhyay, U. (2002). Biological activities and medicinal properties of neem (*Azadirachta indica*). *Current Science*, *82*(11), 1336–1345.

Bonda, S., Mohanty, S., & Nayak, S. K. (2012). Viscoelastic, mechanical, and thermal characterization of fly ash-filled ABS composites and comparison of fly ash surface treatments. *Polymer Composites*, *33*(1), 22–34. doi:10.1002/pc.21242

Bose, T., & Malbrunot, P. (2007). *Hydrogen: facing the energy challenges of the 21st century.* John Libbey Eurotext.

Breslow, R. (1991). Hydrophobic effects on simple organic reactions in water. *Accounts of Chemical Research, 24*(6), 159–164. doi:10.1021/ar00006a001

Buhler, D. R., & Reed, D. J. (1990). *Ethel Browning's toxicity and metabolism of industrial solvents.* Elsevier.

Busto, E., Gotor-Fernández, V., & Gotor, V. (2006). Enantioselective Synthesis of 4-(Dimethylamino)pyridines through a Chemical Oxidation-Enzymatic Reduction Sequence. Application in Asymmetric Catalysis. *Advanced Synthesis & Catalysis, 348*(18), 2626–2632. doi:10.1002/adsc.200600274

Byrne, F. P., Jin, S., Paggiola, G., Petchey, T. H., Clark, J. H., Farmer, T. J., ... Sherwood, J. (2016). Tools and techniques for solvent selection: green solvent selection guides. *Sustainable Chemical Processes, 4*(1), 1-24.

Byrne, C. M., Allen, S. D., Lobkovsky, E. B., & Coates, G. W. (2004). Alternating Copolymerization of Limonene Oxide and Carbon Dioxide. *Journal of the American Chemical Society, 126*(37), 11404–11405. doi:10.1021/ja0472580 PMID:15366863

Bystrzanowska, M., Pena-Pereira, F., Marcinkowski, Ł., & Tobiszewski, M. (2019). How green are ionic liquids?–A multicriteria decision analysis approach. *Ecotoxicology and Environmental Safety, 174*, 455–458. doi:10.1016/j.ecoenv.2019.03.014 PMID:30852310

Camp, J. E. (2018). Bio-available Solvent Cyrene: Synthesis, Derivatization, and Applications. *ChemSusChem, 11*(18), 3048–3055. doi:10.1002/cssc.201801420 PMID:30044553

Capello, C., Fischer, U., & Hungerbühler, K. (2007). What is a green solvent? A comprehensive framework for the environmental assessment of solvents. *Green Chemistry, 9*(9), 927–934. doi:10.1039/b617536h

Carissimi, G., Montalbán, M. G., Fuster, M. G., & Víllora, G. (2022). *Nanoparticles as Drug Delivery Systems, 21st Century Nanostructured Materials - Physics*. Chemistry, Classification, and Emerging Applications in Industry, Biomedicine, and Agriculture. doi:10.5772/intechopen.100253

Cavallo, E., Charreau, H., Cerrutti, P., & Foresti, M. L. (2017). *Yarrowia lipolytica*: A model yeast for citric acid production. *FEMS Yeast Research, 17*(8), fox084. doi:10.1093/femsyr/fox084 PMID:29106510

Cavallo, E., Nobile, M., Cerrutti, P., & Foresti, M. L. (2020). Exploring the production of citric acid with *Yarrowia lipolytica* using corn wet milling products as alternative low-cost fermentation media. *Biochemical Engineering Journal, 155*, 107463. doi:10.1016/j.bej.2019.107463

Chahal, K., & Mishra, R. (2021). Challenges and Future Prospects of Nanotechnology for Antiviral Drugs. *International Journal of Applied Pharmaceutical Sciences and Research.* https://www.sierrajournals.com/index.php/IJAPSR/issue/view/91 doi:10.21477/ijapsr.6.2.2

Chandarana & Chandra. (2021). *Production of Bioplastics from Banana Peels*. Academic Press.

Chand, N., Sharma, P., & Fahim, M. (2010). Correlation of mechanical and tribological properties of organosilane modified cenosphere filled high density polyethylene. *Materials Science and Engineering A*, *527*(21-22), 5873–5878. doi:10.1016/j.msea.2010.06.022

Chandra, R. (1998). Biodegradable polymers. *Progress in Polymer Science*, *23*(7), 1273–1335. doi:10.1016/S0079-6700(97)00039-7

Chaowasakoo, T., & Sombatsompop, N. (2007). Mechanical and morphological properties of fly ash/epoxy composites using conventional thermal and microwave curing methods. *Composites Science and Technology, 67*(11-12), 2282-2291.

Chaudhary, A., Beckman, E. J., & Russell, A. J. (1995). Rational Control of Polymer Molecular Weight and Dispersity during Enzyme-Catalyzed Polyester Synthesis in Supercritical Fluids. *Journal of the American Chemical Society*, *117*(13), 3728–3733. doi:10.1021/ja00118a009

Chen, J., Spear, S. K., Huddleston, J. G., & Rogers, R. D. (2005). Polyethylene glycol and solutions of polyethylene glycol as green reaction media. *Green Chemistry*, *7*(2), 64–82. doi:10.1039/b413546f

Chiesa, P., Consonni, S., Kreutz, T., & Williams, R. (2005). Co-production of hydrogen, electricity and CO2 from coal with commercially ready technology. Part A: Performance and emissions. *International Journal of Hydrogen Energy*, *30*(7), 747–767. doi:10.1016/j.ijhydene.2004.08.002

Chinh, N. T., Mai, T. T., Trang, N. T. T., Giang, N. V., Trung, T. H., Huong, N. T. T., & Hoang, T. (2016). Tensile, electrical properties and morphology of polyethylene/modified fly ash composites using ultraflow. *Vietnam Journal of Chemistry*, *54*(6), 776.

Chysirichote, T. (2020). Valorization of banana peel for citric acid production under solid state fermentation with *Aspergillus niger. Chemical and Biochemical Engineering Quarterly*, *34*(1), 49–57. doi:10.15255/CABEQ.2019.1740

Cigolotti, V., Mcphail, S. J., & Tommasino, M. C. (2018). *HyLAW National Policy Chapter—Italy*. HyLaw Project.

Cintas, P., Tabasso, S., Veselov, V. V., & Cravotto, G. (2020). Alternative reaction conditions: Enabling technologies in solvent-free protocols. *Current Opinion in Green and Sustainable Chemistry*, *21*, 44–49. doi:10.1016/j.cogsc.2019.11.007

Cipriani, G., Di Dio, V., Genduso, F., La Cascia, D., Liga, R., Miceli, R., & Galluzzo, G. R. (2014). Perspective on hydrogen energy carrier and its automotive applications. *International Journal of Hydrogen Energy*, *39*(16), 8482–8494. doi:10.1016/j.ijhydene.2014.03.174

Clarke, C. J., Tu, W. C., Levers, O., Brohl, A., & Hallett, J. P. (2018). Green and sustainable solvents in chemical processes. *Chemical Reviews*, *118*(2), 747–800. doi:10.1021/acs.chemrev.7b00571 PMID:29300087

Cohce, M. K., Dincer, I., & Rosen, M. A. (2010). Thermodynamic analysis of hydrogen production from biomass gasification. *International Journal of Hydrogen Energy, 3*(10), 4970–4980. doi:10.1016/j.ijhydene.2009.08.066

Coppola, G., Gaudio, M. T., Lopresto, C. G., Calabro, V., Curcio, S., & Chakraborty, S. (2021). *Bioplastic from Renewable Biomass: A Facile Solution for a Greener Environment*. Earth Systems and Environment. doi:10.100741748-021-00208-7

Cortes-Clerget, M., Yu, J., Kincaid, J. R., Walde, P., Gallou, F., & Lipshutz, B. H. (2021). Water as the reaction medium in organic chemistry: From our worst enemy to our best friend. *Chemical Science (Cambridge), 12*(12), 4237–4266. doi:10.1039/D0SC06000C PMID:34163692

Cox, K. E., & Williamson, K. D. (Eds.). (1997). *Hydrogen, Its Technology and Implications: Implication of Hydrogen Energy*. CRC Press LLC.

Cvjetko Bubalo, M., Vidović, S., Radojčić Redovniković, I., & Jokić, S. (2015). Green solvents for green technologies. *Journal of Chemical Technology and Biotechnology (Oxford, Oxfordshire), 90*(9), 1631–1639. doi:10.1002/jctb.4668

da Silva, D. A., de Paula, R. C., & Feitosa, J. P. (2007). Graft copolymerisation of acrylamide onto cashew gum. *European Polymer Journal, 43*(6), 2620–2629. doi:10.1016/j.eurpolymj.2007.03.041

da Silva, L. V., Coelho, M. A. Z., da Silva, M. R. S., & Amaral, P. F. F. (2020). Investigation of mitochondrial protein expression profiles of *Yarrowia lipolytica* in response to citric acid production. *Bioprocess and Biosystems Engineering, 43*(9), 1703–1715. doi:10.100700449-020-02363-z PMID:32363419

Dachineni, R. (2016). *Cellular Uptake Mechanisms for Hydrophobically Modified Carbohydrate Polymer Based Nanoparticles*. doi:10.26226/morressier.57d6b2bfd462b8028d88ce64

Dalebrook, A. F., Gan, W., Grasemann, M., Moret, S., & Laurenczy, G. (2013). Hydrogen storage: Beyond conventional methods. *Chemical Communications, 49*(78), 8735–8751. doi:10.1039/c3cc43836h PMID:23964360

Darensbourg, D. J. (2007). Making Plastics from Carbon Dioxide: Salen Metal Complexes as Catalysts for the Production of Polycarbonates from Epoxides and CO2. *Chemical Reviews, 107*(6), 2388–2410. doi:10.1021/cr068363q PMID:17447821

Darvish, H. (2019). Bibliometric Analysis using Bibliometrix an R Package. *Journal of Scientometric Res., 8*(3), 156–160. doi:10.5530/jscires.8.3.32

Das, K., Ray, D., Adhikary, K., Bandyopadhyay, N. R., Mohanty, A. K., & Misra, M. (2010). Development of recycled polypropylene matrix composites reinforced with fly ash. *Journal of Reinforced Plastics and Composites, 29*(4), 510–517. doi:10.1177/0731684408099415

Deepthi, A., & Sathi, V. (2021). Supercritical Fluids as Green Solvents. In *Green Organic Reactions* (pp. 323–338). Springer. doi:10.1007/978-981-33-6897-2_17

Deepthi, M. V., Sharma, M., Sailaja, R. R. N., Anantha, P., Sampathkumaran, P., & Seetharamu, S. (2010). Mechanical and thermal characteristics of high density polyethylene–fly ash cenospheres composites. *Materials & Design*, *31*(4), 2051–2060. doi:10.1016/j.matdes.2009.10.014

Degnan, T., & Shinde, S. L. (2019). Waste-plastic processing provides global challenges and opportunities. *MRS Bulletin*, *44*(06), 436–437. doi:10.1557/mrs.2019.133

Delogu, F., Gorrasi, G., & Sorrentino, A. (2017). Fabrication of polymer nanocomposites via ball milling: Present status and future perspectives. *Progress in Materials Science*, *86*, 75–126. doi:10.1016/j.pmatsci.2017.01.003

Denmark, S. E., & Wynn, T. (2001). Lewis base activation of lewis acids: Catalytic enantioselective allylation and propargylation of aldehydes. *Accounts of Chemical Research*, *2002*(35), 209. doi:10.1021/ar000145a

Deshayes, S., Liagre, M., Loupy, A., Luche, J. L., & Petit, A. (1999). Microwave activation in phase transfer catalysis. *Tetrahedron*, *55*(36), 10851–10870. doi:10.1016/S0040-4020(99)00601-8

Dhillon, G. S., Brar, S. K., Kaur, S., & Verma, M. (2013). Screening of agro-industrial wastes for citric acid bioproduction by *Aspergillus niger* NRRL 2001 through solid state fermentation. *Journal of the Science of Food and Agriculture*, *93*(7), 1560–1567. doi:10.1002/jsfa.5920 PMID:23108761

Dhillon, G. S., Brar, S. K., Verma, M., & Tyagi, R. D. (2011a). Enhanced solid-state citric acid bio-production using apple pomace waste through surface response methodology. *Journal of Applied Microbiology*, *110*(4), 1045–1055. doi:10.1111/j.1365-2672.2011.04962.x PMID:21294819

Dhillon, G. S., Brar, S. K., Verma, M., & Tyagi, R. D. (2011b). Apple pomace ultrafiltration sludge–A novel substrate for fungal bioproduction of citric acid: Optimisation studies. *Food Chemistry*, *128*(4), 864–871. doi:10.1016/j.foodchem.2011.03.107

Djukić-Vuković, A., Mladenović, D., Ivanović, J., Pejin, J., & Mojović, L. (2019). Towards sustainability of lactic acid and poly-lactic acid polymers production. *Renewable & Sustainable Energy Reviews*, *108*, 238–252. doi:10.1016/j.rser.2019.03.050

Dolzhenko, A. V. (2021). Green solvents for eco-friendly synthesis of bioactive heterocycles. In *Green Synthetic Approaches for Biologically Relevant Heterocycles* (pp. 393–470). Elsevier. doi:10.1016/B978-0-12-820792-5.00007-X

Dourou, M., Kancelista, A., Juszczyk, P., Sarris, D., Bellou, S., Triantaphyllidou, I. E., Rywinska, A., Papanikolaou, S., & Aggelis, G. (2016). Bioconversion of olive mill wastewater into high-added value products. *Journal of Cleaner Production*, *139*, 957–969. doi:10.1016/j.jclepro.2016.08.133

Duan, H., Song, G., Qu, S., Dong, X., & Xu, M. (2019). Post-consumer packaging waste from express delivery in China. *Resources, Conservation and Recycling*, *144*, 137–143. doi:10.1016/j.resconrec.2019.01.037

Duke's Phytochemical and Ethnobotanical Databases. (n.d.). *National Germplasm Resources Laboratory.* Available from: http://wwwarsgringov/duke/ plantshtml

Eaves, D. (2004). *Handbook of polymer foams*. Rapra Technology Ltd.

Eberhardt, R., Allmendinger, M., & Rieger, B. (2003). DMAP/Cr(III) Catalyst Ratio: The Decisive Factor for Poly(propylene carbonate) Formation in the Coupling of CO2 and Propylene Oxide. *Macromolecular Rapid Communications*, *24*(2), 194–196. doi:10.1002/marc.200390022

Effenberger, F. B., Couto, R. A., Kiyohara, P. K., Machado, G., Masunaga, S. H., Jardim, R. F., & Rossi, L. M. (2017). Economically attractive route for the preparation of high quality magnetic nanoparticles by the thermal decomposition of iron(III) acetylacetonate. *Nanotechnology*, *28*(11), 115603. Advance online publication. doi:10.1088/1361-6528/aa5ab0 PMID:28192283

El-Holi, M. A., & Al-Delaimy, S. (2003). Citric acid production from whey with sugars and additives by *Aspergillus niger. African Journal of Biotechnology*, *2*(10), 356–359. doi:10.5897/AJB2003.000-1073

El-malek, F. A., Khairy, H., Farag, A., & Omar, S. (2020). The sustainability of microbial bioplastics, production and applications. *International Journal of Biological Macromolecules*, *157*, 319–328. doi:10.1016/j.ijbiomac.2020.04.076 PMID:32315677

El-Samragy, Y. A., Khorshid, M. A., Foda, M. I., & Shehata, A. E. (1996). Effect of fermentation conditions on the production of citric acid from cheese whey by *Aspergillus niger. International Journal of Food Microbiology*, *29*(2-3), 411–416. doi:10.1016/0168-1605(95)00072-0 PMID:8796442

Eskander, S. (2017). *EI-Din H, Saleh M. Environmental Science and Engineering* (Vol. 8). Studium Press.

Ez-Zahraoui, S., Kassab, Z., Ablouh, E. H., Sehaqui, H., Bouhfid, R., Alami, J., & Qaiss, A. E. K. (2021). Effect of fly ash and coupling agent on the structural, morphological, thermal, and mechanical properties of polyamide 6/acrylonitrile-butadiene-styrene blend. *Polymer Composites*.

Faber, K. (2004). *Biotransformations in Organic Chemistry* (5th ed.). Springer-Verlag. doi:10.1007/978-3-642-18537-3

Fang, J., Pan, L., Gu, Q. X., Juengpanich, S., Zheng, J. H., Tong, C. H., Wang, Z. Y., Nan, J. J., & Wang, Y. F. (2020). Scientometric analysis of mTOR signaling pathway in liver disease. *Annals of Translational Medicine*, *8*(4), 93. doi:10.21037/atm.2019.12.110 PMID:32175386

Favis, B. D., & Corroller, P. (2017). *Polymeric material and process for recycling plastic blends*. U.S. Patent 9670344.

Fazlzadeh, M., Rahmani, K., Zarei, A., Abdoallahzadeh, H., Nasiri, F., & Khosravi, R. (2017). A novel green synthesis of zero valent iron nanoparticles (NZVI) using three plant extracts and their efficient application for removal of Cr(VI) from aqueous solutions. *Advanced Powder Technology*, *28*(1), 122–130. doi:10.1016/j.apt.2016.09.003

Federal Highway Administration. (2017). *Department of Transportation's Federal Highway Administration Privacy Policy*. https://www.google.nl/ url?sa = t&source = web & rct = j&url = https:// www.fhwa.dot.gov/ pavement/ recycling/ fach01. Cfm & ved = 2ahU KEwj okpa 7wZn uAhW s7XM BHVQ SAWo QFjA jegQ IKhAB & usg = AOvV aw3p G8U4 gmdq U8iK NvRH 6xR6

Femi-Ola, T. O., Oluyege, J. O., & Gbadebo, A. O. (2009). Citric acid production from pineapple waste. *Continental Journal of Microbiology*, *3*, 1–5.

Feng, G. R., Peng, J. J., Qiu, H. Y., Jiang, J. X., Tao, L., & Lai, G. Q. (2007). Synthesis of novel greener functionalized ionic liquids containing appended hydroxyl. *Synthetic Communications*, *37*(16), 2671–2675. doi:10.1080/00397910701465230

Ferreira, P., Lopes, M., Mota, M., & Belo, I. (2016). Oxygen mass transfer impact on citric acid production by *Yarrowia lipolytica* from crude glycerol. *Biochemical Engineering Journal*, *110*, 35–42. doi:10.1016/j.bej.2016.02.001

Fickers, P., Cheng, H., & Sze Ki Lin, C. (2020). Sugar alcohols and organic acids synthesis in *Yarrowia lipolytica*: Where are we? *Microorganisms*, *8*(4), 574. doi:10.3390/microorganisms8040574 PMID:32326622

Filly, A., Fabiano-Tixier, A. S., Louis, C., Fernandez, X., & Chemat, F. (2016). Water as a green solvent combined with different techniques for extraction of essential oil from lavender flowers. *Comptes Rendus. Chimie*, *19*(6), 707–717. doi:10.1016/j.crci.2016.01.018

Fly ash market size, share & industry analysis, by type (class F and Class C), by application (cement & concretes fills & embarkments, waste stabilization, mining, oil field service and road stabilization and others) and regional forecast, 2020-2027. (2020). https://www.fortunebusinessinsights.com/industry-reports/fly-ash-market-101087-

Fly ash market-growth, trends, covid-19 impact and forecasts 2021-2026. (n.d.). https://www.mordorintelligence.com/industry-reports/fly-ash-market

Folino. (2020). *Biodegradation of Wasted Bioplastics in Natural and Industrial Environments: A Review*. Academic Press.

Francis, R. (2016). *Recycling of Polymers: Methods, Characterization and Applications*. John Wiley & Sons. doi:10.1002/9783527689002

Ganaie, K. A., & Nawachoo, I. A. (2003). Traditional treatment of toothache by Gujjar and Bakerwal tribes of Kashmir in India. *Journal of Economic and Taxonomic Botany*, *27*, 105–107.

Gan, Y. X., Jayatissa, A. H., Yu, Z., Chen, X., & Li, M. (2020). Hydrothermal Synthesis of Nanomaterials. *Journal of Nanomaterials*, *2020*, 1–3. Advance online publication. doi:10.1155/2020/8917013

Gao, X., Hiraoka, T., Ohmagari, S., Tanaka, S., Sheng, Z., Liu, K., Xu, M., Chen, P., & Hokamoto, K. (2019). High-efficiency production of large-size few-layer graphene platelets via pulsed discharge of graphite strips. *Nanomaterials (Basel, Switzerland)*, *9*(12), 1785. Advance online publication. doi:10.3390/nano9121785 PMID:31888135

Gasiorek, E. (2008). Effect of operating conditions on biomass growth during citric acid production by solid-state fermentation. *Chemical Papers*, *62*(2), 141–146. doi:10.247811696-008-0003-5

Geresh, S., Gdalevsky, G. Y., Gilboa, I., Voorspoels, J., Remon, J. P., & Kost, J. (2004). Bioadhesive grafted starch copolymers as platforms for peroral drug delivery: A study of theophylline release. *Journal of Controlled Release*, *94*(2–3), 391–399. doi:10.1016/j.jconrel.2003.10.019 PMID:14744489

Globe News Wire. (2018). https://www.globenewswire.com/ news- release/ 2018/ 10/ 04/ 1601143/ 0/en/ Global- Fly- Ash- Market- Will- Reach- USD- 11-0- Billion- by- 2024- Zion- Market- Research.html

Goh, C. K., Valavan, S. E., Low, T. K., & Tang, L. H. (2016). Effects of different surface modification and contents on municipal solid waste incineration fly ash/epoxy composites. *Waste Management, 58*, 309-315.

Gohatre, O. K., Biswal, M., Mohanty, S., & Nayak, S. K. (2020). Study on thermal, mechanical and morphological properties of recycled poly (vinyl chloride)/fly ash composites. *Polymer International*, *69*(6), 552–563. doi:10.1002/pi.5988

Gohatre, O. K., Biswal, M., Mohanty, S., & Nayak, S. K. (2021). Effect of silane treated fly ash on physico-mechanical, morphological, and thermal properties of recycled poly (vinyl chloride) composites. *Journal of Applied Polymer Science*, *138*(19), 50387. doi:10.1002/app.50387

Golsby, A. (2015). *New Generation Polymer Technology Available to be Used in Plaster and Cementitious Fibre and Non-Fibre Spray On and Grout Systems*. Academic Press.

Gómez-López, P., Puente-Santiago, A., Castro-Beltrán, A., do Nascimento, L. A. S., Balu, A. M., Luque, R., & Alvarado-Beltrán, C. G. (2020). Nanomaterials and catalysis for green chemistry. *Current Opinion in Green and Sustainable Chemistry*, *24*, 48–55. doi:10.1016/j.cogsc.2020.03.001

Groote, P. D., Devaux, J., & Godard, P. (2002). Effect of Benzenesulfonamide Plasticizers on the glass- transition temperature of semicrystalline polydodecamide. *Journal of Polymer Science. Part B, Polymer Physics*, *40*(19), 2208–2218. doi:10.1002/polb.10281

Guhanathan, S., Devi, M. S., & Murugesan, V. (2001). Effect of coupling agents on the mechanical properties of fly ash/polyester particulate composites. *Journal of Applied Polymer Science*, *82*(7), 1755–1760. doi:10.1002/app.2017

Gummadi, J., Kumar, G. V., & Rajesh, G. (2012). Evaluation of flexural properties of fly ash filled polypropylene composites. *International Journal of Modern Engineering Research*, *2*(4), 2584–2590.

Guo, C., Qiao, J., Zhang, S., Ren, X., & Li, M. (2018). Purification of polyphenols from kiwi fruit peel extracts using macroporous resins and high-performance liquid chromatography analysis. *International Journal of Food Science & Technology*, *53*(6), 1486–1493. doi:10.1111/ijfs.13729

Gupta, P. K. (2020). Solvents, Gasses, and Vapors. In *Problem Solving Questions in Toxicology* (pp. 181–193). Springer. doi:10.1007/978-3-030-50409-0_13

Gu, Y., & Jérôme, F. (2010). Glycerol as a sustainable solvent for green chemistry. *Green Chemistry*, *12*(7), 1127–1138. doi:10.1039/c001628d

Häckl, K., & Kunz, W. (2018). Some aspects of green solvents. *Comptes Rendus. Chimie*, *21*(6), 572–580. doi:10.1016/j.crci.2018.03.010

Hamdy, H. S. (2013). Citric acid production by *Aspergillus niger* grown on orange peel medium fortified with cane molasses. *Annals of Microbiology*, *63*(1), 267–278. doi:10.100713213-012-0470-3

Hang, Y. D., Luh, B. S., & Woodams, E. E. (1987). Microbial production of citric acid by solid state fermentation of kiwifruit peel. *Journal of Food Science*, *52*(1), 226–227. doi:10.1111/j.1365-2621.1987.tb14014.x

Hang, Y. D., Splittstoesser, D. F., Woodams, E. E., & Sherman, R. M. (1977). Citric acid fermentation of brewery waste. *Journal of Food Science*, *42*(2), 383–384. doi:10.1111/j.1365-2621.1977.tb01504.x

Hang, Y. D., & Woodams, E. E. (1984). Apple pomace: A potential substrate for citric acid production by *Aspergillus niger. Biotechnology Letters*, *6*(11), 763–764. doi:10.1007/BF00133071

Hang, Y. D., & Woodams, E. E. (1986). Solid state fermentation of apple pomace for citric acid production. *MIRCEN Journal of Applied Microbiology and Biotechnology*, *2*(2), 283–287. doi:10.1007/BF00933494

Hang, Y. D., & Woodams, E. E. (1998). Production of citric acid from corncobs by *Aspergillus niger. Bioresource Technology*, *65*(3), 251–253. doi:10.1016/S0960-8524(98)00015-7

Hang, Y. D., & Woodams, E. E. (2000). Corn husks: A potential substrate for production of citric acid by *Aspergillus niger. Lebensmittel-Wissenschaft + Technologie*, *33*(7), 520–521. doi:10.1006/fstl.2000.0711

Hang, Y. D., & Woodams, E. E. (2001). Enzymatic enhancement of citric acid production by *Aspergillus niger* from corn cobs. *Lebensmittel-Wissenschaft + Technologie*, *34*(7), 484–486. doi:10.1006/fstl.2001.0784

Hansen, B. B., Spittle, S., Chen, B., Poe, D., Zhang, Y., Klein, J. M., Horton, A., Adhikari, L., Zelovich, T., Doherty, B. W., Gurkan, B., Maginn, E. J., Ragauskas, A., Dadmun, M., Zawodzinski, T. A., Baker, G. A., Tuckerman, M. E., Savinell, R. F., & Sangoro, J. R. (2020). Deep eutectic solvents: A review of fundamentals and applications. *Chemical Reviews*, *121*(3), 1232–1285. doi:10.1021/acs.chemrev.0c00385 PMID:33315380

Hartonen, K., & Riekkola, M. L. (2017). Water as the first choice green solvent. In *The Application of Green Solvents in Separation Processes* (pp. 19–55). Elsevier. doi:10.1016/B978-0-12-805297-6.00002-4

Hassan, M. D., Castanha, R. C. G., & Wolfram, D. (2020). Scientometric analysis of global trypanosomiasis research: 1988–2017. *Journal of Infection and Public Health, 13*(4), 514–520. doi:10.1016/j.jiph.2019.10.006 PMID:31831393

Hauk, P., Wencel-Delord, J., Ackermann, L., Walde, P., & Gallou, F. (2021). Organic Synthesis in Aqueous Multiphase Systems-Challenges and Opportunities ahead of Us. *Current Opinion in Colloid & Interface Science*, *56*, 101506. doi:10.1016/j.cocis.2021.101506

Hayashi, T., Kanai, H., Yodoya, S., Oka, M., & Hayashi, T. (2002). Biodegradation of random co-polypeptide hydrogels consisting of N-hydroxypropyl l-glutamine as one component. *European Polymer Journal*, *38*(1), 139–146. doi:10.1016/S0014-3057(01)00137-9

Hebeish, A., Abd El-Thalouth, I., El-Kashouti, M. A., & Abdel-Fattah, S. H. (1979). Graft copolymerization of acrylonitrile onto starch using potassium permanganate as initiator. *Angewandte Makromolekulare Chemie*, *78*(1), 101–108. doi:10.1002/apmc.1979.050780107

Herndon, J. M., & Whiteside, M. (2017). Further evidence of coal fly ash utilization in tropospheric geoengineering: Implications on human and environmental health. *J Geog Environ Earth Sci Intn*, *9*(1), 1–8. doi:10.9734/JGEESI/2017/31417

Hesham, A. E. L., Mostafa, Y. S., & AlSharqi, L. E. O. (2020). Optimization of citric acid production by immobilized cells of novel yeast isolates. *Mycobiology*, *48*(2), 122–132. doi:10.1080/12298093.2020.1726854 PMID:32363040

Hijosa-Valsero, M., Paniagua-García, A. I., & Díez-Antolínez, R. (2017). Biobutanol production from apple pomace: The importance of pretreatment methods on the fermentability of lignocellulosic agro-food wastes. *Applied Microbiology and Biotechnology*, *101*(21), 8041–8052. doi:10.100700253-017-8522-z PMID:28932978

Holz, M., Förster, A., Mauersberger, S., & Barth, G. (2009). Aconitase overexpression changes the product ratio of citric acid production by *Yarrowia lipolytica. Applied Microbiology and Biotechnology*, *81*(6), 1087–1096. doi:10.100700253-008-1725-6 PMID:18850095

Hopewell, J., Dvorak, R., & Kosior, E. (2009). Plastics recycling: Challenges and opportunities. *Philosophical Transactions of the Royal Society of London. Series B, Biological Sciences*, *364*(1526), 2115–2126. doi:10.1098/rstb.2008.0311 PMID:19528059

Howard, G. T. (2002). Biodegradation of polyurethane: A review. *International Biodeterioration & Biodegradation*, *49*(4), 245–252. doi:10.1016/S0964-8305(02)00051-3

Hume, A. S., & Ho, K. (2019). Toxicity of solvents. In *Basic Environmental Toxicology* (pp. 157–184). CRC Press. doi:10.1201/9781315138091-6

Hu, W., Li, W. J., Yang, H. Q., & Chen, J. H. (2019). Current strategies and future prospects for enhancing microbial production of citric acid. *Applied Microbiology and Biotechnology*, *103*(1), 201–209. doi:10.100700253-018-9491-6 PMID:30421107

Huzaisham & Marsi. (2020). *Utilization of banana (Musa paradisiaca) peel as bioplastic for planting bag application*. Academic Press.

Ilyas, R., Sapuan, S., Ibrahim, R., Abral, H., Ishak, M., Zainudin, E., Atikah, M., Mohd Nurazzi, N., Atiqah, A., Ansari, M., Syafri, E., Asrofi, M., Sari, N. H., & Jumaidin, R. (2019). Effect of sugar palm nanofibrillated cellulose concentrations on morphological, mechanical and physical properties of biodegradable films based on agro-waste sugar palm (Arenga pinnata (Wurmb.) Merr) starch. *Journal of Materials Research and Technology*, *8*(5), 4819–4830. doi:10.1016/j.jmrt.2019.08.028

Imandi, S. B., Bandaru, V. V. R., Somalanka, S. R., Bandaru, S. R., & Garapati, H. R. (2008). Application of statistical experimental designs for the optimization of medium constituents for the production of citric acid from pineapple waste. *Bioresource Technology*, *99*(10), 4445–4450. doi:10.1016/j.biortech.2007.08.071 PMID:17936623

Iordache, I., Schitea, D., Floristean, A., Stefănescu, I., & Brahy, N. (2018). INC-DTCI-ICSI RM. Vâlcea participation in hylaw project. *Smart Energy and Sustainable Environment*, *21*(1), 75–82.

Iqbal, J., & Utara, U. (2015). Isolation of *Aspergillus niger* strains from soil and their screening and optimization for enhanced citric acid production using cane molasses as carbon source. *Journal of Applied Environmental and Biological Sciences*, *5*(4), 128–137.

Isroi, A. C., & Panji, T. (2016). Bioplastic production from oil palm empty fruit bunch. *International Conference on Biomass*.

Isroi, A. C., Panji, T., Wibowo, N. A., & Syamsu, K. (2017). Bioplastic production from cellulose of oil palm empty fruit bunch. *IOP Conference Series. Earth and Environmental Science*, *65*, 11. doi:10.1088/1755-1315/65/1/012011

Jabali, K. A., Ashiq, M., Ahmad, S., & Rehman, S. U. (2020). A Bibliometric Analysis of Research Productivity on Diabetes Modeling and Artificial Pancreas 2001 to 2020. *Library Philosophy and Practice (e-journal),* 4305. https://digitalcommons.unl.edu/libphilprac/4305

Jahangirian, H., Lemraski, E. G., Webster, T. J., Rafiee-Moghaddam, R., & Abdollahi, Y. (2017). A review of drug delivery systems based on nanotechnology and green chemistry: Green nanomedicine. *International Journal of Nanomedicine*, *12*, 2957–2978. doi:10.2147/IJN.S127683 PMID:28442906

Jamir, L., Kumar, V., Kaur, J., Kumar, S., & Singh, H. (2021). Composition, valorization and therapeutical potential of molasses: A critical review. *Environmental Technology Reviews*, *10*(1), 131–142. doi:10.1080/21622515.2021.1892203

Javed, S., Asgher, M., Sheikh, M. A., & Nawaz, H. (2010). Strain improvement through UV and chemical mutagenesis for enhanced citric acid production in molasses-based solid state fermentation. *Food Biotechnology*, *24*(2), 165–179. doi:10.1080/08905436.2010.482014

Jianlong, W., Xianghua, W., & Ding, Z. (2000). Production of citric acid from molasses integrated with in-situ product separation by ion-exchange resin adsorption. *Bioresource Technology*, *75*(3), 231–234. doi:10.1016/S0960-8524(00)00067-5

Jimenez-Gonzalez, C. (2019). Life cycle considerations of solvents. *Current Opinion in Green and Sustainable Chemistry*, *18*, 66–71. doi:10.1016/j.cogsc.2019.02.004

Jung, J. H., Lee, J. H., & Shinkai, S. (2011). Functionalized magnetic nanoparticles as chemosensors and adsorbents for toxic metal ions in environmental and biological fields. *Chemical Society Reviews*, *40*(9), 4464–4474. doi:10.1039/c1cs15051k PMID:21607241

Kalia, S., Sharma, A., & Kaith, B. S. (2007). Ab initio study of gas phase and water-assisted tautomerization of maleimide and formamide. *Journal of Chemical Sciences*, *119*(6), 617–624. doi:10.100712039-007-0077-6

Kamal, M., Sharma, C. S., Upadhyaya, P., Verma, V., Pandey, K. N., Kumar, V., & Agrawal, D. D. (2012). Calcium carbonate (CaCO3) nanoparticle filled polypropylene: Effect of particle surface treatment on mechanical, thermal, and morphological performance of composites. *Journal of Applied Polymer Science*, *124*(4), 2649–2656. doi:10.1002/app.35319

Kamzolova, S. V., Fatykhova, A. R., Dedyukhina, E. G., Anastassiadis, S. G., Golovchenko, N. P., & Morgunov, I. G. (2011). Citric acid production by yeast grown on glycerol-containing waste from biodiesel industry. *Food Technology and Biotechnology*, *49*(1), 65–74.

Kamzolova, S. V., Vinokurova, N. G., Lunina, J. N., Zelenkova, N. F., & Morgunov, I. G. (2015). Production of technical-grade sodium citrate from glycerol-containing biodiesel waste by *Yarrowia lipolytica*. *Bioresource Technology*, *193*, 250–255. doi:10.1016/j.biortech.2015.06.092 PMID:26141285

Kang, H. M., Cai, Y. L., & Liu, P. S. (2006). Synthesis, characterization and thermal sensitivity of chitosan-based graft copolymers. *Carbohydrate Research*, *341*(17), 2851–2857. doi:10.1016/j.carres.2006.08.001 PMID:17027942

Kanoujiya & Khanna. (2019). *Production of Bioplastic from banana peels*. Academic Press.

Kareem, S. O., Akpan, I., & Alebiowu, O. O. (2010). Production of citric acid by *Aspergillus niger* using pineapple waste. *Malaysian Journal of Microbiology*, *6*(2), 161–165.

Kareem, S. O., & Rahman, R. A. (2013). Utilization of banana peels for citric acid production by *Aspergillus niger*. *Agriculture and Biology Journal of North America*, *4*(4), 384–387. doi:10.5251/abjna.2013.4.4.384.387

Karthikeyan, A., & Sivakumar, N. (2010). Citric acid production by Koji fermentation using banana peel as a novel substrate. *Bioresource Technology*, *101*(14), 5552–5556. doi:10.1016/j.biortech.2010.02.063 PMID:20219361

Kastiawan, I., Sutantra, I. N., & Sutikno. (2020). Correlation of Holding Time and Bottom Ash Particle Size to Mechanical Properties of Polypropylene Composite. *Key Engineering Materials*, *867*, 172–181. doi:10.4028/www.scientific.net/KEM.867.172

Ke, L., Lu, C., Shen, R., Lu, T., Ma, B., & Hua, Y. (2020). Knowledge Mapping of Drug-Induced Liver Injury: A Scientometric Investigation (2010-2019). *Frontiers in Pharmacology*, *11*, 842. doi:10.3389/fphar.2020.00842 PMID:32581801

Khare, C. P. (2004). Indian Herbal remedies, Rational Western Therapy, Ayurvedic and other traditional Usage, Botany. *Springer–Verlag*, *5*, 6.

Khare, S. K., Jha, K., & Gandhi, A. P. (1995). Citric acid production from okara (soy-residue) by solid-state fermentation. *Bioresource Technology*, *54*(3), 323–325. doi:10.1016/0960-8524(95)00155-7

Khatami, K., Perez-Zabaleta, M., Owusu-Agyeman, I., & Cetecioglu, Z. (2021). Waste to bioplastics: How close are we to sustainable polyhydroxyalkanoates production? *Waste Management (New York, N.Y.)*, *119*, 374–388. doi:10.1016/j.wasman.2020.10.008 PMID:33139190

Khatoon, N., Jamal, A., & Ali, M. I. (2018). Lignin peroxidase isoenzyme: A novel approach to biodegrade the toxic synthetic polymer waste. *Environmental Technology*, *40*(11), 1366–1375. doi:10.1080/09593330.2017.1422550 PMID:29291675

Khoshnoud, P., & Abu-Zahra, N. (2019). The effect of particle size of fly ash (FA) on the interfacial interaction and performance of PVC/FA composites. *Journal of Vinyl and Additive Technology*, *25*(2), 134–143. doi:10.1002/vnl.21633

Khosravi-Darani, K., & Zoghi, A. (2008). Comparison of pretreatment strategies of sugarcane baggase: Experimental design for citric acid production. *Bioresource Technology*, *99*(15), 6986–6993. doi:10.1016/j.biortech.2008.01.024 PMID:18334291

Kiel, H., Guvrin, R., & Henis, Y. (1981). Citric acid fermentation by *Aspergillus niger* on low sugar concentrations and cotton waste. *Applied and Environmental Microbiology*, *42*(1), 1–4. doi:10.1128/aem.42.1.1-4.1981 PMID:16345802

Kim, K. H., Lee, H. Y., & Lee, C. Y. (2015). Pretreatment of sugarcane molasses and citric acid production by *Candida zeylanoides. Microbiology and Biotechnology Letters*, *43*(2), 164–168. doi:10.4014/mbl.1503.03006

Kindermann, N., Cristòfol, L., & Kleij, A. W. (2017). Access to Biorenewable Polycarbonates with Unusual Glass-Transition Temperature (Tg) Modulation. *ACS Catalysis*, *7*(6), 3860–3863. doi:10.1021/acscatal.7b00770

Klibanov, A. M. (2001). Improving enzymes by using them in organic solvents. *Nature*, *409*(6817), 241–246. doi:10.1038/35051719 PMID:11196652

Kohl, B. E. (2018). *H2-Herstellung; voestalpine Beitrag zum HyLaw Workshop*. Voestalpine AG.

Kolicheski, M. B., Soccol, C. R., Martin, B., Medeiros, T., & Rainbault, M. (1997). Citric acid production on three cellulosic supports in solid state fermentation. In S. Roussos, B. K. Lonsane, M. Raimbault, & G. Viniegra-Gonzalez (Eds.), *Advances in solid state fermentation* (pp. 449–462). Kluwer Academic Publishers. doi:10.1007/978-94-017-0661-2_37

Kotov, Y. A. (2003). Electric explosion of wires as a method for preparation of nanopowders. *Journal of Nanoparticle Research*, *5*(5–6), 539–550. doi:10.1023/B:NANO.0000006069.45073.0b

Koutinas, M., Patsalou, M., Stavrinou, S., & Vyrides, I. (2015). High temperature alcoholic fermentation of orange peel by the newly isolated thermotolerant *Pichia kudriavzevii* KVMP 10. *Letters in Applied Microbiology*, *62*(1), 75–83. doi:10.1111/lam.12514 PMID:26510181

Krasik, Y. E., Fedotov, A., Sheftman, D., Efimov, S., Sayapin, A., Gurovich, V. T., Veksler, D., Bazalitski, G., Gleizer, S., Grinenko, A., & Oreshkin, V. I. (2010). Underwater electrical wire explosion. *Plasma Sources Science & Technology*, *19*(3), 034020. Advance online publication. doi:10.1088/0963-0252/19/3/034020

Kreutz, T., Williams, R., Consonni, S., & Chiesa, P. (2005). Co-production of hydrogen, electricity and CO2 from coal with commercially ready technology. Part B: Economic analysis. *International Journal of Hydrogen Energy*, *30*(7), 769–784. doi:10.1016/j.ijhydene.2004.08.001

Kulkarni, M. B., Bambole, V. A., & Mahanwar, P. A. (2014). Effect of particle size of fly ash cenospheres on the properties of acrylonitrile butadiene styrene-filled composites. *Journal of Thermoplastic Composite Materials*, *27*(2), 251–267. doi:10.1177/0892705712443253

Kulkarni, M. B., & Mahanwar, P. A. (2016). Studies on effect of titanate-coupling agent (0.5, 1.5, and 2.5%) on the mechanical, thermal, and morphological properties of fly ash–filled polypropylene composites. *Journal of Thermoplastic Composite Materials*, *29*(3), 344–365. doi:10.1177/0892705713518795

Kulkarni, S. M., & Kishore. (2002). Effects of surface treatments and size of fly ash particles on the compressive properties of epoxy based particulate composites. *Journal of Materials Science*, *37*(20), 4321–4326. doi:10.1023/A:1020648418233

Kumar, D., Jain, V. K., Shanker, G., & Srivastava, A. (2003). Utilisation of fruits waste for citric acid production by solid state fermentation. *Process Biochemistry*, *38*(12), 1725–1729. doi:10.1016/S0032-9592(02)00253-4

Kumar, D., Verma, R., & Bhalla, T. C. (2010). Citric acid production by *Aspergillus niger* van. Tieghem MTCC 281 using waste apple pomace as a substrate. *Journal of Food Science and Technology*, *47*(4), 458–460. doi:10.100713197-010-0077-2 PMID:23572671

Kumari, S. C., Dhand, V., & Padma, P. N. (2021). Green synthesis of metallic nanoparticles: A review. *Nanomaterials (Basel, Switzerland)*, 259–281. doi:10.1016/B978-0-12-822401-4.00022-2

Kumar, P. N., Nagappan, V., & Karthikeyan, C. (2019). Effects of fly ash, calcium carbonate fillers on mechanical, moisture absorption properties in poly vinyl chloride resin. *Materials Today: Proceedings*, *16*, 1219–1225. doi:10.1016/j.matpr.2019.05.217

Kumar, S. S., Swapna, T. S., & Sabu, A. (2018). Coffee husk: A potential agro-industrial residue for bioprocess. In R. Singhania, R. Agarwal, R. Kumar, & R. Sukumaran (Eds.), *Waste to wealth. Energy, environment, and sustainability* (pp. 97–109). Springer. doi:10.1007/978-981-10-7431-8_6

Kumar, V., Kumar, P., & Singh, J. (2020). *Environmental Degradation: Causes and Remediation Strategies, Environmental Degradation: Causes and Remediation Strategies*. doi:10.26832/aesa-2020-edcrs

Kurbanoglu, E. B. (2004). Enhancement of citric acid production with ram horn hydrolysate by *Aspergillus niger. Bioresource Technology*, *92*(1), 97–101. doi:10.1016/j.biortech.2003.08.002 PMID:14643992

Kurbanoglu, E. B., & Kurbanoglu, N. I. (2003). Production of citric acid from ram horn hydrolysate by *Aspergillus niger. Process Biochemistry*, *38*(10), 1421–1424. doi:10.1016/S0032-9592(03)00029-3

Kushwah, K. S., & Verma, D. K. (2022). *Biological Synthesis of Metallic Nanoparticles from Different Plant Species, 21st Century Nanostructured Materials - Physics*. Chemistry, Classification, and Emerging Applications in Industry, Biomedicine, and Agriculture. doi:10.5772/intechopen.101355

Laurent, S., Forge, D., Port, M., Roch, A., Robic, C., Vander Elst, L., & Muller, R. N. (2008). Magnetic iron oxide nanoparticles: Synthesis, stabilization, vectorization, physicochemical characterizations and biological applications. *Chemical Reviews*, *108*(6), 2064–2110. doi:10.1021/cr068445e PMID:18543879

Laycock, B., Nikolić, M., Colwell, J. M., Gauthier, E., Halley, P., Bottle, S., & George, G. (2017). Lifetime prediction of biodegradable polymers. *Progress in Polymer Science*, *71*, 144–189. doi:10.1016/j.progpolymsci.2017.02.004

Leung, A. Y. (1980). *Encyclopeida on common natural ingredients used in food, drugs and cosmetics*. John Willey and Sons.

Linko, Y. Y., Wang, Z. L., & Seppälä, J. (1995). Lipase-catalyzed synthesis of poly(1,4-butyl sebacate) from sebacic acid or its derivatives with 1,4-butanediol. *Journal of Biotechnology*, *40*(2), 133–138. doi:10.1016/0168-1656(95)00039-S

Liu, X., Lv, J., Xu, J., Xia, J., He, A., Zhang, T., Xu, X., & Xu, J. (2018). Effects of osmotic pressure and pH on citric acid and erythritol production from waste cooking oil by *Yarrowia lipolytica. Engineering in Life Sciences*, *18*(6), 344–352. doi:10.1002/elsc.201700114 PMID:32624914

Liu, X., Lv, J., Xu, J., Zhang, T., Deng, Y., & He, J. (2015c). Citric acid production in *Yarrowia lipolytica* SWJ-1b yeast when grown on waste cooking oil. *Applied Biochemistry and Biotechnology*, *175*(5), 2347–2356. doi:10.100712010-014-1430-0 PMID:25488499

Liu, X., Lv, J., Zhang, T., & Deng, Y. (2015a). Citric acid production from hydrolysate of pretreated straw cellulose by *Yarrowia lipolytica* SWJ-1b using batch and fed-batch cultivation. *Preparative Biochemistry & Biotechnology*, *45*(8), 825–835. doi:10.1080/10826068.2014.979203 PMID:25356914

Liu, X., Wang, X., Xu, J., Xia, J., Lv, J., Zhang, T., Wu, Z., Deng, Y., & He, J. (2015b). Citric acid production by *Yarrowia lipolytica* SWJ-1b using corn steep liquor as a source of organic nitrogen and vitamins. *Industrial Crops and Products*, *78*, 154–160. doi:10.1016/j.indcrop.2015.10.029

Li, V. C. (2019a). *Introduction to Engineered Cementitious Composites (ECC), Engineered Cementitious Composites (ECC)*. doi:10.1007/978-3-662-58438-5

Li, Y., Wang, Z., Sun, L., Liu, L., Xu, C., & Kuang, H. (2019). Nanoparticle-based sensors for food contaminants. *Trends in Analytical Chemistry*, *113*, 74–83. doi:10.1016/j.trac.2019.01.012

Li, Z. L., Wang, W., Li, M., Zhang, M. J., Tang, M. J., Su, Y. Y., Liu, Z., Ju, X.-J., Xie, R., & Chu, L. Y. (2018). Facile Fabrication of Bubble-Propelled Micromotors Carrying Nanocatalysts for Water Remediation. *Industrial & Engineering Chemistry Research*, *57*(13), 4562–4570. doi:10.1021/acs.iecr.7b04941

Li, Z., Chen, Z., Yang, N., Wei, K., Ling, Z., Liu, Q., Chen, G., & Ye, B. H. (2021). Trends in research on the carbon footprint of higher education: A bibliometric analysis (2010-2019). *Journal of Cleaner Production*, *289*, 125642. doi:10.1016/j.jclepro.2020.125642

Lomba, L., Giner, B., Bandrés, I., Lafuente, C., & Pino, M. R. (2011). Physicochemical properties of green solvents derived from biomass. *Green Chemistry*, *13*(8), 2062–2070. doi:10.1039/c0gc00853b

Loupy, A. (2004). Solvent-free microwave organic synthesis as an efficient procedure for green chemistry. *Comptes Rendus. Chimie*, *7*(2), 103–112. doi:10.1016/j.crci.2003.10.015

Lu, A. H., Salabas, E. L., & Schüth, F. (2007). Magnetic nanoparticles: Synthesis, protection, functionalization, and application. *Angewandte Chemie International Edition*, *46*(8), 1222–1244. doi:10.1002/anie.200602866 PMID:17278160

Lucas, N., Bienaime, C., Belloy, C., Queneudec, M., Silvestre, F., & Nava-Saucedo, J. E. (2008). Polymer biodegradation: Mechanisms and estimation techniques – A review. *Chemosphere*, *73*(4), 429–442. doi:10.1016/j.chemosphere.2008.06.064 PMID:18723204

Lu, F., & Astruc, D. (2020). Nanocatalysts and other nanomaterials for water remediation from organic pollutants. *Coordination Chemistry Reviews*, *408*, 213180. doi:10.1016/j.ccr.2020.213180

Magangana, T. P., Makunga, N. P., Fawole, O. A., & Opara, U. L. (2020). Processing factors affecting the phytochemical and nutritional properties of pomegranate (*Punica granatum L.*) peel waste: A review. *Molecules (Basel, Switzerland)*, *25*(20), 4690. doi:10.3390/molecules25204690 PMID:33066412

Mahato, N., Sharma, K., Sinha, M., Dhyani, A., Pathak, B., Jang, H., Park, S., Pashikanti, S., & Cho, S. (2021). Biotransformation of citrus waste-I: Production of biofuel and valuable compounds by fermentation. *Processes (Basel, Switzerland)*, *9*(2), 220. doi:10.3390/pr9020220

Malazy, O. T., Ramezani, A., Atlasi, R., Larijani, B., & Abdollahi, M. (2016). Scientometric study of academic publications on antioxidative herbal medicines in type 2 diabetes mellitus. *Journal of Diabetes and Metabolic Disorders*, *15*(1), 48. doi:10.118640200-016-0273-3 PMID:27785446

Malik, M. A., Wani, M. Y., & Hashim, M. A. (2012). Microemulsion method: A novel route to synthesize organic and inorganic nanomaterials. 1st Nano Update. *Arabian Journal of Chemistry*, *5*(4), 397–417. doi:10.1016/j.arabjc.2010.09.027

Malolan, R., Gopinath, K. P., Vo, D. V. N., Jayaraman, R. S., Adithya, S., Ajay, P. S., & Arun, J. (2021). Green ionic liquids and deep eutectic solvents for desulphurization, denitrification, biomass, biodiesel, bioethanol and hydrogen fuels: A review. *Environmental Chemistry Letters*, *19*(2), 1001–1023. doi:10.100710311-020-01113-7

Manimaran. (2016). *Production of Biodegradable plastic from banana peel.* Academic Press.

Maruyama, E. (n.d.). Nanotechnology research promotion at RIKEN - past, present and future. *Proceedings of the 2nd IEEE Conference on Nanotechnology*. 10.1109/NANO.2002.1032269

Masruri, M., Azhar, A. Z., Rosyada, I., & Febrianto, A. (2019). The effect of kaffir lime (*Citrus hystrix* DC) essential oil on bioplastic derived from cassava peel waste. *Journal of Physics: Conference Series*, *1374*(1), 012015. doi:10.1088/1742-6596/1374/1/012015

Mas-Tur, A., Roig-Tierno, N., Sarin, S., Haon, C., Sego, T., Belkhouja, M., Porter, A., & Merigo, J. M. (2021). Co-citation, bibliographic coupling and leading authors, institutions and countries in the 50 years of Technological Forecasting and Social Change. *Technological Forecasting and Social Change*, *165*, 120487. doi:10.1016/j.techfore.2020.120487

Masuhiro, T., Shafiul, I., Takayuki, A., Alessandra, B., Giuliano, F., & … . (2005).. . *AUTEX Research Journal*, *5*, 40.

Matlack, A. (2021). *Introduction to Green Chemistry* (2nd ed.). T&F India.

Maulida, M. S., & Tarigan, P. (2016). Production of starch based bioplastic from cassava peel reinforced with microcrystalline cellulose avicel PH101 using sorbitol as plasticizer. *Journal of Physics: Conference Series*, *710*, 012012. doi:10.1088/1742-6596/710/1/012012

McCarty, R. D., Cox, K. E., & Williamson, K. D. (2019). Hydrogen: Its Technology and Implications: Hydrogen Properties. *International Journal of Innovative Research in the Humanities*, *1*(1), 22–28.

McNaught, A. D., & Wilkinson, A. (1997). *Compendium of Chemical Terminology (IUPAC Chemical Data)* (2nd ed.). Wiley.

Medeirosgarciaalcântara, J., Distante, F., Storti, G., Moscatelli, D., Morbidelli, M., & Sponchioni, M. (2020). Current trends in the production of biodegradable bioplastics: The case of polyhydroxyalkanoates. *Biotechnology Advances*, *42*, 107582. doi:10.1016/j.biotechadv.2020.107582 PMID:32621947

Mehyar, G. F., Delaimy, K. S., & Ibrahim, S. A. (2005). Citric acid production by *Aspergillus niger* using date-based medium fortified with whey and additives. *Food Biotechnology*, *19*(2), 137–144. doi:10.1081/FBT-200063458

Mishra, S., & Shimpi, N. G. (2005). *Comparison of nano CaCO 3 and flyash filled with styrene butadiene rubber on mechanical and thermal properties*. Academic Press.

Mohammadzadeh, M. (2009). *Characterization of Recycled Thermoplastic Polymers* [Master's Thesis]. University of Borås, Borås, Sweden.

Monrroy, M., Rueda, L., Aparicio, A. L., & García, J. R. (2019). Fermentation of *Musa paradisiaca* peels to produce citric acid. *Journal of Chemistry*, *2019*, 1–8. doi:10.1155/2019/8356712

Mores, S., de Souza Vandenberghe, L. P., Júnior, A. I. M., de Carvalho, J. C., de Mello, A. F. M., Pandey, A., & Soccol, C. R. (2020). Citric acid bioproduction and downstream processing: Status, opportunities, and challenges. *Bioresource Technology*, 124426. PMID:33249260

Mostafa, A. M., Mwafy, E. A., & Toghan, A. (2021). ZnO nanoparticles decorated carbon nanotubes via pulsed laser ablation method for degradation of methylene blue dyes. *Colloids and Surfaces. A, Physicochemical and Engineering Aspects*, *627*(July), 127204. doi:10.1016/j.colsurfa.2021.127204

Mueller, N. C., & Nowack, B. (2010). Nanoparticles for Remediation: Solving Big Problems with Little Particles. *Elements*, *6*(6), 395–400. doi:10.2113/gselements.6.6.395

Mülhaupt, R. (2012). Green Polymer Chemistry and Bio-based Plastics: Dreams and Reality. *Macromolecular Chemistry and Physics*, *214*(2), 159–174. doi:10.1002/macp.201200439

Müller, R. J., Kleeberg, I., & Deckwer, W. D. (2001). Biodegradation of polyesters containing aromatic constituents. *Journal of Biotechnology*, *86*(2), 87–95. doi:10.1016/S0168-1656(00)00407-7 PMID:11245897

Nadkarni, K. M. (1976). *Indian Materia Medica* (3rd ed.). PopularPrakashan.

Nagendiran, S., Badghaish, A., Hussein, I. A., Shuaib, A. N., Furquan, S. A., & Al-Mehthel, M. H. (2016). Epoxy/oil fly ash composites prepared through in situ polymerization: Enhancement of thermal and mechanical properties. *Polymer Composites*, *37*(2), 512–522. doi:10.1002/pc.23207

Nakano, R., Ito, S., & Nozaki, K. (2014). Copolymerization of carbon dioxide and butadiene via a lactone intermediate. *Nature Chemistry*, *6*(4), 325–331. doi:10.1038/nchem.1882 PMID:24651200

Narancic, T., Cerrone, F., Beagan, N., & O'Connor, K. E. (2020). Recent Advances in Bioplastics: Application and Biodegradation. *Polymers*, *12*(4), 920. doi:10.3390/polym12040920 PMID:32326661

Nath, D. C. D., Bandyopadhyay, S., Yu, A., Zeng, Q., Das, T., Blackburn, D., & White, C. (2009). Structure–property interface correlation of fly ash–isotactic polypropylene composites. *Journal of Materials Science*, *44*(22), 6078–6089. doi:10.100710853-009-3839-3

Nath, D., & Banerjee, P. (2013). Green nanotechnology - A new hope for medical biology. *Environmental Toxicology and Pharmacology*, *36*(3), 997–1014. doi:10.1016/j.etap.2013.09.002 PMID:24095717

Nayak, A., & Bhushan, B. (2019). An overview of the recent trends on the waste valorization techniques for food wastes. *Journal of Environmental Management*, *233*, 352–370. doi:10.1016/j.jenvman.2018.12.041 PMID:30590265

Nayak, B., & Singh, R. (2001). Synthesis and characterization of grafted hydroxypropyl guar gum by ceric ion induced initiation. *European Polymer Journal*, *37*(8), 1655–1666. doi:10.1016/S0014-3057(01)00035-0

Nayak, P. L. (1999). Biodegradable Polymers: Opportunities and Challenges. *Journal of Macromolecular Science, Part C: Polymer Reviews*, *39*(3), 481–505. doi:10.1081/MC-100101425

Ndlovu, N., Banda, L., Muitire, C., Munyengwa, N., & Mayaya, T. (2020). Bibliometric and Scientometric Analysis of Scopus-Indexed Literature on Hemp (Cannabis Sativa) Breeding. International Journal of Plant & Soil Science, 32(17), 29-43.

Nichols, C., Moore, T., & Griffith, S. (2001). *Thermoplastic Polymers with Improved Infrared Reheat Properties*. U.S. Patent 09973436.

Norouzi, N. (2021a). *Analysis of legal considerations and challenges of wastewater recycling management.* 5th International Congress of Developing Agriculture, Natural Resources, Environment and Tourism of Iran, Shiraz, Iran.

Norouzi, N. (2021b). Post-COVID-19 and globalization of oil and natural gas trade: Challenges, opportunities, lessons, regulations, and strategies. *International Journal of Energy Research*, *45*(10), 14338–14356. doi:10.1002/er.6762 PMID:34219899

Norouzi, N. (2021c). An Overview on the renewable hydrogen market. *International Journal of Energy Studies*, *6*(1), 67–94.

Norouzi, N., & Ataei, E. (2021a). Covid-19 Crisis and Environmental law: Opportunities and challenges. *Hasanuddin Law Review*, *7*(1), 46–60. doi:10.20956/halrev.v7i1.2772

Norouzi, N., & Ataei, E. (2021b). Environmental Protection Regulations in the Light of Public Law and Social Obligations. *Research Journal of Ecology and Environmental Sciences*, *1*(1), 1–16.

Norouzi, N., Khanmohammadi, H. U., & Ataei, E. (2021). The Law in the Face of the COVID-19 Pandemic: Early Lessons from Uruguay. *Hasanuddin Law Review*, *7*(2), 75–88. doi:10.20956/halrev.v7i2.2827

Nugroho, A. W., Prasetyo, M. K. P., & Budiyantoro, C. (2020). Effect of Fly Ash on the Mechanical Properties of Polyvinyl Chloride-Fly Ash Composite. In *Proceedings of the 6th International Conference and Exhibition on Sustainable Energy and Advanced Materials* (pp. 667-674). Springer. 10.1007/978-981-15-4481-1_63

Nyika, J. M. (2022). *Nanotechnology and Its Applications in Environmental Remediation, Research Anthology on Emerging Techniques in Environmental Remediation*. doi:10.4018/978-1-6684-3714-8.ch004

O'Hagan, D., & Zaidi, N. A. (1994). Enzyme-catalysed condensation polymerization of 11-hydroxyundecanoic acid with lipase from Candida cylindracea. *Polymer*, *35*(16), 3576–3578. doi:10.1016/0032-3861(94)90928-8

Obileke, K., Onyeaka, H., Omoregbe, O., Makaka, G., Nwokolo, N., & Mukumba, P. (2020). Bioenergy from Bio-waste: A bibliometric analysis of the trend in scientific research from 1998-2018. *Biomass Conversion and Biorefinery*. Advance online publication. doi:10.100713399-020-00832-9

Ochiai, B., Hatano, Y., & Endo, T. (2008). Fixing Carbon Dioxide Concurrently with Radical Polymerization for Utilizing Carbon Dioxide by Low-Energy Cost. *Macromolecules*, *41*(24), 9937–9939. doi:10.1021/ma801960q

Ojeda, T., Freitas, A., Birck, K., Dalmolin, E., Jacques, R., Bento, F., & Camargo, F. (2011). Degradability of linear polyolefins under natural weathering. *Polymer Degradation & Stability*, *96*(4), 703–707. doi:10.1016/j.polymdegradstab.2010.12.004

Oklu, N. K., Matsinha, L. C., & Makhubela, B. C. (2019). *Bio-solvents: synthesis, industrial production and applications*. Solvents, Ionic Liquids and Solvent Effects.

Okuhara, T. (2002). Water-tolerant solid acid catalysts. *Chemical reviews, 102*(10), 3641-3666.36. *Chemical Communications (Cambridge)*, *268*. Advance online publication. doi:10.1039/b108851n

Olah, G. A., Goeppert, A., & Prakash, G. K. S. (2008). Chemical Recycling of Carbon Dioxide to Methanol and Dimethyl Ether: From Greenhouse Gas to Renewable, Environmentally Carbon Neutral Fuels and Synthetic Hydrocarbons. *The Journal of Organic Chemistry*, *74*(2), 487–498. doi:10.1021/jo801260f PMID:19063591

Otto, S., Bertoncin, F., & Engberts, J. B. (1996). Lewis acid catalysis of a Diels– Alder reaction in water. *Journal of the American Chemical Society*, *118*(33), 7702–7707. doi:10.1021/ja960318k

Özdal, M., & Kurbanoglu, E. B. (2019). Citric acid production by *Aspergillus niger* from agro-industrial by-products: Molasses and chicken feather peptone. *Waste and Biomass Valorization*, *10*(3), 631–640. doi:10.100712649-018-0240-y

Pacheco-Fernández, I., & Pino, V. (2019). Green solvents in analytical chemistry. *Current Opinion in Green and Sustainable Chemistry*, *18*, 42–50. doi:10.1016/j.cogsc.2018.12.010

Palit, S. (2019a). *Nanotechnology Research, Industrial Ecology, and the Visionary Future, Chemical Nanoscience and Nanotechnology*. doi:10.1201/9780429398254-7

Palit, S. (2019b). *Nanotechnology Research, Green Engineering, and Sustainability: A Vision for the Future, Chemical Nanoscience and Nanotechnology*. doi:10.1201/9780429398254-8

Pandev, M., Lucchese, P., Mansilla, C., Duigou, A. L., Abrashev, B., & Vladikova, D. (2017). Hydrogen Economy: The future for a sustainable and green society. *Izvestiia po Himiia*, *49*, 84–92.

Pandey, A., Soccol, C. R., Nigam, P., Brand, D., Mohan, R., & Roussos, S. (2000). Biotechnological potential of coffee pulp and coffee husk for bioprocesses. *Biochemical Engineering Journal*, *6*(2), 153–162. doi:10.1016/S1369-703X(00)00084-X PMID:10959086

Papadaki, E., & Mantzouridou, F. T. (2019). Citric acid production from the integration of Spanish-style green olive processing wastewaters with white grape pomace by *Aspergillus niger*. *Bioresource Technology*, *280*, 59–69. doi:10.1016/j.biortech.2019.01.139 PMID:30754006

Papagianni, M. (2007). Advances in citric acid fermentation by *Aspergillus niger*: Biochemical aspects, membrane transport and modeling. *Biotechnology Advances*, *25*(3), 244–263. doi:10.1016/j.biotechadv.2007.01.002 PMID:17337335

Papanikolaou, S., Galiotou-Panayotou, M., Fakas, S., Komaitis, M., & Aggelis, G. (2008). Citric acid production by *Yarrowia lipolytica* cultivated on olive-mill wastewater-based media. *Bioresource Technology*, *99*(7), 2419–2428. doi:10.1016/j.biortech.2007.05.005 PMID:17604163

Parameswaran, B. (2009). Sugarcane bagasse. *Biotechnology for Agro-Industrial Residues Utilisation*, 239-252.

Pardo, S. G., Bernal, C., Ares, A., Abad, M. J., & Cano, J. (2010). Rheological, thermal, and mechanical characterization of fly ash-thermoplastic composites with different coupling agents. *Polymer Composites*, *31*(10), 1722–1730. doi:10.1002/pc.20962

Park, H. M., Li, X., Jin, C. Z., Park, C. Y., Cho, W. J., & Ha, C.-S. (2002). Preparation and properties of biodegradable thermoplastic starch/clay hybrids. *Macromolecular Materials and Engineering*, *287*(8), 553–558. doi:10.1002/1439-2054(20020801)287:8<553::AID-MAME553>3.0.CO;2-3

Parvaiz, M. R., Mohanty, S., Nayak, S. K., & Mahanwar, P. A. (2011). Effect of surface modification of fly ash on the mechanical, thermal, electrical and morphological properties of polyetheretherketone composites. *Materials Science and Engineering A*, *528*(13-14), 4277–4286. doi:10.1016/j.msea.2011.01.026

Pazouki, M., Felse, P. A., Sinha, J., & Panda, T. (2000). Comparative studies on citric acid production by *Aspergillus niger* and *Candida lipolytica* using molasses and glucose. *Bioprocess Engineering*, *22*(4), 353–361. doi:10.1007/PL00009115

PD CEN/TR 15351. (2006). BSI.

Pena-Pereira, F., & Tobiszewski, M. (Eds.). (2017). *The application of green solvents in separation processes*. Elsevier.

Peng, Y. H., Shih, Y. H., Lai, Y. C., Liu, Y. Z., Liu, Y. T., & Lin, N. C. (2014). Degradation of polyurethane by bacterium isolated from soil and assessment of polyurethanolytic activity of a Pseudomonas putida strain. *Environmental Science and Pollution Research International, 21*(16), 9529–9537. doi:10.100711356-014-2647-8 PMID:24633845

Pepè Sciarria, T., Batlle-Vilanova, P., Colombo, B., Scaglia, B., Balaguer, M. D., Colprim, J., Puig, S., & Adani, F. (2018). Bio-electrorecycling of carbon dioxide into bioplastics. *Green Chemistry, 20*(17), 4058–4066. doi:10.1039/C8GC01771A

Perea, L.N., Gaviria, D., & Redondo, M.I. (2020). Bioeconomy: Bibliometric analysis from 2006 to 2019. *Revista ESPACIOS, 41*(43).

Pratt, A. (2014). Environmental applications of magnetic nanoparticles. In *Frontiers of Nanoscience* (pp. 259–307). Elsevier Ltd. doi:10.1016/B978-0-08-098353-0.00007-5

Properties of Polymers: Their Correlation with Chemical Structure; Their Numerical Estimation and Prediction from Additive Group Contributions. (2014). Elsevier Science & Technology.

Prüss-Ustün, A., Wolf, J., Bartram, J., Clasen, T., Cumming, O., Freeman, M. C., Gordon, B., Hunter, P. R., Medlicott, K., & Johnston, R. (2019). Burden of disease from inadequate water, sanitation and hygiene for selected adverse health outcomes: An updated analysis with a focus on low- and middle-income countries. *International Journal of Hygiene and Environmental Health, 222*(5), 765–777. doi:10.1016/j.ijheh.2019.05.004 PMID:31088724

PT Online. (n.d.). https://www.ptonline.com/articles/fly-ash-filler-stages-a-comeback

Pukánszky, B., & Móczó, J. (2004, August). Morphology and properties of particulate filled polymers. *Macromolecular Symposia, 214*(1), 115–134. doi:10.1002/masy.200451009

Qin, F., Du, J., Gao, J., Liu, G., Song, Y., Yang, A., Wang, H., Ding, Y., & Wang, Q. (2020). Bibliometric Profile of Global Microplastics Research from 2004 to 2019. *International Journal of Environmental Research and Public Health, 17*(16), 5639. doi:10.3390/ijerph17165639 PMID:32764298

Rabinovitch, E. B., Isner, J. D., Sidor, J. A., & Wiedl, D. J. (1997). Effect of extrusion conditions on rigid PVC foam. *Journal of Vinyl and Additive Technology, 3*(3), 210–215. doi:10.1002/vnl.10193

Rahaman, M. S., Ahmad, S., Ansari, K. M. N., & Abdel-Magid, I. M. (2020). Scientometrics Profile of the Banasthali Vidyapith: A Deemed University of Rajasthan, India. *Library Philosophy and Practice (e-journal).* 4329. https://digitalcommons.unl.edu/libphilprac/4329

Rahmat, S. (2022). *Effects, uptake, and translocation of selenium-based nanoparticles in plants, Toxicity of Nanoparticles in Plants.* doi:10.1016/B978-0-323-90774-3.00006-4

Ramachandra, Y. L., Narayanamurthy, G., Jois, S., Chavan, A., & Satwadi, P. R. (2013). Production of citric acid in basal coffee husk medium by *Aspergillus niger* under solid state fermentation. *Advances in Biological Research (Faisalabad), 7*(6), 234–240.

Ramakrishnan, N., Sharma, S., Gupta, A., & Alashwal, B. Y. (2018). Keratin based bioplastic film from chicken feathers and its characterization. *International Journal of Biological Macromolecules*, *111*, 352–358. doi:10.1016/j.ijbiomac.2018.01.037 PMID:29320725

Raquez, J. M., Bourgeois, A., Jacobs, H., Degée, P., Alexandre, M., & Dubois, P. (2011). Oxidative degradations of oxodegradable LDPE enhanced with thermoplastic pea starch: Thermo-mechanical properties, morphology, and UV-ageing studies. *Journal of Applied Polymer Science*, *122*(1), 489–496. doi:10.1002/app.34190

Ratti, R. (2014). Ionic liquids: Synthesis and applications in catalysis. *Adv. Chem*, *2014*(3), 1–16. doi:10.1155/2014/729842

Reihmann, M., & Ritter, H. (2006). Synthesis of Phenol Polymers Using Peroxidases. *Enzyme-Catalyzed Synthesis of Polymers*, 1–49.

Revelle, R., & Suess, H. E. (1957). Carbon Dioxide Exchange Between Atmosphere and Ocean and the Question of an Increase of Atmospheric CO2 during the Past Decades. *Tellus*, *9*(1), 18–27. doi:10.3402/tellusa.v9i1.9075

Rivas, B., Torrado, A., Torre, P., Converti, A., & Domínguez, J. M. (2008). Submerged citric acid fermentation on orange peel autohydrolysate. *Journal of Agricultural and Food Chemistry*, *56*(7), 2380–2387. doi:10.1021/jf073388r PMID:18321055

Roobini, U., & Gnanavel, D. G. (2018). Citric acid Production from whey by *Aspergillus niger. IOSR Journal of Engineering*, *8*(9), 65–71.

Rossow, M. (2003). *Fly Ash Facts for Highway Engineers*. New York NY Continuing Education and Development.

Roukas, T., & Kotzekidou, P. (1986). Production of citric acid from brewery wastes by surface fermentation using *Aspergillus niger. Journal of Food Science*, *51*(1), 225–228. doi:10.1111/j.1365-2621.1986.tb10876.x

Roukas, T., & Kotzekidou, P. (1987). Influence of some trace metals and stimulants on citric acid production from brewery wastes by *Aspergillus niger. Enzyme and Microbial Technology*, *9*(5), 291–294. doi:10.1016/0141-0229(87)90006-8

Roukas, T., & Kotzekidou, P. (2020). Pomegranate peel waste: A new substrate for citric acid production by *Aspergillus niger* in solid-state fermentation under non-aseptic conditions. *Environmental Science and Pollution Research International*, *27*(12), 13105–13113. doi:10.100711356-020-07928-9 PMID:32016865

Ruiz-Rosero, J., Ramirez-Gonzalez, G., & Viveros-Delgado, J. (n.d.). *Software survey: ScientoPy, a scientometric tool for topics trend analysis in scientific publications*. Academic Press.

Rymowicz, W., Fatykhova, A. R., Kamzolova, S. V., Rywińska, A., & Morgunov, I. G. (2010). Citric acid production from glycerol-containing waste of biodiesel industry by *Yarrowia lipolytica* in batch, repeated batch, and cell recycle regimes. *Applied Microbiology and Biotechnology*, *87*(3), 971–979. doi:10.100700253-010-2561-z PMID:20376633

Rymowicz, W., Rywińska, A., Żarowska, B., & Juszczyk, P. (2006). Citric acid production from raw glycerol by acetate mutants of *Yarrowia lipolytica. Chemical Papers*, *60*(5), 391–394. doi:10.247811696-006-0071-3

Rywińska, A., Rymowicz, W., Zarowska, B., & Wojtatowicz, M. (2009). Biosynthesis of citric acid from glycerol by acetate mutants of *Yarrowia lipolytica* in fed-batch fermentation. *Food Technology and Biotechnology*, *47*(1), 1–6.

Rzechonek, D. A., Dobrowolski, A., Rymowicz, W., & Mirończuk, A. M. (2019). Aseptic production of citric and isocitric acid from crude glycerol by genetically modified *Yarrowia lipolytica. Bioresource Technology*, *271*, 340–344. doi:10.1016/j.biortech.2018.09.118 PMID:30292133

Sadak, O., Hackney, R., Sundramoorthy, A. K., Yilmaz, G., & Gunasekaran, S. (2020). Azo dye-functionalized magnetic Fe3O4/polyacrylic acid nanoadsorbent for removal of lead (II) ions. *Environmental Nanotechnology, Monitoring & Management*, *14*(October), 100380. doi:10.1016/j.enmm.2020.100380

Saharan, B.S., & Ankita, S. D. (2012). Bioplastics- for Sustainable Development: A Review. *International Journal of Microbial Resource Technology*, *1*(1), 11–23.

Saini, P. K., Fiorani, G., Mathers, R. T., & Williams, C. K. (2017). Zinc versus Magnesium: Orthogonal Catalyst Reactivity in Selective Polymerizations of Epoxides, Bio-derived Anhydrides and Carbon Dioxide. *Chemistry (Weinheim an der Bergstrasse, Germany)*, *23*(18), 4260–4265. doi:10.1002/chem.201605690 PMID:28295663

Salatin, S., Dizaj, S. M., & Khosroushahi, A. Y. (2015). Effect of the surface modification, size, and shape on cellular uptake of nanoparticles. Cell Biology International, 39(8), 881-890. doi:10.1002/cbin.10459

Saleem, A., Hussain, A., Chaudhary, A., Iqtedar, M., Javid, A., & Akram, A. M. (2020). Acid hydrolysis optimization of pomegranate peels waste using response surface methodology for ethanol production. *Biomass Conversion and Biorefinery*, 1–12.

Saleh, N. J., & Mustafa, S. N. (2011). A study of some mechanical, thermal and physical properties of polymer blend with Iraqi kaolin filler. *Eng. & Tech. Journal*, *29*(11), 2114–2132.

Santaniello, E., Ferraboschi, P., & Grisenti, P. (1993). Lipase-catalyzed transesterification in organic solvents: Applications to the preparation of enantiomerically pure compounds. *Enzyme and Microbial Technology*, *15*(5), 367–382. doi:10.1016/0141-0229(93)90123-J

Sapna, D. K. (2018). *Biodegradable polymer-based nanoadsorbents for environmental remediation, New Polymer Nanocomposites for Environmental Remediation*. doi:10.1016/B978-0-12-811033-1.00012-3

Saran, A., Boddula, R., & Sharan, S. R. (2017). *Biological Nanoparticles: Optical and Photothermal Properties, Fungal Biology,Fungal Nanotechnology*. doi:10.1007/978-3-319-68424-6_8

Sardon, H., & Dove, A. P. (2018). Plastics recycling with a difference. *Science*, *360*(6387), 380–381. doi:10.1126cience.aat4997 PMID:29700253

Sarkar, A., Sengupta, S., & Sen, S. (2019). Nanoparticles for Soil Remediation. In *Nanoscience and Biotechnology for Environmental Applications* (pp. 249–262). doi:10.1007/978-3-319-97922-9_9

Sarris, D., Galiotou-Panayotou, M., Koutinas, A. A., Komaitis, M., & Papanikolaou, S. (2011). Citric acid, biomass and cellular lipid production by *Yarrowia lipolytica* strains cultivated on olive mill wastewater-based media. *Journal of Chemical Technology and Biotechnology (Oxford, Oxfordshire)*, *86*(11), 1439–1448. doi:10.1002/jctb.2658

Sarris, D., Rapti, A., Papafotis, N., Koutinas, A. A., & Papanikolaou, S. (2019). Production of added-value chemical compounds through bioconversions of olive-mill wastewaters blended with crude glycerol by a *Yarrowia lipolytica* strain. *Molecules (Basel, Switzerland)*, *24*(2), 222. doi:10.3390/molecules24020222 PMID:30634450

Sarris, D., Stoforos, N. G., Mallouchos, A., Kookos, I. K., Koutinas, A. A., Aggelis, G., & Papanikolaou, S. (2017). Production of added-value metabolites by *Yarrowia lipolytica* growing in olive mill wastewater-based media under aseptic and non-aseptic conditions. *Engineering in Life Sciences*, *17*(6), 695–709. doi:10.1002/elsc.201600225 PMID:32624815

Sartika, M., Lubis, M., Harahap, M. B., Afrida, E., & Ginting, M. H. S. (2018). Production of bioplastic from avocado seed starch as matrix and microcrystalline cellulose from sugar palm fibers with Schweizer's reagent as solvent. *Asian Journal of Chemistry*, *30*(5), 1051–1056. doi:10.14233/ajchem.2018.21155

Sashiwa, H., Fukuda, R., Okura, T., Sato, S., & Nakayama, A. (2018). Microbial Degradation Behavior in Seawater of Polyester Blends Containing Poly(3-hydroxybutyrate-co-3-hydroxyhexanoate) (PHBHHx). *Marine Drugs*, *16*(1), 34. doi:10.3390/md16010034 PMID:29342118

Sassi, G., Ruggeri, B., Specchia, V., & Gianetto, A. (1991). Citric acid production by *A. niger* with banana extract. *Bioresource Technology*, *37*(3), 259–269. doi:10.1016/0960-8524(91)90193-N

Satapathy, B. K., Das, A., & Patnaik, A. (2011). Ductile-to-brittle transition in cenosphere-filled polypropylene composites. *Journal of Materials Science*, *46*(6), 1963–1974. doi:10.100710853-010-5032-0

Satyanarayana, K. G., Arizaga, G. G., & Wypych, F. (2009). Biodegradable composites based on lignocellulosic fibers—An overview. *Progress in Polymer Science*, *34*(9), 982–1021. doi:10.1016/j.progpolymsci.2008.12.002

Saumya, K. M., Alam, M., & Hussain, A. (2016). Effect of fly ash particle and vegetable oil on the mechanical properties of fly ash-vegetable oil reinforced hard PVC plastic. *Int Res J Eng Technol*, *3*, 833–839.

Savitskaya, T., Kimlenka, I., Lu, Y., Hrynshpan, D., Sarkisov, V., Yu, J., & Wang, L. (2021). Principle of Green Chemistry. In *Green Chemistry* (pp. 1–14). Springer.

Sayın Börekçi, B., Kaban, G., & Kaya, M. (2020). Citric acid production of yeasts: An overview. *The Eurobiotech Journal*, *5*(2), 79–91. doi:10.2478/ebtj-2021-0012

Schitea, D., Floristean, A., & Brahy, N. (2018). INC-DTCI-ICSI RM. Vâlcea participation in hylaw project. *Progress of Cryogenics & Isotopes Separation, 21*(1).

Scott, G. (2002). Why Degradable Polymers? *Degradable Polymers*, 1–15.

Searle, N. D., McGreer, M., & Zielnik, A. (2010). *Weathering of Polymeric Materials. Encyclopedia of Polymer Science and Technology*. Published.

Secinaro, S., Brescia, V., Calandra, D., & Biancone, P. (2020). Employing bibliometric analysis to identify suitable business models for electric cars. *Journal of Cleaner Production*, *264*, 121503. Advance online publication. doi:10.1016/j.jclepro.2020.121503

Selvavinayagam, A. V., Muthu, M., & Boopalan, E. (2018). Scientometrics, Techniques, Sources And Their Key Points To Analysis Of Lis Research: An Overview. *International Journal of Science & Technology, 8*(1). www.ijst.co.in

Sen, G., Singh, R. P., & Pal, S. (2010). Microwave-initiated synthesis of polyacrylamide grafted sodium alginate: Synthesis and characterization. *Journal of Applied Polymer Science*, *115*(1), 63–71. doi:10.1002/app.30596

Sengupta, S., Pal, K., Ray, D., & Mukhopadhyay, A. (2011). Furfuryl palmitate coated fly ash used as filler in recycled polypropylene matrix composites. *Composites. Part B, Engineering*, *42*(7), 1834–1839. doi:10.1016/j.compositesb.2011.06.021

Şen, S., & Nugay, N. (2000). Uncured and cured state properties of fly ash filled unsaturated polyester composites. *Journal of Applied Polymer Science*, *77*(5), 1128–1136. doi:10.1002/1097-4628(20000801)77:5<1128::AID-APP21>3.0.CO;2-D

Shah, M., Rajhans, S., Pandya, H.A., & Mankand, A.U. (2021). Bioplastic for future: A review then and now. *World Journal of Advanced Research and Reviews, 9*(2), 56-67. doi:10.30574/wjarr.2021.9.2.0054

Shah, T. V., & Vasava, D. V. (2019a). A glimpse of biodegradable polymers and their biomedical applications. *E-Polymers*, *19*(1), 385–410. doi:10.1515/epoly-2019-0041

Shalini, R., & Gupta, D. K. (2010). Utilization of pomace from apple processing industries: A review. *Journal of Food Science and Technology*, *47*(4), 365–371. doi:10.100713197-010-0061-x PMID:23572655

Shankaranand, V. S., & Lonsane, B. K. (1994). Coffee husk: An inexpensive substrate for production of citric acid by *Aspergillus niger* in a solid-state fermentation system. *World Journal of Microbiology & Biotechnology*, *10*(2), 165–168. doi:10.1007/BF00360879 PMID:24420939

Shanmugasundaram, K. R., Seethapathy, P. G., & Shanmugasundaram, E. R. (1987). Anna Pavala Sindhooram—An antiatherosclerotic Indian drug. *Journal of Ethnopharmacology*, *7*(3), 247–265. doi:10.1016/0378-8741(83)90001-6 PMID:6876850

Sharma, M. K., & Shah, D. O. (1985). *Introduction to Macro- and Microemulsions*. doi:10.1021/bk-1985-0272.ch001

Sharma, A. K., & Mahanwar, P. A. (2010). Effect of particle size of fly ash on recycled poly (ethylene terephthalate)/fly ash composites. *International Journal of Plastics Technology*, *14*(1), 53–64. doi:10.100712588-010-0006-2

Sharma, N., Sahota, P. P., & Singh, M. P. (2021). Organic acid production from agricultural waste. In B. K. Kashyap, M. K. Solanki, D. V. Kamboj, & A. K. Pandey (Eds.), *Waste to Energy: Prospects and Applications* (pp. 415–437). Springer.

Sheldon, R. A. (2019). The greening of solvents: Towards sustainable organic synthesis. *Current Opinion in Green and Sustainable Chemistry*, *18*, 13–19. doi:10.1016/j.cogsc.2018.11.006

Shojaosadati, S. A., & Babaeipour, V. (2002). Citric acid production from apple pomace in multi-layer packed bed solid-state bioreactor. *Process Biochemistry*, *37*(8), 909–914. doi:10.1016/S0032-9592(01)00294-1

Show, P. L., Oladele, K. O., Siew, Q. Y., Aziz Zakry, F. A., Lan, J. C. W., & Ling, T. C. (2015). Overview of citric acid production from *Aspergillus niger*. *Frontiers in Life Science*, *8*(3), 271–283. doi:10.1080/21553769.2015.1033653

Shubham, P., & Tiwari, S. K. (2013). Effect of fly ash concentration and its surface modification on fiber reinforced epoxy composite's mechanical properties. *IJSER*, *4*, 1173–1180.

Sidauruk, M. G. E., Hutauruk, K. S. N., Martgrita, M. M., & Manurung, A. (2019). Citric acid production from toba banana peel (*Musa acuminata* colla) through submerged fermentation using *Aspergillus niger*. *Microbiology Indonesia*, *13*(4), 2–2.

Singh, B., & Dhiman, M. (2021). *Carbon aerogels for environmental remediation, Advances in Aerogel Composites for Environmental Remediation*. doi:10.1016/B978-0-12-820732-1.00012-6

Singh, G. B., & Subramaniam, K. V. (2018). Characterization of Indian fly ashes using different experimental techniques. *Indian Concrete Journal*, *92*(3), 10–23.

Singh, N., Hui, D., Singh, R., Ahuja, I., Feo, L., & Fraternali, F. (2017). Recycling of plastic solid waste: A state of art review and future applications. *Composites. Part B, Engineering*, *115*, 409–422. doi:10.1016/j.compositesb.2016.09.013

Singh, V. V. (2022). *Green nanotechnology for environmental remediation, Sustainable Nanotechnology for Environmental Remediation*. doi:10.1016/B978-0-12-824547-7.00017-5

Singh, V., Tiwari, A., Pandey, S., & Singh, S. K. (2007). Peroxydisulfate initiated synthesis of potato starch-graft-poly(acrylonitrile) under microwave irradiation. *Express Polymer Letters*, *1*(1), 51–58. doi:10.3144/expresspolymlett.2007.10

Soccol, C. R., Vandenberghe, L. P., Rodrigues, C., & Pandey, A. (2006). New perspectives for citric acid production and application. *Food Technology and Biotechnology*, *44*(2).

Soh, S. H., & Lee, L. Y. (2019). Microencapsulation and nanoencapsulation using supercritical fluid (SCF) techniques. *Pharmaceutics*, *11*(1), 21. doi:10.3390/pharmaceutics11010021 PMID:30621309

Sreenivasulu, K., Satyaveni, M., Rajwardanreddy, M., Sowjanya, M. S., Nama, S., & Baburao, C. (2013). A Review On Magnetic Nano Particles. *International Journal of Current Pharmaceutical & Clinical Research, 3*(2), 45–49.

Srivastava, N., Srivastava, M., Ramteke, P. W., & Mishra, P. K. (2019). Solid-state fermentation strategy for microbial metabolites production: An overview. *New and Future Developments in Microbial Biotechnology and Bioengineering*, 345-354.

Srivastava, N., Srivastava, M., Ramteke, P., & Mishra, P. (2019). Solid-state fermentation strategy for microbial metabolites production: An overview. In V. K. Gupta & A. Pandey (Eds.), *New and future developments in microbial biotechnology and bioengineering* (pp. 345–354). Elsevier. doi:10.1016/B978-0-444-63504-4.00023-2

Sriyapai, P., Chansiri, K., & Sriyapai, T. (2018). Isolation and Characterization of Polyester-Based Plastics-Degrading Bacteria from Compost Soils. *Microbiology, 87*(2), 290–300. doi:10.1134/S0026261718020157

Stenesh, J. (1998). The citric acid cycle. In *Biochemistry* (pp. 273–291). Springer. doi:10.1007/978-1-4757-9427-4_11

Stevens, E. S. (2002). *Green Plastics* (1st ed., Vol. 1). Amsterdam University Press. doi:10.1515/9780691214177

Sudarsana, D., & Baba, M. S. (2019). Global nuclear fuel research during 2000 to 2017: A scientometric analysis. *Annals of Library and Information Studies, 66*, 85–93.

Suffo, M., Mata, M. D. L., & Molina, S. (2020). A sugar-beet waste based thermoplastic agro-composite as substitute for raw materials. *Journal of Cleaner Production, 257*, 120382. doi:10.1016/j.jclepro.2020.120382

Sukla, M., Regar, R. L., & Jangid, B. L. (2012). Henna (*Lawsonia inermis* L.) Cultivation: Available agri-enterprise in arid fringes of western Rajasthan. *Den News, 14*(2).

Sultan & Johari. (2017). *The Development of Banana Peel/Corn Starch Bioplastic Film: A Preliminary Study*. Academic Press.

Sun, T., Li, G., Ning, T. Y., Zhang, Z. M., Mi, Q. H., & Lal, R. (2018). Suitability of mulching with biodegradable film to moderate soil temperature and moisture and to increase photosynthesis and yield in peanut. *Agricultural Water Management, 208*, 214–223. doi:10.1016/j.agwat.2018.06.027

Surabhi, S. (2017). Fly ash in India: Generation vis-a-vis utilization and global perspective. *Int J Appl Chem, 13*(1), 29–52.

Sushma, S. P., & Kumar, A. K. (2014). Hardness and tensile testing of PVC and fly ash composite. *Advance research and innovations in mechanical, material science, industrial engineering and management-ICARMMIEM*, 273-277.

Swanepoel, J. C., & Strydom, C. A. (2002). Utilisation of fly ash in a geopolymeric material. *Applied Geochemistry, 17*(8), 1143–1148. doi:10.1016/S0883-2927(02)00005-7

Talele, G. S., & Shahare, H. V. (2021). Green Solvents for Pharmaceuticals. In *Natural Pharmaceuticals and Green Microbial Technology* (pp. 1–9). Apple Academic Press. doi:10.1201/9781003003229-1

Tamura, M., Ito, K., Honda, M., Nakagawa, Y., Sugimoto, H., & Tomishige, K. (2016). Direct Copolymerization of CO2 and Diols. *Scientific Reports*, *6*(1), 24038. doi:10.1038rep24038 PMID:27075987

Tang, Z. X., Claveau, D., Corcuff, R., Belkacemi, K., & Arul, J. (2008). Preparation of nano-CaO using thermal-decomposition method. *Materials Letters*, *62*(14), 2096–2098. doi:10.1016/j.matlet.2007.11.053

Tebby, C., Mombelli, E., Pandard, P., & Péry, A. R. (2011). Exploring an ecotoxicity database with the OECD (Q) SAR Toolbox and DRAGON descriptors in order to prioritise testing on algae, daphnids, and fish. *The Science of the Total Environment*, *409*(18), 3334–3343. doi:10.1016/j.scitotenv.2011.05.029 PMID:21684579

Thomas, N. L., Eastupy, J. R., & Quirk, P. (2004). *Rigid PVC foam. Formulating for sustainability, Blowing agent and foaming process*. Academic Press.

Tocháček, J., & Vrátníčková, Z. (2014). Polymer life-time prediction: The role of temperature in UV accelerated ageing of polypropylene and its copolymers. *Polymer Testing*, *36*, 82–87. doi:10.1016/j.polymertesting.2014.03.019

Tokiwa, Y., Calabia, B., Ugwu, C., & Aiba, S. (2009). Biodegradability of Plastics. *International Journal of Molecular Sciences*, *10*(9), 3722–3742. doi:10.3390/ijms10093722 PMID:19865515

Torrado, A. M., Cortés, S., Salgado, J. M., Max, B., Rodríguez, N., Bibbins, B. P., Converti, A., & Domínguez, J. M. (2011). Citric acid production from orange peel wastes by solid-state fermentation. *Brazilian Journal of Microbiology*, *42*(1), 394–409. doi:10.1590/S1517-83822011000100049 PMID:24031646

Tosta, M. R., Prates, L. L., Christensen, D. A., & Yu, P. (2019). Biodegradation kinetics by microorganisms, enzymatic biodigestion, and fractionation of protein in seeds of cool-climate-adapted oats: Comparison among oat varieties, between milling-type and feed-type oats, and with barley grain. *Journal of Cereal Science*, *89*, 102814. doi:10.1016/j.jcs.2019.102814

Tran, C. T., & Mitchell, D. A. (1995). Pineapple waste-a novel substrate for citric acid production by solid-state fermentation. *Biotechnology Letters*, *17*(10), 1107–1110. doi:10.1007/BF00143111

Tran, C. T., Sly, L. I., & Mitchell, D. A. (1998). Selection of a strain of *Aspergillus* for the production of citric acid from pineapple waste in solid-state fermentation. *World Journal of Microbiology & Biotechnology*, *14*(3), 399–404. doi:10.1023/A:1008821413239

Tripathy, T., & Singh, R. P. (2001). Characterization of polyacrylamide-grafted sodium alginate: A novel polymeric flocculant. *Journal of Applied Polymer Science*, *81*(13), 3296–3308. doi:10.1002/app.1786

UNESCO. (2021). *Valuing water*. United Nations Educational, Scientific and Cultural Organization. doi:10.4324/9781315627250-3

United Nations. (n.d.). The United Nations World Water Development Report 2021. *Valuing Water. Author.*

Uyama, H., & Kobayashi, S. (2006). Enzymatic Synthesis of Polyesters via Polycondensation. *Advances in Polymer Science*, *194*, 133–158. doi:10.1007/12_031

Uzma, N., Khaja Mohinuddin Salar, B. M., Kumar, B. S., Aziz, N., David, M. A., & Reddy, V. D. (2008). Impact of organic solvents and environmental pollutants on the physiological function in petrol filling workers. *International Journal of Environmental Research and Public Health*, *5*(3), 139–146. doi:10.3390/ijerph5030139 PMID:19139531

Vahdat, A., Bahrami, H., Ansari, N., & Ziaie, F. (2007). Radiation grafting of styrene onto polypropylene fibres by a 10MeV electron beam. *Radiation Physics and Chemistry*, *76*(5), 787–793. doi:10.1016/j.radphyschem.2006.05.009

Valavanidis, A., Fiotakis, K., & Vlachogianni, T. (2008). Airborne particulate matter and human health: Toxicological assessment and importance of size and composition of particles for oxidative damage and carcinogenic mechanisms. *Journal of Environmental Science and Health. Part C, Environmental Carcinogenesis & Ecotoxicology Reviews*, *26*(4), 339–362. doi:10.1080/10590500802494538 PMID:19034792

Van der Merwe, E. M., Mathebula, C. L., & Prinsloo, L. C. (2014). Characterization of the surface and physical properties of South African coal fly ash modified by sodium lauryl sulphate (SLS) for applications in PVC composites. *Powder Technology*, *266*, 70–78. doi:10.1016/j.powtec.2014.06.008

Vandenberghe, L. P., Soccol, C. R., Pandey, A., & Lebeault, J. M. (2000). Solid-state fermentation for the synthesis of citric acid by *Aspergillus niger. Bioresource Technology*, *74*(2), 175–178. doi:10.1016/S0960-8524(99)00107-8

Vega-Castro, O., Contreras-Calderon, J., León, E., Segura, A., Arias, M., Pérez, L., & Sobral, P. J. (2016). Characterization of a polyhydroxyalkanoate obtained from pineapple peel waste using Ralsthonia eutropha. *Journal of Biotechnology*, *231*, 232–238. doi:10.1016/j.jbiotec.2016.06.018 PMID:27316828

Vendruscolo, F., Albuquerque, P. M., Streit, F., Esposito, E., & Ninow, J. L. (2008). Apple pomace: A versatile substrate for biotechnological applications. *Critical Reviews in Biotechnology*, *28*(1), 1–12. doi:10.1080/07388550801913840 PMID:18322854

Vengosh, A., Cowan, E. A., Coyte, R. M., Kondash, A. J., Wang, Z., Brandt, J. E., & Dwyer, G. S. (2019). Evidence for unmonitored coal ash spills in Sutton Lake, North Carolina: Implications for contamination of lake ecosystems. *The Science of the Total Environment*, *686*, 1090–1103. doi:10.1016/j.scitotenv.2019.05.188 PMID:31200305

Venzhik, Y., Sokolov, A., Sokolov, O., Moshkov, I., & Dykman, L. (2022). *Effects, uptake, and translocation of Au-based nanoparticles in plant, Toxicity of Nanoparticles in Plants*. doi:10.1016/B978-0-323-90774-3.00016-7

Verma, M. L. (2017). *Fungus-Mediated Bioleaching of Metallic Nanoparticles from Agro-industrial By-Products, Fungal Biology, Fungal Nanotechnology*. doi:10.1007/978-3-319-68424-6_5

Virot, M., Tomao, V., Ginies, C., & Chemat, F. (2008). Total lipid extraction of food using d-limonene as an alternative to n-hexane. *Chromatographia*, *68*(3), 311–313. doi:10.136510337-008-0696-1

Wagare, D. S., Shirsath, S. E., Shaikh, M., & Netankar, P. (2021). Sustainable solvents in chemical synthesis: A review. *Environmental Chemistry Letters*, *19*(4), 1–20. doi:10.100710311-020-01176-6

Wang, J. P., Chen, Y. Z., Zhang, S. J., & Yu, H. Q. (2008). A chitosan-based flocculant prepared with gamma-irradiation-induced grafting. *Bioresource Technology*, *99*(9), 3397–3402. doi:10.1016/j.biortech.2007.08.014 PMID:17888656

Wang, J., Qin, A., & Tang, B. Z. (2020). Multicomponent Polymerizations Involving Green Monomers. *Macromolecular Rapid Communications*, *42*(6), 2000547. doi:10.1002/marc.202000547 PMID:33314433

Wang, M., Liu, P., Gu, Z., Cheng, H., & Li, X. (2019). A Scientometric Review of Resource Recycling Industry. *International Journal of Environmental Research and Public Health*, *16*(23), 4654. doi:10.3390/ijerph16234654 PMID:31766728

Wang, Y., Wang, Q., Zhu, R., Yang, C., Chen, Z., Bai, Y., Li, M., & Zhai, X. (2016). Trends of spinal tuberculosis research (1994-2015) A bibliometric study. *Medicine*, *95*(38), e4923. doi:10.1097/MD.0000000000004923 PMID:27661044

Wang, Z. H., Lu, Y. L., Liu, J., Dang, Z. M., Zhang, L. Q., & Wang, W. (2011). Preparation of nanoalumina/EPDM composites with good performance in thermal conductivity and mechanical properties. *Polymers for Advanced Technologies*, *22*(12), 2302–2310. doi:10.1002/pat.1761

Westminster City Council | Commercial Waste Services. (2020, July 6). *Commercial Waste*. https://cleanstreets.westminster.gov.uk/

White, J. R., & Turnbull, A. (1994). Weathering of polymers: Mechanisms of degradation and stabilization, testing strategies and modelling. *Journal of Materials Science*, *29*(3), 584–613. doi:10.1007/BF00445969

Wiesbrock, F., Hoogenboom, R., & Schubert, U. S. (2004). Microwave-Assisted Polymer Synthesis: State-of-the-Art and Future Perspectives. *Macromolecular Rapid Communications*, *25*(20), 1739–1764. doi:10.1002/marc.200400313

Wong, C. H., & Whitesides, G. M. (1994). *Enzymes in Synthetic Organic Chemistry*. Pergamon.

Wong, K. Y., & Truss, R. W. (1994). Effect of flyash content and coupling agent on the mechanical properties of flyash-filled polypropylene. *Composites Science and Technology*, *52*(3), 361–368. doi:10.1016/0266-3538(94)90170-8

Wu, J., Liang, Q., Yu, X., Lü, Q. F., Ma, L., Qin, X., Chen, G., & Li, B. (2021). Deep Eutectic Solvents for Boosting Electrochemical Energy Storage and Conversion: A Review and Perspective. *Advanced Functional Materials*, *31*(22), 2011102. doi:10.1002/adfm.202011102

Xing, H., Wang, T., Zhou, Z., & Dai, Y. (2005). Novel Brønsted-acidic ionic liquids for esterifications. *Industrial & Engineering Chemistry Research*, *44*(11), 4147–4150. doi:10.1021/ie0488703

Xiong, T., & Li, S. (2022). *Effects, uptake, and translocation of Cu-based nanoparticles in plants, Toxicity of Nanoparticles in Plants*. doi:10.1016/B978-0-323-90774-3.00012-X

Xuan, Y., Huajin, T., & Zhiwen, Q. (2019, October). Effect of mixing time of coupling agent on properties of 3D printing prototype of fly ash filled ABS. *IOP Conference Series. Materials Science and Engineering*, *631*(3), 032025. doi:10.1088/1757-899X/631/3/032025

Xue, C., Nan, H., Lu, Y., Chen, H., Zhao, C., & Xu, S. (2021). Effects of inorganic-organic surface modification on the mechanical and thermal properties of poly (vinyl chloride) composites reinforced with fly-ash. *Polymer Composites*, *42*(4), 1867–1877. doi:10.1002/pc.25942

Yadegary, M., Hamidi, A., Alavi, S. A., Khodaverdi, E., Yahaghi, H., Sattari, S., Bagherpour, G., & Yahaghi, E. (2013). Citric acid production from sugarcane bagasse through solid state fermentation method using *Aspergillus niger* mold and optimization of citric acid production by Taguchi method. *Jundishapur Journal of Microbiology*, *6*(9). Advance online publication. doi:10.5812/jjm.7625

Yalcin, S. K. (2012). Enhancing citric acid production of *Yarrowia lipolytica* by mutagenesis and using natural media containing carrot juice and celery byproducts. *Food Science and Biotechnology*, *21*(3), 867–874. doi:10.100710068-012-0112-3

Yalcin, S. K., Bozdemir, M. T., & Ozbas, Z. Y. (2009). Utilization of whey and grape must for citric acid production by two *Yarrowia lipolytica* strains. *Food Biotechnology*, *23*(3), 266–283. doi:10.1080/08905430903106860

Yalcin, S. K., Bozdemir, M. T., & Ozbas, Z. Y. (2010). Citric acid production by yeasts: Fermentation conditions, process optimization and strain improvement. *Current Research. Technology and Education Topics in Applied Microbiology and Microbial Biotechnology*, *9*, 1374–1382.

Yamada-Onodera, K., Mukumoto, H., Katsuyaya, Y., Saiganji, A., & Tani, Y. (2001). Degradation of polyethylene by a fungus, Penicillium simplicissimum YK. *Polymer Degradation & Stability*, *72*(2), 323–327. doi:10.1016/S0141-3910(01)00027-1

Yang, Y. F., Gai, G. S., Cai, Z. F., & Chen, Q. R. (2006). Surface modification of purified fly ash and application in polymer. *Journal of Hazardous Materials*, *133*(1-3), 276–282. doi:10.1016/j.jhazmat.2005.10.028 PMID:16314041

Yan, W. T., Lu, S., Yang, Y. D., Ning, W. Y., Cai, Y., Hu, X. M., Zhang, Q., & Xiong, K. (2021). Research trends, hot spots and prospects for necroptosis in the field of neuroscience. *Neural Regeneration Research*, *16*(8), 1628–1637. doi:10.4103/1673-5374.303032 PMID:33433494

Yao, N., Zhang, P., Song, L., Kang, M., Lu, Z., & Zheng, R. (2013). Stearic acid coating on circulating fluidized bed combustion fly ashes and its effect on the mechanical performance of polymer composites. *Applied Surface Science*, *279*, 109–115. doi:10.1016/j.apsusc.2013.04.045

Yaradoddi. (2016). *Biodegradable plastic production from fruit waste material and its sustainable usage for green applications*. Academic Press.

Yaseen, G., Ahmad, M., Zafar, M., Akram, A., Sultana, S., Kilic, O., & Sonmez, G. D. (2021). Current status of solvents used in the pharmaceutical industry. In *Green Sustainable Process for Chemical and Environmental Engineering and Science* (pp. 195–219). Elsevier. doi:10.1016/B978-0-12-821885-3.00004-9

Yazici Guvenc, S., Alan, B., Adar, E., & Bilgili, M. S. (2017). The impact of nanoparticles on aerobic degradation of municipal solid waste. *Waste Management & Research*, *35*(4), 426–436. doi:10.1177/0734242X17695884 PMID:28367755

Yilmaz, F., Lee, D. J., Song, J. W., Hong, H. S., Son, H. T., Yoon, J. S., & Hong, S. J. (2013). Fabrication of cobalt nano-particles by pulsed wire evaporation method in nitrogen atmosphere. *Powder Technology*, *235*, 1047–1052. doi:10.1016/j.powtec.2012.10.024

Yin, J., Teng, Y., Meng, X., Ge, Y., & Zhang, X. (2020). Influence of fly ash on the curing characteristics of an epoxy resin. *Polymers & Polymer Composites*, *28*(8-9), 579–588. doi:10.1177/0967391119894083

Young, P. J., Harper, A. B., Huntingford, C., Paul, N. D., Morgenstern, O., Newman, P. A., Oman, L. D., Madronich, S., & Garcia, R. R. (2021). The Montreal Protocol protects the terrestrial carbon sink. *Nature*, *596*(7872), 384–388. doi:10.103841586-021-03737-3 PMID:34408332

Zhang, F., Zheng, B., Zhang, J., Huang, X., Liu, H., Guo, S., & Zhang, J. (2010). Horseradish Peroxidase Immobilized on Graphene Oxide: Physical Properties and Applications in Phenolic Compound Removal. *The Journal of Physical Chemistry C*, *114*(18), 8469–8473. doi:10.1021/jp101073b

Zhang, H., Xu, J., Su, X., Bao, J., Wang, K., & Mao, Z. (2017). Citric acid production by recycling its wastewater treated with anaerobic digestion and nanofiltration. *Process Biochemistry*, *58*, 245–251. doi:10.1016/j.procbio.2017.04.022

Zhang, L., Zhu, Y., Zhang, J., Zeng, G., Dong, H., Cao, W., Fang, W., Cheng, Y., Wang, Y., & Ning, Q. (2019). Impacts of iron oxide nanoparticles on organic matter degradation and microbial enzyme activities during agricultural waste composting. *Waste Management (New York, N.Y.)*, *95*, 289–297. doi:10.1016/j.wasman.2019.06.025 PMID:31351614

Zhang, S., Jagtap, S. S., Deewan, A., & Rao, C. V. (2019). pH selectively regulates citric acid and lipid production in *Yarrowia lipolytica* W29 during nitrogen-limited growth on glucose. *Journal of Biotechnology, 290*, 10–15. doi:10.1016/j.jbiotec.2018.10.012 PMID:30496777

Zhang, W., & Yuan, H. (2019). A Bibliometric Analysis of Energy Performance Contracting Research from 2008 to 2018. *Sustainability, 11*(13), 3548. doi:10.3390u11133548

Zhang, X., Zhou, L., Pi, H., Guo, S., & Fu, J. (2014). Performance of layered double hydroxides intercalated by a UV stabilizer in accelerated weathering and thermal stabilization of PVC. *Polymer Degradation & Stability, 102*, 204–211. doi:10.1016/j.polymdegradstab.2014.01.003

Zhao, L., Tang, Z. Y., & Zou, X. (2019). Mapping the Knowledge Domain of Smart-City Research: A Bibliometric and Scientometric Analysis. *Sustainability, 11*(23), 6648. doi:10.3390u11236648

Zhou, Y. S., Liu, S., Liu, J. H., Sun, J. L., & Zhou, M. J. (2012). Rheological and Mechanical Properties of Fly Ash Microspheres Modified PVC. *Plastics*.

Zhuang, X. Y., Chen, L., Komarneni, S., Zhou, C. H., Tong, D. S., Yang, H. M., Yu, W. H., & Wang, H. (2016). Fly ash-based geopolymer: Clean production, properties and applications. *Journal of Cleaner Production, 125*, 253–267. doi:10.1016/j.jclepro.2016.03.019

Zhu, F., Ma, S., Liu, T., & Deng, X. (2018). Green synthesis of nano zero-valent iron/Cu by green tea to remove hexavalent chromium from groundwater. *Journal of Cleaner Production, 174*, 184–190. doi:10.1016/j.jclepro.2017.10.302

Zimmerman, J. B., Anastas, P. T., Erythropel, H. C., & Leitner, W. (2020). Designing for a green chemistry future. *Science, 367*(6476), 397–400. doi:10.1126cience.aay3060 PMID:31974246

Zotta, T., Solieri, L., Iacumin, L., Picozzi, C., & Gullo, M. (2020). Valorization of cheese whey using microbial fermentations. *Applied Microbiology and Biotechnology, 104*(7), 2749–2764. doi:10.100700253-020-10408-2 PMID:32009200

Related References

To continue our tradition of advancing academic research, we have compiled a list of recommended IGI Global readings. These references will provide additional information and guidance to further enrich your knowledge and assist you with your own research and future publications.

Ajiboye, O. E., & Yusuff, O. S. (2017). Foreign Land Acquisition: Food Security and Food Chains – The Nigerian Experience. In I. Management Association (Ed.), Natural Resources Management: Concepts, Methodologies, Tools, and Applications (pp. 1524-1545). Hershey, PA: IGI Global. https://doi.org/ doi:10.4018/978-1-5225-0803-8.ch072

Alapiki, H. E., & Amadi, L. A. (2018). Sustainable Food Consumption in the Neoliberal Order: Challenges and Policy Implications. In A. Obayelu (Ed.), *Food Systems Sustainability and Environmental Policies in Modern Economies* (pp. 90–123). Hershey, PA: IGI Global. doi:10.4018/978-1-5225-3631-4.ch005

Altaş, A. (2018). Geographical Information System Applications Utilized in Museums in Turkey Within the Scope of the Cultural Heritage Tourism: A Case Study of Mobile Application of Müze Asist. In S. Chaudhuri & N. Ray (Eds.), *GIS Applications in the Tourism and Hospitality Industry* (pp. 42–60). Hershey, PA: IGI Global. doi:10.4018/978-1-5225-5088-4.ch002

Andreea, I. R. (2018). Beyond Macroeconomics of Food and Nutrition Security. *International Journal of Sustainable Economies Management*, *7*(1), 13–22. doi:10.4018/IJSEM.2018010102

Anwar, J. (2017). Reproductive and Mental Health during Natural Disaster: Implications and Issues for Women in Developing Nations – A Case Example. In I. Management Association (Ed.), Gaming and Technology Addiction: Breakthroughs in Research and Practice (pp. 446-472). Hershey, PA: IGI Global. https://doi.org/ doi:10.4018/978-1-5225-0778-9.ch021

Awadh, H., Aksissou, M., Benhardouze, W., Darasi, F., & Snaiki, J. (2018). Socioeconomic Status of Artisanal Fishers in the West Part of Moroccan Mediterranean. *International Journal of Social Ecology and Sustainable Development, 9*(1), 40–52. doi:10.4018/IJSESD.2018010104

Aye, G. C., & Haruna, R. F. (2018). Effect of Climate Change on Crop Productivity and Prices in Benue State, Nigeria: Implications for Food Security. In V. Erokhin (Ed.), *Establishing Food Security and Alternatives to International Trade in Emerging Economies* (pp. 244–268). Hershey, PA: IGI Global. doi:10.4018/978-1-5225-2733-6.ch012

Azizan, S. A., & Suki, N. M. (2017). Consumers' Intentions to Purchase Organic Food Products. In T. Esakki (Ed.), *Green Marketing and Environmental Responsibility in Modern Corporations* (pp. 68–82). Hershey, PA: IGI Global. doi:10.4018/978-1-5225-2331-4.ch005

Barakabitze, A. A., Fue, K. G., Kitindi, E. J., & Sanga, C. A. (2017). Developing a Framework for Next Generation Integrated Agro Food-Advisory Systems in Developing Countries. In I. Management Association (Ed.), Agri-Food Supply Chain Management: Breakthroughs in Research and Practice (pp. 47-67). Hershey, PA: IGI Global. doi:10.4018/978-1-5225-1629-3.ch004

Beachcroft-Shaw, H., & Ellis, D. (2018). Using Successful Cases to Promote Environmental Sustainability: A Social Marketing Approach. In I. Management Association (Ed.), Sustainable Development: Concepts, Methodologies, Tools, and Applications (pp. 936-953). Hershey, PA: IGI Global. doi:10.4018/978-1-5225-3817-2.ch042

Behnassi, M., Kahime, K., Boussaa, S., Boumezzough, A., & Messouli, M. (2017). Infectious Diseases and Climate Vulnerability in Morocco: Governance and Adaptation Options. In I. Management Association (Ed.), Public Health and Welfare: Concepts, Methodologies, Tools, and Applications (pp. 91-109). Hershey, PA: IGI Global. https://doi.org/ doi:10.4018/978-1-5225-1674-3.ch005

Bekele, F., & Bekele, I. (2017). Social and Environmental Impacts on Agricultural Development. In W. Ganpat, R. Dyer, & W. Isaac (Eds.), *Agricultural Development and Food Security in Developing Nations* (pp. 21–56). Hershey, PA: IGI Global. doi:10.4018/978-1-5225-0942-4.ch002

Benaouda, A., & García-Peñalvo, F. J. (2018). Towards an Intelligent System for the Territorial Planning: Agricultural Case. In F. García-Peñalvo (Ed.), *Global Implications of Emerging Technology Trends* (pp. 158–178). Hershey, PA: IGI Global. doi:10.4018/978-1-5225-4944-4.ch010

Bhaskar, A., Rao, G. B., & Vencatesan, J. (2017). Characterization and Management Concerns of Water Resources around Pallikaranai Marsh, South Chennai. In P. Rao & Y. Patil (Eds.), *Reconsidering the Impact of Climate Change on Global Water Supply, Use, and Management* (pp. 102–121). Hershey, PA: IGI Global. doi:10.4018/978-1-5225-1046-8.ch007

Bhyan, P., Shrivastava, B., & Kumar, N. (2022). Requisite Sustainable Development Contemplating Buildings: Economic and Environmental Sustainability. In A. Hussain, K. Tiwari, & A. Gupta (Eds.), *Addressing Environmental Challenges Through Spatial Planning* (pp. 269–288). IGI Global. https://doi.org/10.4018/978-1-7998-8331-9.ch014

Bogataj, D., & Drobne, D. (2017). Control of Perishable Goods in Cold Logistic Chains by Bionanosensors. In S. Joo (Ed.), *Applying Nanotechnology for Environmental Sustainability* (pp. 376–402). Hershey, PA: IGI Global. doi:10.4018/978-1-5225-0585-3.ch016

Bogataj, D., & Drobne, D. (2017). Control of Perishable Goods in Cold Logistic Chains by Bionanosensors. In I. Management Association (Ed.), Materials Science and Engineering: Concepts, Methodologies, Tools, and Applications (pp. 471-497). Hershey, PA: IGI Global. https://doi.org/ doi:10.4018/978-1-5225-1798-6.ch019

Bogueva, D., & Marinova, D. (2018). What Is More Important: Perception of Masculinity or Personal Health and the Environment? In D. Bogueva, D. Marinova, & T. Raphaely (Eds.), *Handbook of Research on Social Marketing and Its Influence on Animal Origin Food Product Consumption* (pp. 148–162). Hershey, PA: IGI Global. doi:10.4018/978-1-5225-4757-0.ch010

Bouzid, M. (2017). Waterborne Diseases and Climate Change: Impact and Implications. In M. Bouzid (Ed.), *Examining the Role of Environmental Change on Emerging Infectious Diseases and Pandemics* (pp. 89–108). Hershey, PA: IGI Global. doi:10.4018/978-1-5225-0553-2.ch004

Bowles, D. C. (2017). Climate Change-Associated Conflict and Infectious Disease. In M. Bouzid (Ed.), *Examining the Role of Environmental Change on Emerging Infectious Diseases and Pandemics* (pp. 68–88). Hershey, PA: IGI Global. doi:10.4018/978-1-5225-0553-2.ch003

Buck, J. J., & Lowry, R. K. (2017). Oceanographic Data Management: Quills and Free Text to the Digital Age and "Big Data". In P. Diviacco, A. Leadbetter, & H. Glaves (Eds.), *Oceanographic and Marine Cross-Domain Data Management for Sustainable Development* (pp. 1–22). Hershey, PA: IGI Global. doi:10.4018/978-1-5225-0700-0.ch001

Buse, C. G. (2017). Are Climate Change Adaptation Policies a Game Changer?: A Case Study of Perspectives from Public Health Officials in Ontario, Canada. In M. Bouzid (Ed.), *Examining the Role of Environmental Change on Emerging Infectious Diseases and Pandemics* (pp. 230–257). Hershey, PA: IGI Global. doi:10.4018/978-1-5225-0553-2.ch010

Calderon, F. A., Giolo, E. G., Frau, C. D., Rengel, M. G., Rodriguez, H., Tornello, M., ... Gallucci, R. (2018). Seismic Microzonation and Site Effects Detection Through Microtremors Measures: A Review. In N. Ceryan (Ed.), *Handbook of Research on Trends and Digital Advances in Engineering Geology* (pp. 326–349). Hershey, PA: IGI Global. doi:10.4018/978-1-5225-2709-1.ch009

Carfì, D., Donato, A., & Panuccio, D. (2018). A Game Theory Coopetitive Perspective for Sustainability of Global Feeding: Agreements Among Vegan and Non-Vegan Food Firms. In I. Management Association (Ed.), Game Theory: Breakthroughs in Research and Practice (pp. 71-104). Hershey, PA: IGI Global. https://doi.org/doi:10.4018/978-1-5225-2594-3.ch004

Castagnolo, V. (2018). Analyzing, Classifying, Safeguarding: Drawing for the Borgo Murattiano Neighbourhood of Bari. In G. Carlone, N. Martinelli, & F. Rotondo (Eds.), *Designing Grid Cities for Optimized Urban Development and Planning* (pp. 93–108). Hershey, PA: IGI Global. doi:10.4018/978-1-5225-3613-0.ch006

Chekima, B. (2018). The Dilemma of Purchase Intention: A Conceptual Framework for Understanding Actual Consumption of Organic Food. *International Journal of Sustainable Economies Management*, *7*(2), 1–13. doi:10.4018/IJSEM.2018040101

Chen, Y. (2017). Sustainable Supply Chains and International Soft Landings: A Case of Wetland Entrepreneurship. In B. Christiansen & F. Kasarcı (Eds.), *Corporate Espionage, Geopolitics, and Diplomacy Issues in International Business* (pp. 232–247). Hershey, PA: IGI Global. doi:10.4018/978-1-5225-1031-4.ch013

Çıtak, L., Akel, V., & Ersoy, E. (2018). Investors' Reactions to the Announcement of New Constituents of BIST Sustainability Index: An Analysis by Event Study and Mean-Median Tests. In M. Risso & S. Testarmata (Eds.), *Value Sharing for Sustainable and Inclusive Development* (pp. 270–289). Hershey, PA: IGI Global. doi:10.4018/978-1-5225-3147-0.ch012

D'Aleo, V., D'Aleo, F., & Bonanno, R. (2018). New Food Industries Toward a New Level of Sustainable Supply: Success Stories, Business Models, and Strategies. In V. Erokhin (Ed.), *Establishing Food Security and Alternatives to International Trade in Emerging Economies* (pp. 74–97). Hershey, PA: IGI Global. doi:10.4018/978-1-5225-2733-6.ch004

Dagevos, H., & Reinders, M. J. (2018). Flexitarianism and Social Marketing: Reflections on Eating Meat in Moderation. In D. Bogueva, D. Marinova, & T. Raphaely (Eds.), *Handbook of Research on Social Marketing and Its Influence on Animal Origin Food Product Consumption* (pp. 105–120). Hershey, PA: IGI Global. doi:10.4018/978-1-5225-4757-0.ch007

Danisman, G. O. (2022). What Drives Eco-Design Innovations in European SMEs? In U. Akkucuk (Ed.), *Disruptive Technologies and Eco-Innovation for Sustainable Development* (pp. 191–206). IGI Global. https://doi.org/10.4018/978-1-7998-8900-7.ch011

Deenapanray, P. N., & Ramma, I. (2017). Adaptations to Climate Change and Climate Variability in the Agriculture Sector in Mauritius: Lessons from a Technical Needs Assessment. In I. Management Association (Ed.), Natural Resources Management: Concepts, Methodologies, Tools, and Applications (pp. 655-680). Hershey, PA: IGI Global. https://doi.org/ doi:10.4018/978-1-5225-0803-8.ch030

Deenapanray, P. N., & Ramma, I. (2017). Adaptations to Climate Change and Climate Variability in the Agriculture Sector in Mauritius: Lessons from a Technical Needs Assessment. In I. Management Association (Ed.), Natural Resources Management: Concepts, Methodologies, Tools, and Applications (pp. 655-680). Hershey, PA: IGI Global. https://doi.org/ doi:10.4018/978-1-5225-0803-8.ch030

Deshpande, S., Basu, S. K., Li, X., & Chen, X. (2017). Smart, Innovative and Intelligent Technologies Used in Drug Designing. In I. Management Association (Ed.), Pharmaceutical Sciences: Breakthroughs in Research and Practice (pp. 1175-1191). Hershey, PA: IGI Global. https://doi.org/ doi:10.4018/978-1-5225-1762-7.ch045

Dlamini, P. N. (2017). Use of Information and Communication Technologies Tools to Capture, Store, and Disseminate Indigenous Knowledge: A Literature Review. In P. Ngulube (Ed.), *Handbook of Research on Theoretical Perspectives on Indigenous Knowledge Systems in Developing Countries* (pp. 225–247). Hershey, PA: IGI Global. doi:10.4018/978-1-5225-0833-5.ch010

Dolejsova, M., & Kera, D. (2017). The Fermentation GutHub Project and the Internet of Microbes. In S. Konomi & G. Roussos (Eds.), *Enriching Urban Spaces with Ambient Computing, the Internet of Things, and Smart City Design* (pp. 25–46). Hershey, PA: IGI Global. doi:10.4018/978-1-5225-0827-4.ch002

Dolunay, O. (2018). A Paradigm Shift: Empowering Farmers to Eliminate the Waste in the Form of Fresh Water and Energy Through the Implementation of 4R+T. In I. Management Association (Ed.), Sustainable Development: Concepts, Methodologies, Tools, and Applications (pp. 882-892). Hershey, PA: IGI Global. https://doi.org/doi:10.4018/978-1-5225-3817-2.ch039

Dube, P., Heijman, W. J., Ihle, R., & Ochieng, J. (2018). The Potential of Traditional Leafy Vegetables for Improving Food Security in Africa. In V. Erokhin (Ed.), *Establishing Food Security and Alternatives to International Trade in Emerging Economies* (pp. 220–243). Hershey, PA: IGI Global. doi:10.4018/978-1-5225-2733-6.ch011

Duruji, M. M., & Urenma, D. F. (2017). The Environmentalism and Politics of Climate Change: A Study of the Process of Global Convergence through UNFCCC Conferences. In I. Management Association (Ed.), *Natural Resources Management: Concepts, Methodologies, Tools, and Applications* (pp. 77-108). Hershey, PA: IGI Global. 10.4018/978-1-5225-0803-8.ch004

Dutta, U. (2017). Agro-Geoinformatics, Potato Cultivation, and Climate Change. In S. Londhe (Ed.), *Sustainable Potato Production and the Impact of Climate Change* (pp. 247–271). Hershey, PA: IGI Global. doi:10.4018/978-1-5225-1715-3.ch012

Edirisinghe, R., Stranieri, A., & Wickramasinghe, N. (2017). A Taxonomy for mHealth. In N. Wickramasinghe (Ed.), *Handbook of Research on Healthcare Administration and Management* (pp. 596–615). Hershey, PA: IGI Global. doi:10.4018/978-1-5225-0920-2.ch036

Ekpeni, N. M., & Ayeni, A. O. (2018). Global Natural Hazard and Disaster Vulnerability Management. In A. Eneanya (Ed.), *Handbook of Research on Environmental Policies for Emergency Management and Public Safety* (pp. 83–104). Hershey, PA: IGI Global. doi:10.4018/978-1-5225-3194-4.ch005

Ene, C., Voica, M. C., & Panait, M. (2017). Green Investments and Food Security: Opportunities and Future Directions in the Context of Sustainable Development. In M. Mieila (Ed.), *Measuring Sustainable Development and Green Investments in Contemporary Economies* (pp. 163–200). Hershey, PA: IGI Global. doi:10.4018/978-1-5225-2081-8.ch007

Escamilla, I., Ruíz, M. T., Ibarra, M. M., Soto, V. L., Quintero, R., & Guzmán, G. (2018). Geocoding Tweets Based on Semantic Web and Ontologies. In M. Lytras, N. Aljohani, E. Damiani, & K. Chui (Eds.), *Innovations, Developments, and Applications of Semantic Web and Information Systems* (pp. 372–392). Hershey, PA: IGI Global. doi:10.4018/978-1-5225-5042-6.ch014

Escribano, A. J. (2018). Marketing Strategies for Trendy Animal Products: Sustainability as a Core. In F. Quoquab, R. Thurasamy, & J. Mohammad (Eds.), *Driving Green Consumerism Through Strategic Sustainability Marketing* (pp. 169–203). Hershey, PA: IGI Global. doi:10.4018/978-1-5225-2912-5.ch010

Eudoxie, G., & Roopnarine, R. (2017). Climate Change Adaptation and Disaster Risk Management in the Caribbean. In W. Ganpat & W. Isaac (Eds.), *Environmental Sustainability and Climate Change Adaptation Strategies* (pp. 97–125). Hershey, PA: IGI Global. doi:10.4018/978-1-5225-1607-1.ch004

Farmer, L. S. (2017). Data Analytics for Strategic Management: Getting the Right Data. In V. Wang (Ed.), *Encyclopedia of Strategic Leadership and Management* (pp. 810–822). Hershey, PA: IGI Global. doi:10.4018/978-1-5225-1049-9.ch056

Fattal, L. R. (2017). Catastrophe: An Uncanny Catalyst for Creativity. In R. Shin (Ed.), *Convergence of Contemporary Art, Visual Culture, and Global Civic Engagement* (pp. 244–262). Hershey, PA: IGI Global. doi:10.4018/978-1-5225-1665-1.ch014

Forti, I. (2017). A Cross Reading of Landscape through Digital Landscape Models: The Case of Southern Garda. In A. Ippolito (Ed.), *Handbook of Research on Emerging Technologies for Architectural and Archaeological Heritage* (pp. 532–561). Hershey, PA: IGI Global. doi:10.4018/978-1-5225-0675-1.ch018

Gharbi, A., De Runz, C., & Akdag, H. (2017). Urban Development Modelling: A Survey. In S. Faiz & K. Mahmoudi (Eds.), *Handbook of Research on Geographic Information Systems Applications and Advancements* (pp. 96–124). Hershey, PA: IGI Global. doi:10.4018/978-1-5225-0937-0.ch004

Ghosh, I., & Ghoshal, I. (2018). Implications of Trade Liberalization for Food Security Under the ASEAN-India Strategic Partnership: A Gravity Model Approach. In V. Erokhin (Ed.), *Establishing Food Security and Alternatives to International Trade in Emerging Economies* (pp. 98–118). Hershey, PA: IGI Global. doi:10.4018/978-1-5225-2733-6.ch005

Glaves, H. M. (2017). Developing a Common Global Framework for Marine Data Management. In P. Diviacco, A. Leadbetter, & H. Glaves (Eds.), *Oceanographic and Marine Cross-Domain Data Management for Sustainable Development* (pp. 47–68). Hershey, PA: IGI Global. doi:10.4018/978-1-5225-0700-0.ch003

Godulla, A., & Wolf, C. (2018). Future of Food: Transmedia Strategies of National Geographic. In R. Gambarato & G. Alzamora (Eds.), *Exploring Transmedia Journalism in the Digital Age* (pp. 162–182). Hershey, PA: IGI Global. doi:10.4018/978-1-5225-3781-6.ch010

Gomes, P. P. (2018). Food and Environment: A Review on the Sustainability of Six Different Dietary Patterns. In A. Obayelu (Ed.), *Food Systems Sustainability and Environmental Policies in Modern Economies* (pp. 15–31). Hershey, PA: IGI Global. doi:10.4018/978-1-5225-3631-4.ch002

Gonzalez-Feliu, J. (2018). Sustainability Evaluation of Green Urban Logistics Systems: Literature Overview and Proposed Framework. In A. Paul, D. Bhattacharyya, & S. Anand (Eds.), *Green Initiatives for Business Sustainability and Value Creation* (pp. 103–134). Hershey, PA: IGI Global. doi:10.4018/978-1-5225-2662-9.ch005

Goodland, R. (2017). A Fresh Look at Livestock Greenhouse Gas Emissions and Mitigation. In I. Management Association (Ed.), Natural Resources Management: Concepts, Methodologies, Tools, and Applications (pp. 124-139). Hershey, PA: IGI Global. doi:10.4018/978-1-5225-0803-8.ch006

Goundar, S., & Appana, S. (2018). Mainstreaming Development Policies for Climate Change in Fiji: A Policy Gap Analysis and the Role of ICTs. In I. Management Association (Ed.), Sustainable Development: Concepts, Methodologies, Tools, and Applications (pp. 402-432). Hershey, PA: IGI Global. doi:10.4018/978-1-5225-3817-2.ch020

Granell-Canut, C., & Aguilar-Moreno, E. (2018). Geospatial Influence in Science Mapping. In M. Khosrow-Pour, D.B.A. (Ed.), Encyclopedia of Information Science and Technology, Fourth Edition (pp. 3473-3483). Hershey, PA: IGI Global. doi:10.4018/978-1-5225-2255-3.ch302

Grigelis, A., Blažauskas, N., Gelumbauskaitė, L. Ž., Gulbinskas, S., Suzdalev, S., & Ferrarin, C. (2017). Marine Environment Data Management Related to the Human Activity in the South-Eastern Baltic Sea (The Lithuanian Sector). In P. Diviacco, A. Leadbetter, & H. Glaves (Eds.), *Oceanographic and Marine Cross-Domain Data Management for Sustainable Development* (pp. 282–302). Hershey, PA: IGI Global. doi:10.4018/978-1-5225-0700-0.ch012

Guma, I. P., Rwashana, A. S., & Oyo, B. (2018). Food Security Policy Analysis Using System Dynamics: The Case of Uganda. *International Journal of Information Technologies and Systems Approach*, *11*(1), 72–90. doi:10.4018/IJITSA.2018010104

Guma, I. P., Rwashana, A. S., & Oyo, B. (2018). Food Security Indicators for Subsistence Farmers Sustainability: A System Dynamics Approach. *International Journal of System Dynamics Applications*, *7*(1), 45–64. doi:10.4018/IJSDA.2018010103

Gupta, P., & Goyal, S. (2017). Wildlife Habitat Evaluation. In A. Santra & S. Mitra (Eds.), *Remote Sensing Techniques and GIS Applications in Earth and Environmental Studies* (pp. 258–264). Hershey, PA: IGI Global. doi:10.4018/978-1-5225-1814-3.ch013

Hanson, T., & Hildebrand, E. (2018). GPS Travel Diaries in Rural Transportation Research: A Focus on Older Drivers. In I. Management Association (Ed.), Intelligent Transportation and Planning: Breakthroughs in Research and Practice (pp. 609-625). Hershey, PA: IGI Global. doi:10.4018/978-1-5225-5210-9.ch027

Hartman, M. B. (2017). Research-Based Climate Change Public Education Programs. In I. Management Association (Ed.), Natural Resources Management: Concepts, Methodologies, Tools, and Applications (pp. 992-1003). Hershey, PA: IGI Global. doi:10.4018/978-1-5225-0803-8.ch046

Hashim, N. (2017). Zanzibari Seaweed: Global Climate Change and the Promise of Adaptation. In I. Management Association (Ed.), Natural Resources Management: Concepts, Methodologies, Tools, and Applications (pp. 365-391). Hershey, PA: IGI Global. https://doi.org/ doi:10.4018/978-1-5225-0803-8.ch019

Herrera, J. E., Argüello, L. V., Gonzalez-Feliu, J., & Jaimes, W. A. (2017). Decision Support System Design Requirements, Information Management, and Urban Logistics Efficiency: Case Study of Bogotá, Colombia. In G. Jamil, A. Soares, & C. Pessoa (Eds.), *Handbook of Research on Information Management for Effective Logistics and Supply Chains* (pp. 223–238). Hershey, PA: IGI Global. doi:10.4018/978-1-5225-0973-8.ch012

Huizinga, T., Ayanso, A., Smoor, M., & Wronski, T. (2017). Exploring Insurance and Natural Disaster Tweets Using Text Analytics. *International Journal of Business Analytics*, *4*(1), 1–17. doi:10.4018/IJBAN.2017010101

Hung, K., Kalantari, M., & Rajabifard, A. (2017). An Integrated Method for Assessing the Text Content Quality of Volunteered Geographic Information in Disaster Management. *International Journal of Information Systems for Crisis Response and Management*, *9*(2), 1–17. doi:10.4018/IJISCRAM.2017040101

Husnain, A., & Avdic, A. (2018). Identifying the Contemporary Status of E-Service Sustainability Research. In I. Management Association (Ed.), Sustainable Development: Concepts, Methodologies, Tools, and Applications (pp. 467-485). Hershey, PA: IGI Global. https://doi.org/ doi:10.4018/978-1-5225-3817-2.ch022

Iarossi, M. P., & Ferro, L. (2017). "The Past is Never Dead. It's Not Even Past": Virtual Archaeological Promenade. In A. Ippolito & M. Cigola (Eds.), *Handbook of Research on Emerging Technologies for Digital Preservation and Information Modeling* (pp. 228–255). Hershey, PA: IGI Global. doi:10.4018/978-1-5225-0680-5.ch010

Ignjatijević, S., & Cvijanović, D. (2018). Analysis of Serbian Honey Production and Exports. In *Exploring the Global Competitiveness of Agri-Food Sectors and Serbia's Dominant Presence: Emerging Research and Opportunities* (pp. 109–139). Hershey, PA: IGI Global. doi:10.4018/978-1-5225-2762-6.ch005

Jana, S. K., & Karmakar, A. K. (2017). Globalization, Governance, and Food Security: The Case of BRICS. In I. Management Association (Ed.), Natural Resources Management: Concepts, Methodologies, Tools, and Applications (pp. 692-712). Hershey, PA: IGI Global. https://doi.org/ doi:10.4018/978-1-5225-0803-8.ch032

Jana, S. K., & Karmakar, A. K. (2017). Food Security in Asia: Is There Convergence? In I. Management Association (Ed.), Natural Resources Management: Concepts, Methodologies, Tools, and Applications (pp. 109-123). Hershey, PA: IGI Global. https://doi.org/ doi:10.4018/978-1-5225-0803-8.ch005

John, J., & Kumar, S. (2018). A Locational Decision Making Framework for Shipbreaking Under Multiple Criteria. In I. Management Association (Ed.), Operations and Service Management: Concepts, Methodologies, Tools, and Applications (pp. 504-527). Hershey, PA: IGI Global. doi:10.4018/978-1-5225-3909-4.ch024

John, J., & Srivastava, R. K. (2018). Decision Insights for Shipbreaking using Environmental Impact Assessment: Review and Perspectives. *International Journal of Strategic Decision Sciences*, *9*(1), 45–62. doi:10.4018/IJSDS.2018010104

Joshi, Y., & Rahman, Z. (2018). Determinants of Sustainable Consumption Behaviour: Review and Conceptual Framework. In A. Paul, D. Bhattacharyya, & S. Anand (Eds.), *Green Initiatives for Business Sustainability and Value Creation* (pp. 239–262). Hershey, PA: IGI Global. doi:10.4018/978-1-5225-2662-9.ch011

Juma, D. W., Reuben, M., Wang, H., & Li, F. (2018). Adaptive Coevolution: Realigning the Water Governance Regime to the Changing Climate. In I. Management Association (Ed.), Hydrology and Water Resource Management: Breakthroughs in Research and Practice (pp. 346-357). Hershey, PA: IGI Global. https://doi.org/ doi:10.4018/978-1-5225-3427-3.ch014

K., S., & Tripathy, B. K. (2018). Neighborhood Rough-Sets-Based Spatial Data Analytics. In M. Khosrow-Pour, D.B.A. (Ed.), *Encyclopedia of Information Science and Technology, Fourth Edition* (pp. 1835-1844). Hershey, PA: IGI Global. https://doi.org/ doi:10.4018/978-1-5225-2255-3.ch160

Kabir, F. (2018). Towards a More Gender-Inclusive Climate Change Policy. In N. Mahtab, T. Haque, I. Khan, M. Islam, & I. Wahid (Eds.), *Handbook of Research on Women's Issues and Rights in the Developing World* (pp. 354–369). Hershey, PA: IGI Global. doi:10.4018/978-1-5225-3018-3.ch022

Kabir, F. (2018). Towards a More Gender-Inclusive Climate Change Policy. In I. Management Association (Ed.), Climate Change and Environmental Concerns: Breakthroughs in Research and Practice (pp. 525-540). Hershey, PA: IGI Global. https://doi.org/ doi:10.4018/978-1-5225-5487-5.ch027

Kanyamuka, J. S., Jumbe, C. B., & Ricker-Gilbert, J. (2018). Making Agricultural Input Subsidies More Effective and Profitable in Africa: The Role of Complementary Interventions. In A. Obayelu (Ed.), *Food Systems Sustainability and Environmental Policies in Modern Economies* (pp. 172–187). Hershey, PA: IGI Global. doi:10.4018/978-1-5225-3631-4.ch008

Karimi, H., & Gholamrezafahimi, F. (2017). Study of Integrated Coastal Zone Management and Its Environmental Effects: A Case of Iran. In R. Singh, A. Singh, & V. Srivastava (Eds.), *Environmental Issues Surrounding Human Overpopulation* (pp. 64–88). Hershey, PA: IGI Global. doi:10.4018/978-1-5225-1683-5.ch004

Kaya, I. R., Hutabarat, J., & Bambang, A. N. (2018). "Sasi": A New Path to Sustain Seaweed Farming From Up-Stream to Down-Stream in Kotania Bay, Molucass. *International Journal of Social Ecology and Sustainable Development*, *9*(2), 28–36. doi:10.4018/IJSESD.2018040103

Khader, V. (2018). Technologies for Food, Health, Livelihood, and Nutrition Security. In I. Management Association (Ed.), Food Science and Nutrition: Breakthroughs in Research and Practice (pp. 94-112). Hershey, PA: IGI Global. https://doi.org/ doi:10.4018/978-1-5225-5207-9.ch005

Kocadağlı, A. Y. (2017). The Temporal and Spatial Development of Organic Agriculture in Turkey. In W. Ganpat, R. Dyer, & W. Isaac (Eds.), *Agricultural Development and Food Security in Developing Nations* (pp. 130–156). Hershey, PA: IGI Global. doi:10.4018/978-1-5225-0942-4.ch006

Koundouri, P., Giannouli, A., & Souliotis, I. (2017). An Integrated Approach for Sustainable Environmental and Socio-Economic Development Using Offshore Infrastructure. In I. Management Association (Ed.), Renewable and Alternative Energy: Concepts, Methodologies, Tools, and Applications (pp. 1581-1601). Hershey, PA: IGI Global. doi:10.4018/978-1-5225-1671-2.ch056

Kumar, A., & Dash, M. K. (2017). Sustainability and Future Generation Infrastructure on Digital Platform: A Study of Generation Y. In N. Ray (Ed.), *Business Infrastructure for Sustainability in Developing Economies* (pp. 124–142). Hershey, PA: IGI Global. doi:10.4018/978-1-5225-2041-2.ch007

Kumar, A., Mukherjee, A. B., & Krishna, A. P. (2017). Application of Conventional Data Mining Techniques and Web Mining to Aid Disaster Management. In A. Kumar (Ed.), *Web Usage Mining Techniques and Applications Across Industries* (pp. 138–167). Hershey, PA: IGI Global. doi:10.4018/978-1-5225-0613-3.ch006

Kumar, C. P. (2017). Impact of Climate Change on Groundwater Resources. In I. Management Association (Ed.), Natural Resources Management: Concepts, Methodologies, Tools, and Applications (pp. 1094-1120). Hershey, PA: IGI Global. doi:10.4018/978-1-5225-0803-8.ch052

Kumari, S., & Patil, Y. (2017). Achieving Climate Smart Agriculture with a Sustainable Use of Water: A Conceptual Framework for Sustaining the Use of Water for Agriculture in the Era of Climate Change. In P. Rao & Y. Patil (Eds.), *Reconsidering the Impact of Climate Change on Global Water Supply, Use, and Management* (pp. 122–143). Hershey, PA: IGI Global. doi:10.4018/978-1-5225-1046-8.ch008

Kumari, S., & Patil, Y. (2018). Achieving Climate Smart Agriculture With a Sustainable Use of Water: A Conceptual Framework for Sustaining the Use of Water for Agriculture in the Era of Climate Change. In I. Management Association (Ed.), Climate Change and Environmental Concerns: Breakthroughs in Research and Practice (pp. 111-133). Hershey, PA: IGI Global. doi:10.4018/978-1-5225-5487-5.ch006

Kursah, M. B. (2017). Least-Cost Pipeline using Geographic Information System: The Limit to Technicalities. *International Journal of Applied Geospatial Research*, *8*(3), 1–15. doi:10.4018/ijagr.2017070101

Lahiri, S., Ghosh, D., & Bhakta, J. N. (2017). Role of Microbes in Eco-Remediation of Perturbed Aquatic Ecosystem. In J. Bhakta (Ed.), *Handbook of Research on Inventive Bioremediation Techniques* (pp. 70–107). Hershey, PA: IGI Global. doi:10.4018/978-1-5225-2325-3.ch004

Lallo, C. H., Smalling, S., Facey, A., & Hughes, M. (2018). The Impact of Climate Change on Small Ruminant Performance in Caribbean Communities. In I. Management Association (Ed.), Climate Change and Environmental Concerns: Breakthroughs in Research and Practice (pp. 193-218). Hershey, PA: IGI Global. doi:10.4018/978-1-5225-5487-5.ch010

Laurini, R. (2017). Nature of Geographic Knowledge Bases. In S. Faiz & K. Mahmoudi (Eds.), *Handbook of Research on Geographic Information Systems Applications and Advancements* (pp. 29–60). Hershey, PA: IGI Global. doi:10.4018/978-1-5225-0937-0.ch002

Lawrence, J., Simpson, L., & Piggott, A. (2017). Protected Agriculture: A Climate Change Adaptation for Food and Nutrition Security. In I. Management Association (Ed.), Natural Resources Management: Concepts, Methodologies, Tools, and Applications (pp. 140-158). Hershey, PA: IGI Global. doi:10.4018/978-1-5225-0803-8.ch007

Leadbetter, A., Cheatham, M., Shepherd, A., & Thomas, R. (2017). Linked Ocean Data 2.0. In P. Diviacco, A. Leadbetter, & H. Glaves (Eds.), *Oceanographic and Marine Cross-Domain Data Management for Sustainable Development* (pp. 69–99). Hershey, PA: IGI Global. doi:10.4018/978-1-5225-0700-0.ch004

Lucas, M. R., Rego, C., Vieira, C., & Vieira, I. (2017). Proximity and Cooperation for Innovative Regional Development: The Case of the Science and Technology Park of Alentejo. In L. Carvalho (Ed.), *Handbook of Research on Entrepreneurial Development and Innovation Within Smart Cities* (pp. 199–228). Hershey, PA: IGI Global. doi:10.4018/978-1-5225-1978-2.ch010

Ma, X., Beaulieu, S. E., Fu, L., Fox, P., Di Stefano, M., & West, P. (2017). Documenting Provenance for Reproducible Marine Ecosystem Assessment in Open Science. In P. Diviacco, A. Leadbetter, & H. Glaves (Eds.), *Oceanographic and Marine Cross-Domain Data Management for Sustainable Development* (pp. 100–126). Hershey, PA: IGI Global. doi:10.4018/978-1-5225-0700-0.ch005

Mabe, L. K., & Oladele, O. I. (2017). Application of Information Communication Technologies for Agricultural Development through Extension Services: A Review. In T. Tossy (Ed.), *Information Technology Integration for Socio-Economic Development* (pp. 52–101). Hershey, PA: IGI Global. doi:10.4018/978-1-5225-0539-6.ch003

Malomo, B. I. (2018). A Review of Psychological Resilience as a Response to Natural Hazards in Nigeria. In A. Eneanya (Ed.), *Handbook of Research on Environmental Policies for Emergency Management and Public Safety* (pp. 147–165). Hershey, PA: IGI Global. doi:10.4018/978-1-5225-3194-4.ch008

Manchiraju, S. (2018). Predicting Behavioral Intentions Toward Sustainable Fashion Consumption: A Comparison of Attitude-Behavior and Value-Behavior Consistency Models. In I. Management Association (Ed.), Fashion and Textiles: Breakthroughs in Research and Practice (pp. 1-21). Hershey, PA: IGI Global. doi:10.4018/978-1-5225-3432-7.ch001

Manzella, G. M., Bartolini, R., Bustaffa, F., D'Angelo, P., De Mattei, M., Frontini, F., ... Spada, A. (2017). Semantic Search Engine for Data Management and Sustainable Development: Marine Planning Service Platform. In P. Diviacco, A. Leadbetter, & H. Glaves (Eds.), *Oceanographic and Marine Cross-Domain Data Management for Sustainable Development* (pp. 127–154). Hershey, PA: IGI Global. doi:10.4018/978-1-5225-0700-0.ch006

Mbonigaba, J. (2018). Comparing the Effects of Unsustainable Production and Consumption of Food on Health and Policy Across Developed and Less Developed Countries. In A. Obayelu (Ed.), *Food Systems Sustainability and Environmental Policies in Modern Economies* (pp. 124–158). Hershey, PA: IGI Global. doi:10.4018/978-1-5225-3631-4.ch006

McKeown, A. E. (2017). Nurses, Healthcare, and Environmental Pollution and Solutions: Breaking the Cycle of Harm. In I. Management Association (Ed.), Natural Resources Management: Concepts, Methodologies, Tools, and Applications (pp. 392-415). Hershey, PA: IGI Global. https://doi.org/ doi:10.4018/978-1-5225-0803-8.ch020

Mili, B., Barua, A., & Katyaini, S. (2017). Climate Change and Adaptation through the Lens of Capability Approach: A Case Study from Darjeeling, Eastern Himalaya. In I. Management Association (Ed.), Natural Resources Management: Concepts, Methodologies, Tools, and Applications (pp. 1351-1365). Hershey, PA: IGI Global. https://doi.org/ doi:10.4018/978-1-5225-0803-8.ch064

Mir, S. A., Shah, M. A., Mir, M. M., & Iqbal, U. (2017). New Horizons of Nanotechnology in Agriculture and Food Processing Industry. In B. Nayak, A. Nanda, & M. Bhat (Eds.), *Integrating Biologically-Inspired Nanotechnology into Medical Practice* (pp. 230–258). Hershey, PA: IGI Global. doi:10.4018/978-1-5225-0610-2.ch009

Moallem, M., Sterrett, W. L., Gordon, C. R., Sukhera, S. M., Mahmood, A., & Bashir, A. (2021). An Investigation of the Effects of Integrating Computing and Project- or Problem-Based Learning in the Context of Environmental Sciences: A Case of Pakistani STEM Teachers. In S. Schroth & J. Daniels (Eds.), *Building STEM Skills Through Environmental Education* (pp. 49–89). IGI Global. https://doi.org/10.4018/978-1-7998-2711-5.ch003

Mujere, N., & Moyce, W. (2017). Climate Change Impacts on Surface Water Quality. In W. Ganpat & W. Isaac (Eds.), *Environmental Sustainability and Climate Change Adaptation Strategies* (pp. 322–340). Hershey, PA: IGI Global. doi:10.4018/978-1-5225-1607-1.ch012

Mukherjee, A. B., Krishna, A. P., & Patel, N. (2018). Geospatial Technology for Urban Sciences. In *Geospatial Technologies in Urban System Development: Emerging Research and Opportunities* (pp. 99–120). Hershey, PA: IGI Global. doi:10.4018/978-1-5225-3683-3.ch005

Nagarajan, S. K., & Sangaiah, A. K. (2017). Vegetation Index: Ideas, Methods, Influences, and Trends. In N. Kumar, A. Sangaiah, M. Arun, & S. Anand (Eds.), *Advanced Image Processing Techniques and Applications* (pp. 347–386). Hershey, PA: IGI Global. doi:10.4018/978-1-5225-2053-5.ch016

Naraine, L., & Meehan, K. (2017). Strengthening Food Security with Sustainable Practices by Smallholder Farmers in Lesser Developed Economies. In W. Ganpat, R. Dyer, & W. Isaac (Eds.), *Agricultural Development and Food Security in Developing Nations* (pp. 57–81). Hershey, PA: IGI Global. doi:10.4018/978-1-5225-0942-4.ch003

Naraine, L., & Meehan, K. (2017). Strengthening Food Security with Sustainable Practices by Smallholder Farmers in Lesser Developed Economies. In W. Ganpat, R. Dyer, & W. Isaac (Eds.), *Agricultural Development and Food Security in Developing Nations* (pp. 57–81). Hershey, PA: IGI Global. doi:10.4018/978-1-5225-0942-4.ch003

Naraine, L., & Meehan, K. (2017). Strengthening Food Security with Sustainable Practices by Smallholder Farmers in Lesser Developed Economies. In W. Ganpat, R. Dyer, & W. Isaac (Eds.), *Agricultural Development and Food Security in Developing Nations* (pp. 57–81). Hershey, PA: IGI Global. doi:10.4018/978-1-5225-0942-4.ch003

Nikolaou, K., Tsakiridou, E., Anastasiadis, F., & Mattas, K. (2017). Exploring Alternative Distribution Channels of Agricultural Products. *International Journal of Food and Beverage Manufacturing and Business Models, 2*(2), 36–66. doi:10.4018/IJFBMBM.2017070103

Nishat, K. J., & Rahman, M. S. (2018). Disaster, Vulnerability, and Violence Against Women: Global Findings and a Research Agenda for Bangladesh. In N. Mahtab, T. Haque, I. Khan, M. Islam, & I. Wahid (Eds.), *Handbook of Research on Women's Issues and Rights in the Developing World* (pp. 235–250). Hershey, PA: IGI Global. doi:10.4018/978-1-5225-3018-3.ch014

O'Hara, S., Jones, D., & Trobman, H. B. (2018). Building an Urban Food System Through UDC Food Hubs. In A. Burtin, J. Fleming, & P. Hampton-Garland (Eds.), *Changing Urban Landscapes Through Public Higher Education* (pp. 116–143). Hershey, PA: IGI Global. doi:10.4018/978-1-5225-3454-9.ch006

Obayelu, A. E. (2018). Integrating Environment, Food Systems, and Sustainability in Feeding the Growing Population in Developing Countries. In A. Obayelu (Ed.), *Food Systems Sustainability and Environmental Policies in Modern Economies* (pp. 1–14). Hershey, PA: IGI Global. doi:10.4018/978-1-5225-3631-4.ch001

Othman, R., Nath, N., & Laswad, F. (2018). Environmental Reporting and Accounting: Sustainability Hybridisation. In G. Azevedo, J. da Silva Oliveira, R. Marques, & A. Ferreira (Eds.), *Handbook of Research on Modernization and Accountability in Public Sector Management* (pp. 130–158). Hershey, PA: IGI Global. doi:10.4018/978-1-5225-3731-1.ch007

Ouadi, A., & Zitouni, A. (2021). Phasor Measurement Improvement Using Digital Filter in a Smart Grid. In A. Recioui & H. Bentarzi (Eds.), *Optimizing and Measuring Smart Grid Operation and Control* (pp. 100–117). IGI Global. https://doi.org/10.4018/978-1-7998-4027-5.ch005

Padigala, B. S. (2018). Traditional Water Management System for Climate Change Adaptation in Mountain Ecosystems. In I. Management Association (Ed.), Climate Change and Environmental Concerns: Breakthroughs in Research and Practice (pp. 630-655). Hershey, PA: IGI Global. doi:10.4018/978-1-5225-5487-5.ch033

Panda, C. K. (2018). Mobile Phone Usage in Agricultural Extension in India: The Current and Future Perspective. In F. Mtenzi, G. Oreku, D. Lupiana, & J. Yonazi (Eds.), *Mobile Technologies and Socio-Economic Development in Emerging Nations* (pp. 1–21). Hershey, PA: IGI Global. doi:10.4018/978-1-5225-4029-8.ch001

Pandian, S. L., Yarrakula, K., & Chaudhury, P. (2018). GIS-Based Decision Support System for Village Level: A Case Study in Andhra Pradesh. In S. Chaudhuri & N. Ray (Eds.), *GIS Applications in the Tourism and Hospitality Industry* (pp. 275–295). Hershey, PA: IGI Global. doi:10.4018/978-1-5225-5088-4.ch012

Quaranta, G., & Salvia, R. (2017). Social-Based Product Innovation and Governance in The Milk Sector: The Case of Carciocacio and Innonatura. In T. Tarnanidis, M. Vlachopoulou, & J. Papathanasiou (Eds.), *Driving Agribusiness With Technology Innovations* (pp. 293–310). Hershey, PA: IGI Global. doi:10.4018/978-1-5225-2107-5.ch015

Rahman, M. K., Schmidlin, T. W., Munro-Stasiuk, M. J., & Curtis, A. (2017). Geospatial Analysis of Land Loss, Land Cover Change, and Landuse Patterns of Kutubdia Island, Bangladesh. *International Journal of Applied Geospatial Research*, *8*(2), 45–60. doi:10.4018/IJAGR.2017040104

Rajack-Talley, T. A. (2017). Agriculture, Trade Liberalization and Poverty in the ACP Countries. In W. Ganpat, R. Dyer, & W. Isaac (Eds.), *Agricultural Development and Food Security in Developing Nations* (pp. 1–20). Hershey, PA: IGI Global. doi:10.4018/978-1-5225-0942-4.ch001

Rajamanickam, S. (2018). Exploring Landscapes in Regional Convergence: Environment and Sustainable Development in South Asia. In I. Management Association (Ed.), Sustainable Development: Concepts, Methodologies, Tools, and Applications (pp. 1051-1087). Hershey, PA: IGI Global. https://doi.org/ doi:10.4018/978-1-5225-3817-2.ch047

Rizvi, S. M., & Dearden, A. (2018). KHETI: ICT Solution for Agriculture Extension and Its Replication in Open and Distance Learning. In U. Pandey & V. Indrakanti (Eds.), *Open and Distance Learning Initiatives for Sustainable Development* (pp. 163–174). Hershey, PA: IGI Global. doi:10.4018/978-1-5225-2621-6.ch008

Roşu, L., & Macarov, L. I. (2017). Management of Drought and Floods in the Dobrogea Region. In I. Management Association (Ed.), Agri-Food Supply Chain Management: Breakthroughs in Research and Practice (pp. 372-403). Hershey, PA: IGI Global. https://doi.org/ doi:10.4018/978-1-5225-1629-3.ch016

Rouzbehani, K., & Rouzbehani, S. (2018). Mapping Women's World: GIS and the Case of Breast Cancer in the US. *International Journal of Public Health Management and Ethics*, *3*(1), 14–25. doi:10.4018/IJPHME.2018010102

Roy, D. (2018). Success Factors of Adoption of Mobile Applications in Rural India: Effect of Service Characteristics on Conceptual Model. In M. Khosrow-Pour, D.B.A. (Ed.), Green Computing Strategies for Competitive Advantage and Business Sustainability (pp. 211-238). Hershey, PA: IGI Global. https://doi.org/ doi:10.4018/978-1-5225-5017-4.ch010

Sajeva, M., Lemon, M., & Sahota, P. S. (2017). Governance for Food Security: A Framework for Social Learning and Scenario Building. *International Journal of Food and Beverage Manufacturing and Business Models*, *2*(2), 67–84. doi:10.4018/ IJFBMBM.2017070104

Sambhanthan, A., & Potdar, V. (2017). A Study of the Parameters Impacting Sustainability in Information Technology Organizations. *International Journal of Knowledge-Based Organizations*, *7*(3), 27–39. doi:10.4018/IJKBO.2017070103

Sanga, C., Kalungwizi, V. J., & Msuya, C. P. (2017). Bridging Gender Gaps in Provision of Agricultural Extension Service Using ICT: Experiences from Sokoine University of Agriculture (SUA) Farmer Voice Radio (FVR) Project in Tanzania. In I. Management Association (Ed.), Discrimination and Diversity: Concepts, Methodologies, Tools, and Applications (pp. 682-697). Hershey, PA: IGI Global. https://doi.org/ doi:10.4018/978-1-5225-1933-1.ch031

Santra, A., & Mitra, D. (2017). Role of Remote Sensing in Potential Fishing Zone Forecast. In A. Santra & S. Mitra (Eds.), *Remote Sensing Techniques and GIS Applications in Earth and Environmental Studies* (pp. 243–257). Hershey, PA: IGI Global. doi:10.4018/978-1-5225-1814-3.ch012

Schaap, D. (2017). SeaDataNet: Towards a Pan-European Infrastructure for Marine and Ocean Data Management. In P. Diviacco, A. Leadbetter, & H. Glaves (Eds.), *Oceanographic and Marine Cross-Domain Data Management for Sustainable Development* (pp. 155–177). Hershey, PA: IGI Global. doi:10.4018/978-1-5225-0700-0.ch007

Seckin-Celik, T. (2017). Sustainability Reporting and Sustainability in the Turkish Business Context. In U. Akkucuk (Ed.), *Ethics and Sustainability in Global Supply Chain Management* (pp. 115–132). Hershey, PA: IGI Global. doi:10.4018/978-1-5225-2036-8.ch006

Segbefia, A. Y., Barnes, V. R., Akpalu, L. A., & Mensah, M. (2018). Environmental Location Assessment for Seaweed Cultivation in Ghana: A Spatial Multi-Criteria Approach. *International Journal of Applied Geospatial Research*, *9*(1), 51–64. doi:10.4018/IJAGR.2018010104

Sen, Y. (2018). How to Manage Sustainability: A Framework for Corporate Sustainability Tools. In I. Management Association (Ed.), Sustainable Development: Concepts, Methodologies, Tools, and Applications (pp. 568-589). Hershey, PA: IGI Global. doi:10.4018/978-1-5225-3817-2.ch026

Shamshiry, E., Abdulai, A. M., Mokhtar, M. B., & Komoo, I. (2015). Regional Landfill Site Selection with GIS and Analytical Hierarchy Process Techniques: A Case Study of Langkawi Island, Malaysia. In P. Thomas, M. Srihari, & S. Kaur (Eds.), *Handbook of Research on Cultural and Economic Impacts of the Information Society* (pp. 248–282). Hershey, PA: IGI Global. doi:10.4018/978-1-4666-8598-7.ch011

Sharma, Y. K., Mangla, S. K., Patil, P. P., & Uniyal, S. (2018). Analyzing Sustainable Food Supply Chain Management Challenges in India. In M. Ram & J. Davim (Eds.), *Soft Computing Techniques and Applications in Mechanical Engineering* (pp. 162–180). Hershey, PA: IGI Global. doi:10.4018/978-1-5225-3035-0.ch008

Silvestrelli, P. (2018). The Impact of Events: To Which Extent Are They Sustainable for Tourist Destinations? Some Evidences From Expo Milano 2015. In M. Risso & S. Testarmata (Eds.), *Value Sharing for Sustainable and Inclusive Development* (pp. 185–204). Hershey, PA: IGI Global. doi:10.4018/978-1-5225-3147-0.ch008

Silvius, G. (2018). Sustainability Evaluation of IT/IS Projects. In I. Management Association (Ed.), Sustainable Development: Concepts, Methodologies, Tools, and Applications (pp. 26-40). Hershey, PA: IGI Global. https://doi.org/ doi:10.4018/978-1-5225-3817-2.ch002

Singh, R., Srivastava, P., Singh, P., Upadhyay, S., & Raghubanshi, A. S. (2017). Human Overpopulation and Food Security: Challenges for the Agriculture Sustainability. In R. Singh, A. Singh, & V. Srivastava (Eds.), *Environmental Issues Surrounding Human Overpopulation* (pp. 12–39). Hershey, PA: IGI Global. doi:10.4018/978-1-5225-1683-5.ch002

Singh, R., Srivastava, P., Singh, P., Upadhyay, S., & Raghubanshi, A. S. (2017). Human Overpopulation and Food Security: Challenges for the Agriculture Sustainability. In R. Singh, A. Singh, & V. Srivastava (Eds.), *Environmental Issues Surrounding Human Overpopulation* (pp. 12–39). Hershey, PA: IGI Global. doi:10.4018/978-1-5225-1683-5.ch002

Srivastava, N. (2017). Climate Change Mitigation: Collective Efforts and Responsibly. In I. Management Association (Ed.), Natural Resources Management: Concepts, Methodologies, Tools, and Applications (pp. 64-76). Hershey, PA: IGI Global. https://doi.org/ doi:10.4018/978-1-5225-0803-8.ch003

Stanganelli, M., & Gerundo, C. (2017). Understanding the Role of Urban Morphology and Green Areas Configuration During Heat Waves. *International Journal of Agricultural and Environmental Information Systems*, *8*(2), 50–64. doi:10.4018/IJAEIS.2017040104

Stewart, M. K., Hagood, D., & Ching, C. C. (2017). Virtual Games and Real-World Communities: Environments that Constrain and Enable Physical Activity in Games for Health. *International Journal of Game-Based Learning*, *7*(1), 1–19. doi:10.4018/IJGBL.2017010101

Stone, R. J. (2017). Modelling the Frequency of Tropical Cyclones in the Lower Caribbean Region. In W. Ganpat & W. Isaac (Eds.), *Environmental Sustainability and Climate Change Adaptation Strategies* (pp. 341–349). Hershey, PA: IGI Global. doi:10.4018/978-1-5225-1607-1.ch013

Syed, A., & Jabeen, U. A. (2018). Climate Change Impact on Agriculture and Food Security. In A. Eneanya (Ed.), *Handbook of Research on Environmental Policies for Emergency Management and Public Safety* (pp. 223–237). Hershey, PA: IGI Global. doi:10.4018/978-1-5225-3194-4.ch012

Tam, G. C. (2017). The Global View of Sustainability. In *Managerial Strategies and Green Solutions for Project Sustainability* (pp. 1–24). Hershey, PA: IGI Global. doi:10.4018/978-1-5225-2371-0.ch001

Tam, G. C. (2017). Understanding Project Sustainability. In *Managerial Strategies and Green Solutions for Project Sustainability* (pp. 110–139). Hershey, PA: IGI Global. doi:10.4018/978-1-5225-2371-0.ch005

Tam, G. C. (2017). Perspectives on Sustainability. In *Managerial Strategies and Green Solutions for Project Sustainability* (pp. 53–76). Hershey, PA: IGI Global. doi:10.4018/978-1-5225-2371-0.ch003

Tang, M., & Karunanithi, A. T. (2018). Visual Logic Maps (vLms). In *Advanced Concept Maps in STEM Education: Emerging Research and Opportunities* (pp. 108–149). Hershey, PA: IGI Global. doi:10.4018/978-1-5225-2184-6.ch005

Taşçıoğlu, M., & Yener, D. (2018). The Value and Scope of GIS in Marketing and Tourism Management. In S. Chaudhuri & N. Ray (Eds.), *GIS Applications in the Tourism and Hospitality Industry* (pp. 189–211). Hershey, PA: IGI Global. doi:10.4018/978-1-5225-5088-4.ch009

Thanh Tung, B. (2021). Pharmacology and Therapeutic Applications of Resveratrol. In A. Hussain & S. Behl (Eds.), *Treating Endocrine and Metabolic Disorders With Herbal Medicines* (pp. 321–333). IGI Global. https://doi.org/10.4018/978-1-7998-4808-0.ch014

Tianming, G. (2018). Food Security and Rural Development on Emerging Markets of Northeast Asia: Cases of Chinese North and Russian Far East. In V. Erokhin (Ed.), *Establishing Food Security and Alternatives to International Trade in Emerging Economies* (pp. 155–176). Hershey, PA: IGI Global. doi:10.4018/978-1-5225-2733-6.ch008

Tiftikçigil, B. Y., Yaşgül, Y. S., & Güriş, B. (2017). Sustainability of Foreign Trade Deficit in Energy: The Case of Turkey. In N. Ray (Ed.), *Business Infrastructure for Sustainability in Developing Economies* (pp. 94–109). Hershey, PA: IGI Global. doi:10.4018/978-1-5225-2041-2.ch005

Tiwari, S., Vaish, B., & Singh, P. (2018). Population and Global Food Security: Issues Related to Climate Change. In I. Management Association (Ed.), Climate Change and Environmental Concerns: Breakthroughs in Research and Practice (pp. 41-64). Hershey, PA: IGI Global. https://doi.org/ doi:10.4018/978-1-5225-5487-5.ch003

Toujani, A., & Achour, H. (2018). A Data Mining Framework for Forest Fire Mapping. In I. Management Association (Ed.), Information Retrieval and Management: Concepts, Methodologies, Tools, and Applications (pp. 771-794). Hershey, PA: IGI Global. doi:10.4018/978-1-5225-5191-1.ch033

Tripathy, B., & K., S. B. (2017). Rough Fuzzy Set Theory and Neighbourhood Approximation Based Modelling for Spatial Epidemiology. In I. Management Association (Ed.), *Public Health and Welfare: Concepts, Methodologies, Tools, and Applications* (pp. 1257-1268). Hershey, PA: IGI Global. https://doi.org/ doi:10.4018/978-1-5225-1674-3.ch058

Trukhachev, A. (2017). New Approaches to Regional Branding through Green Production and Utilization of Existing Natural Advantages. In I. Management Association (Ed.), Advertising and Branding: Concepts, Methodologies, Tools, and Applications (pp. 1758-1778). Hershey, PA: IGI Global. doi:10.4018/978-1-5225-1793-1.ch081

Tsobanoglou, G. O., & Vlachopoulou, E. I. (2017). Social-Ecological Systems in Local Fisheries Communities. In G. Korres, E. Kourliouros, & M. Michailidis (Eds.), *Handbook of Research on Policies and Practices for Sustainable Economic Growth and Regional Development* (pp. 306–316). Hershey, PA: IGI Global. doi:10.4018/978-1-5225-2458-8.ch026

Tuydes-Yaman, H., & Karatas, P. (2018). Evaluation of Walkability and Pedestrian Level of Service. In I. Management Association (Ed.), Intelligent Transportation and Planning: Breakthroughs in Research and Practice (pp. 264-291). Hershey, PA: IGI Global. doi:10.4018/978-1-5225-5210-9.ch012

Uddin, S., Chakravorty, S., Ray, A., & Sherpa, K. S. (2018). Optimal Location of Sub-Station Using Q-GIS and Multi-Criteria Decision Making Approach. *International Journal of Decision Support System Technology*, *10*(2), 65–79. doi:10.4018/IJDSST.2018040104

Uddin, S., Chakravorty, S., Sherpa, K. S., & Ray, A. (2018). Power Distribution System Planning Using Q-GIS. *International Journal of Energy Optimization and Engineering*, *7*(2), 61–75. doi:10.4018/IJEOE.2018040103

Uzun, F. V. (2018). Natural Resources Management. In A. Eneanya (Ed.), *Handbook of Research on Environmental Policies for Emergency Management and Public Safety* (pp. 1–21). Hershey, PA: IGI Global. doi:10.4018/978-1-5225-3194-4.ch001

V, M., Agrawal, R., Sharma, V., & T.N., K. (2018). Supply Chain Social Sustainability and Manufacturing. In I. Management Association (Ed.), *Technology Adoption and Social Issues: Concepts, Methodologies, Tools, and Applications* (pp. 226-252). Hershey, PA: IGI Global. https://doi.org/ doi:10.4018/978-1-5225-5201-7.ch011

van der Vliet-Bakker, J. M. (2017). Environmentally Forced Migration and Human Rights. In C. Akrivopoulou (Ed.), *Defending Human Rights and Democracy in the Era of Globalization* (pp. 146–180). Hershey, PA: IGI Global. doi:10.4018/978-1-5225-0723-9.ch007

Vargas-Hernández, J. G. (2022). A Comprehensive Entrepreneurship Model for the Internationalization of Green Innovation Businesses. In U. Akkucuk (Ed.), *Disruptive Technologies and Eco-Innovation for Sustainable Development* (pp. 131–149). IGI Global. https://doi.org/10.4018/978-1-7998-8900-7.ch008

Vaskov, A. G., Lin, Z. Y., Tyagunov, M. G., Shestopalova, T. A., & Deryugina, G. V. (2018). Design of Renewable Sources GIS for ASEAN Countries. In V. Kharchenko & P. Vasant (Eds.), *Handbook of Research on Renewable Energy and Electric Resources for Sustainable Rural Development* (pp. 1–25). Hershey, PA: IGI Global. doi:10.4018/978-1-5225-3867-7.ch001

Vázquez, D. G., & Gil, M. T. (2017). Sustainability in Smart Cities: The Case of Vitoria-Gasteiz (Spain) – A Commitment to a New Urban Paradigm. In L. Carvalho (Ed.), *Handbook of Research on Entrepreneurial Development and Innovation Within Smart Cities* (pp. 248–268). Hershey, PA: IGI Global. doi:10.4018/978-1-5225-1978-2.ch012

Wahab, I. N., & Soonthodu, S. (2018). Geographical Information System in Eco-Tourism. In S. Chaudhuri & N. Ray (Eds.), *GIS Applications in the Tourism and Hospitality Industry* (pp. 61–75). Hershey, PA: IGI Global. doi:10.4018/978-1-5225-5088-4.ch003

Weiss-Randall, D. (2018). Cultivating Environmental Justice. In *Utilizing Innovative Technologies to Address the Public Health Impact of Climate Change: Emerging Research and Opportunities* (pp. 110–143). Hershey, PA: IGI Global. doi:10.4018/978-1-5225-3414-3.ch004

Weiss-Randall, D. (2018). Cultivating Resilience. In *Utilizing Innovative Technologies to Address the Public Health Impact of Climate Change: Emerging Research and Opportunities* (pp. 204–235). Hershey, PA: IGI Global. doi:10.4018/978-1-5225-3414-3.ch007

Weiss-Randall, D. (2018). Climate Change Solutions: Where Do We Go From Here? In *Utilizing Innovative Technologies to Address the Public Health Impact of Climate Change: Emerging Research and Opportunities* (pp. 236–268). Hershey, PA: IGI Global. doi:10.4018/978-1-5225-3414-3.ch008

Whyte, K. P., List, M., Stone, J. V., Grooms, D., Gasteyer, S., Thompson, P. B., . . . Bouri, H. (2018). Uberveillance, Standards, and Anticipation: A Case Study on Nanobiosensors in U.S. Cattle. In I. Management Association (Ed.), Biomedical Engineering: Concepts, Methodologies, Tools, and Applications (pp. 577-596). Hershey, PA: IGI Global. https://doi.org/ doi:10.4018/978-1-5225-3158-6.ch025

Wulff, E. (2017). Data and Operational Oceanography: A Review in Support of Responsible Fisheries and Aquaculture. In P. Diviacco, A. Leadbetter, & H. Glaves (Eds.), *Oceanographic and Marine Cross-Domain Data Management for Sustainable Development* (pp. 303–324). Hershey, PA: IGI Global. doi:10.4018/978-1-5225-0700-0.ch013

Yener, D. (2017). Geographic Information Systems and Its Applications in Marketing Literature. In S. Faiz & K. Mahmoudi (Eds.), *Handbook of Research on Geographic Information Systems Applications and Advancements* (pp. 158–172). Hershey, PA: IGI Global. doi:10.4018/978-1-5225-0937-0.ch006

Yu, K., Liu, Y., & Sharma, A. (2021). Analyze the Effectiveness of the Algorithm for Agricultural Product Delivery Vehicle Routing Problem Based on Mathematical Model. *International Journal of Agricultural and Environmental Information Systems*, *12*(3), 26–38. https://doi.org/10.4018/IJAEIS.2021070103

Zhou, X., Sharma, A., & Mohindru, V. (2021). Research on Linear Programming Algorithm for Mathematical Model of Agricultural Machinery Allocation. *International Journal of Agricultural and Environmental Information Systems*, *12*(3), 1–12. https://doi.org/10.4018/IJAEIS.2021070101

About the Contributors

Kavita Shakya Chahal is an Assistant Professor in the Department of Botany, Government College Bichhua, Chhindwara, M.P. She obtained M.Sc & Ph.D in Microbiology; and has cleared CSIR-NET in Life Sciences and ASRB-NET in Agricultural Microbiology. She is an Editorial Board Member of American Journal of Life Sciences; Editorial Board Member of International Journal of Current Science Research & Review; Active RMS member (ID: 116866) of International Journal of Creative Research Thoughts; and Member of Indian Academy of Sciences, Delhi. She has to her credit 18 Research papers, 7 Book chapters and 3 Books as an Editor.

* * *

Rupak Chakravarty is currently serving as a Faculty member at the Department of Library and Information Science, Panjab University, Chandigarh since 2006. His areas of research include Open Access, Library Technologies and Technology-enabled Library Services, Plagiarism, Reference Management Software, Cloud Technologies, Informetrics, etc. He has successfully guided 4 Research Scholars and currently 8 scholars are pursuing Ph.D. under his supervision. He has authored more than 90 research publications published in national and international journals and has delivered more than 120 invited talks at Human Resource Development Centres, Skill development Training programmes, capacity building workshops, conferences/seminars and Webinars. He has a passionate interest in free and open source software (FOSS), cloud applications and Educational Technology (ET) and keeps on experimenting with them. He has also served as Associate Editor of the Directory of Open Access Journals (DOAJ) for three consecutive years. He is currently the Chief Editor of International Journal of Information Studies and Libraries (IJISL).

Omdeo Gohatre is currently working in CIPET: IPT, Ahmedabad as an Assistant Professor. I have completed a Ph.D. specialization in Polymer and Plastics Technology from the Biju Patnaik University of Technology, Rourkela (Odisha). During my research work, I have particularly focused on the Recovery and recycling of vinyl chloride from electronic waste (e-waste) particularly utilization and recycling of waste wires and cables insulation by means of the mechanical recycling process. In this regard, I have excellency in the relative study of polymer composites and blends by perfoming Mechanical, thermal, morphological analysis. Expert in polymer recycling and its effects on final value-added products. Hands-on experience in FTIR, mechanical recycling equipment, electrostatic separator machine. Apart from that I have published 08 research articles in peer blinded international journals and 03 book chapters in national and international journals.

Jaidev K. is working as research fellow at Laboratory for Advanced Research in Polymeric Materials:CIPET, Bhubaneswar (Odisha). Expert in plastic waste management and recycled polymer blends.

Ankur Khapre is a PhD student in the University of Hyderabad. He has done his MSc from Devi Ahilya University Indore. Worked in the Chemistry lab in IIT BHU, Varanasi and Participated in International Summer Internship Programme, Osaka University, Japan.

Kashmira Majhi is working as lecturer at the Department of Plastic Technology, CIPET: IPT, Ahmedabad.

Nima Norouzi does research in techno-economic, techno-environmental, techno-social, and legal aspects of Energy and natural resources economics.

Payal Panwar has done her MSc from Devi Ahilya University, Indore. Now she is enrolled in PhD chemistry at Birla Institute of Technology Pilani, Rajasthan.

Subhaprada Sahoo is a student of M.Sc. Polymer Science. Currently she is working on waste management and recycling of plastic project. Published 01 research article in peer blinded international journal.

Bilge Sayın Börekçi is a full-time assistant professor in the Department of Gastronomy and Culinary Arts, University of Ardahan, Turkey. She graduated in Food Engineering from the Faculty of Agriculture of Atatürk University, in 2011. She gained her MSc degree at Ankara University in 2013. Her MSc thesis is about microbial alkaline protease production which is entitled "Determination of the specifications of alkaline proteases of some anaerobic bacteria". In 2020, she completed her PhD study entitled "Determination of citric acid production capacity of autochthonous *Candida zeylanoides* strains". She has experience in the field of microbial productions, process optimization and food safety.

Anuradha Singh (Assistant professor in Department of Chemistry, S S Khanna Girls' Degree College, University of Allahabad) recently completed a research project entitled "Design and Development of bi-functional Reverse Transcriptase Inhibitors of HIV Replication and Their Enhanced Cellular Uptake Using Myristoylated Derivatives" funded by the Department of Science and Technology (DST), New Delhi as Principal Investigator. Presently she is working as Principal Investigator in a start-up project funded by University Grants Commission, New Delhi. Her current research interest is focused around developing multifunctional anti-HIV, anti-HCV, anticancer and antibacterial agents. She is well versed in synthetic chemistry, computer-aided drug design and other techniques used in the field of medicinal chemistry. She has been awarded with gold medal during her post graduate studies and Swarnajayanti Fellowships by National Academy of Sciences, India. Dr. Singh has published more than 25 research papers/book chapters, in addition to several articles, in journals/ books of national and international repute. She also edited three books.

Kavya Singh is a Research Scholar in the Chemistry Department at SS Khanna Girls Degree College (A constituent college of University of Allahabad, Allahabad). Kavya received a Bachelor's Degree in Chemistry (Honours) from Dayalbagh Educational Institute, Agra and a Master's Degree in Chemistry from Dr. BR Ambedkar University, Agra.

Swastika Singh is an Assistant Professor in the Department of Chemistry at S.S. Khanna Girls Degree College. She obtained undergraduate, postgraduate and DPhil degrees from the University of Allahabad. She has over 06 years of teaching experience (both UG and PG levels). She has published many research publications in National and International journals with high impact factors. Dr Singh has participated in various national and international symposia.

Sunil S. Suresh is working as independent Researcher at Cochin, Kerala (India). Expert in polymer recycling and characterization analysis.

Index

A

Aliphatic Polymer 77
Analgesic 200
anti-inflammatory 200, 202
Aspergillus niger 44, 64-75
Autohydrolysis 57, 76

B

banana peel 49, 64-65, 69, 74, 129, 132, 134, 138-140
bibliometric analysis 142-146, 148, 169-172
Bibliometrix 145, 147, 169, 172
biodegradable 45, 49, 57, 77, 81-82, 89-92, 95-101, 119, 123, 129-131, 133, 136, 138-140, 143, 153, 172, 199
biodegradation 82-83, 92, 97-99, 102, 131, 134, 136, 139, 154, 169-170, 172, 184
biomass 36, 39, 43-44, 47, 51, 59, 61, 66, 71, 73, 76, 79, 82, 92, 117, 119, 126-127, 130, 136, 140, 144, 169, 171-172
bioplastics 129-133, 135-136, 139-140, 142-146, 149, 153, 155, 160, 164-165, 167-170
bioreactor 43, 48-49, 53, 60, 74, 184
Bioremediation 50, 76

C

cluster analysis 142
composites 1-2, 7, 9-14, 16-23, 90, 101, 185, 193, 197-199
Compound Annual Growth Rate 45, 76
Copolymerization 83-86, 96-97, 99, 102

D

deep eutectic solvents 120-121, 126-128

E

Energy Law 24, 37
Environmental law 37
Environmental Management 71, 174
Environmental Remediation 45, 190-191, 197-199

F

FDA 45, 76
fermentation 43-46, 48-76, 91, 143
fly ash (FA) 1-2, 19

G

Green Law 24
green solvents 104-105, 109, 114, 120, 122, 125-127
green synthesis 95, 181, 185-187, 189, 192

H

hepatoprotective 200
hydrogen economy 25, 31-32, 35, 37-41
Hydrogen Law 24, 26, 40
Hydrogen Policy 41-42
hydrogen production 33-36, 38, 41
hydrogen safety 39

hydrogen storage 32, 36, 39, 42
hydrogen technologies 29, 31, 39-40
HyLaw 24, 26, 35-37, 40
hypoglycemic 200

I

immune stimulant 200
interfacial-interaction 8
ionic liquids 104, 109, 114-117, 120-121, 124-125, 127-128

L

Lawsonia inermis 200-203, 207, 209-210

M

magnetic nanoparticles 174-177, 179, 181-184, 186-187
mechanical properties 2, 7, 9-16, 18-23, 131, 133, 139, 153, 155
mutation 50

N

nanoparticles 174-199
Nanotechnology 126, 174, 186-188, 190-191, 193-194, 196-199

O

organic acid 43-46, 63, 74

P

polymer matrix 1-2, 4, 7-16
Poly-Olefins 77
Public Perception of Hydrogen 41

R

Remediation 45, 182-184, 187-188, 190-192, 197-199
Renewable Energy Law 37
RStudio 147, 173
Rwa 129

S

Scientometrics 142-143, 171-173
secondary metabolite 200, 204
solid-state fermentation 43-44, 66, 69, 72-75
Supercritical fluids 96, 104, 121-122, 125

T

thermoplastic 9, 11-12, 14, 19, 91, 93, 99-101, 131
Tofu 55, 76

U

United States Hydrogen Policy 42

V

Volumetric Productivity 52, 76
VOSviewer 147, 150, 153, 155-156, 158, 162, 165, 173

W

Waste Management 18, 76, 130, 140, 144, 183-185, 188-189

Y

Yarrowia lipolytica 44-45, 64-68, 70-73, 75

www.ingramcontent.com/pod-product-compliance
Lightning Source LLC
LaVergne TN
LVHW061539290926
849162LV00004B/848